Post-Processing Techniques for Metal-Based Additive Manufacturing

This book shares insights on post-processing techniques adopted to achieve precision-grade surfaces of additive manufactured metals including material characterization techniques and the identified material properties.

Post-processes are discussed from support structure removal and heat treatment to the material removal processes including hybrid manufacturing. Also discussed are case studies on unique applications of additive manufactured metals as an exemplary of the considerations taken during post-processing design and selection.

- Addresses the critical aspect of post-processing for metal additive manufacturing
- Provides systematic introduction of pertinent materials
- Demonstrates post-process technique selection with the enhanced understanding of material characterization methods and evaluation
- Includes in-depth validation of ultra-precision machining technology
- Reviews precision fabrication of industrial-grade titanium alloys, steels, and aluminum alloys, with additive manufacturing technology

The book is aimed at researchers, professionals, and graduate students in advanced manufacturing, additive manufacturing, machining, and materials processing.

Post-Processing Techniques for Metal-Based Additive Manufacturing

Towards Precision Fabrication

Hao Wang, Yan Jin Lee, Yuchao Bai, and
Jiong Zhang

CRC Press
Taylor & Francis Group
Boca Raton London New York

CRC Press is an imprint of the
Taylor & Francis Group, an **informa** business

Designed cover image: © Hao Wang, Yan Jin Lee, Yuchao Bai, and Jiong Zhang

First edition published 2024
by CRC Press
6000 Broken Sound Parkway NW, Suite 300, Boca Raton, FL 33487-2742

and by CRC Press
4 Park Square, Milton Park, Abingdon, Oxon, OX14 4RN

CRC Press is an imprint of Taylor & Francis Group, LLC

Library of Congress Cataloging-in-Publication Data
Names: Wang, Hao (Mechanical engineering teacher), author.
Title: Post-processing techniques for metal-based additive manufacturing :
towards precision fabrication / Hao Wang, Yan Jin Lee, Yuchao Bai, and Jiong Zhang.
Description: First edition. | Boca Raton, FL : CRC Press, 2024. |
Includes bibliographical references and index.
Identifiers: LCCN 2023010863 | ISBN 9781032224473 (hbk) |
ISBN 9781032224480 (pbk) | ISBN 9781003272601 (ebk)
Subjects: LCSH: Additive manufacturing. | Metals–Finishing.
Classification: LCC TS183.25 .W36 2024 |
DDC 621.9/88–dc23/eng/20230501
LC record available at https://lccn.loc.gov/202301086

ISBN: 978-1-032-22447-3 (hbk)
ISBN: 978-1-032-22448-0 (pbk)
ISBN: 978-1-003-27260-1 (ebk)

DOI: 10.1201/9781003272601

Typeset in Times
by Newgen Publishing UK

Contents

Preface

The innovation of additive manufacturing (AM) technology has disrupted the manufacturing industry with the unconventional advantage of achieving near-net-shape products with complex geometries and a great allowance for customization. The technological advances in this area of manufacturing have also given AM a giant leap to the forefront of the precision engineering industry with the capabilities of producing intricate microstructures with relatively good accuracy. Major efforts have been spent on identifying new materials to be manufactured by AM technology and optimizing process parameters to achieve industrial-grade mechanical properties and form accuracy. However, the aspect of post-processing for AM is often briefly mentioned without an in-depth discussion of its critical involvement in the chain of processes to achieve high-precision quality products.

This book is dedicated to providing a comprehensive overview of the developments in metal-based AM and the insights on post-processing techniques adopted to accommodate unique material characteristics and achieve precision-grade surfaces of AM metals. These materials include aluminum-based alloys, titanium alloys, steels, superalloys, high-entropy alloys, and others. Discussions on post-processing will include the systematic workflow that considers a spectrum of procedures to produce precision products, which include support structure removal, heat treatment, and material removal processes for surface finishing. Readers can expect to learn the material-specific challenges that hinder surface quality and the various material removal processes to achieve precision-graded surface finishing. Case studies will also enrich the discussion and instill critical thinking on the exemplary considerations taken during post-processing workflow design and selection for precision manufacturing with AM technology. The established understanding of post-processing procedures will lay the foundation for the development of the hybridization of multi-material, multi-structural, and multi-functional manufacturing.

Hao Wang
National University of Singapore, Singapore

About the Authors

Hao Wang (PhD, MSME, MASME) is an Alexander von Humboldt fellow and Assistant Professor at the Department of Mechanical Engineering in National University of Singapore. He was awarded the Japan Society for the Promotion of Science (JSPS) fellowship in 2020. He has a persistent dedication to the research in microcutting theory and the physics of machining processes. His research covers system and process development, design and implementation of machining and polishing systems, material characterization, among others. His research themes are centralized on bridging the gap between additive manufacturing and ultra-precision machining technology. He has published over 110 peer-reviewed journal articles. He is also an Associate Editor for the *Journal of Intelligent Manufacturing*.

Yan Jin Lee is Research Fellow in the Department of Mechanical Engineering at the National University of Singapore. His research focuses on novel solutions to augment the ultra-precision machining process to produce optical-grade surface finishing with nanoscopic roughness. His PhD research revealed novel physical effects of coatings on the ultra-precision machining of brittle single crystals, which had earned him the NUS President's Graduate Fellowship and the Heidenhain Scholarship recognized by the European Society for Precision Engineering and Nanotechnology. He sits on the International Scientific Committee for the European Society for Precision Engineering and Nanotechnology. Presently, he characterizes the machinability of anisotropic brittle materials and aerospace metals by incorporating nano-/microscopic surface effects in microcutting and the finite element method numerical modeling. His experience in assessing the machinability of various metals, such as maraging steels and aluminum alloys, is beneficial to provide a perspective from conventional approaches compared to post-processes for additive manufacturing.

Yuchao Bai is Senior Research Fellow at the Department of Mechanical Engineering in National University of Singapore. He received his Bachelor's degree and PhD from South China University of Technology in 2013 and 2018, respectively. He has been engaged in the development of additive manufacturing technology and manufacturing processes since 2013. His main research interests include laser metal additive manufacturing and post-processing, multi-material manufacturing, ultra-precision post-processing, and hybrid manufacturing processing for comprehensive research in the field. He is the topic editor of the journal *Coatings*, and the author of more than 40 peer-reviewed scientific papers on metal additive manufacturing and post-processing, such as Additive Manufacturing, Journal of Materials Processing Technology and Materials and Design. He has co-authored "Powder-bed laser melting additive manufacturing technology" (Associate Editor, Chinese Edition, 2021, National Defense Industry Press). His expertise and innovation in the development of additive manufacturing machines, hybrid additive/subtractive machine and process design have yielded a number of patents, where more than 50 were successfully patented by China and Japan, and four are pending Patent Cooperation Treaty (PCT). Several successful patents have been applied in commercial 3D printers. His patent on additive/milling hybrid manufacturing technology obtained the Silver Patent Award and the national Invention-Entrepreneurship Gold Award. He is also serving as a reviewer for more than a dozen international SCI-indexed journals and listed in Top 2% Scientists Worldwide 2022.

Jiong Zhang is Research Fellow at Department of Mechanical Engineering, National University of Singapore (NUS). He obtained his PhD from the Department of Mechanical Engineering at NUS and an MS (mechanical engineering) from Shanghai Jiao Tong University and a BA (mechanical engineering) from Zhejiang University. His research focuses on post-processing for additive manufacturing, internal polishing, and ultraprecision machining. He is particularly interested in developing and modeling post-processing techniques for additively manufactured components, including support removal and surface quality enhancement. He has published more than 20 peer-reviewed journal papers in top-tiered journals such as *Additive Manufacturing, Rapid Prototyping Journal, International Journal of Machine Tools and Manufacture, Materials and Design, Precision Engineering,* and others. He has also published nine conference papers and a book chapter. In addition, he has been a member of euspen's International Scientific Committee since 2021 and is an active reviewer for several SCI-indexed journals.

1 Introduction

1.1 PRECISION MANUFACTURING

Modern-day designs of products and devices are becoming smaller and more complex to accommodate the increasing demand for functionality within confined spaces and efficient weight requirements. As these components continue to become tinnier, the capabilities of precision manufacturing processes become increasingly vital to produce these more complex geometries and microstructures. The continuous development of these components drives research in precision engineering that encompasses both aspects of manufacturing and metrology (Pagano 2014a). The evolution of either aspect cannot progress without the other. The level of precision of manufactured parts can only be affirmed by precise measurements. On the flipside, the precision of metrological systems is enabled by the accuracy of the manufacturing processes to produce them. For instance, stylus tips used in contact measurements must be true to their intended geometry (e.g., tip radius and roundness) for accurate sizing of manufactured parts, while precision manufacturing is also required to fabricate the tip to its intended geometry. Thus, the relationship, which is synonymous with the "chicken and egg" dilemma of which came first, drives the need for continuous research and development in the field of precision engineering. This book will cover manufacturing technologies with a specific focus on the application of these technologies in precision post-processing for additive manufacturing (AM). It is also important to understand the available technology and demand for precision manufacturing.

Precision is most often differentiated into different levels. The extremities of these technologies are defined as the ultra-precision level which can be thought of as technological limits of each manufacturing process regardless of the material removal mechanisms (Brinksmeier 2014). There are also other definitions of ultra-precision that affix a value of <10 nm to the positioning accuracy and resolution of these systems (Pagano 2014b). This differentiates the niche area of ultra-precision from regular precision resolutions that are defined as <10 μm. These values, however, carry no meaning without an association with either the length scale of features producible on the manufactured part or the machine tool itself.

DOI: 10.1201/9781003272601-1

TABLE 1.1
Definition of Machine Tool Performance Measures

Measure	Description
Accuracy	Maximum deviation of the actual position from the ideal position
Repeatability	Variation of the actual position from the designated position after multiple attempts of repositioning
Resolution	Smallest increment of motion along a designated axis

Precision engineering from the perspective of the machine tool is the ratio between the performance and the travel range of the tool, which is defined within limits of 1 to 10^6 of these ratios (Leach and Carmignato 2020). The performance of a machine tool is dictated by its accuracy, repeatability, and resolution as listed in Table 1.1. The purpose of these definitions translates to the achievable accuracy of the manufactured components. A commonality of nanoscopic features on the surface can be found on each precision product in the form of designed sub-micrometric structures or surface roughness. The development of these surfaces and structures with dimensions less than 100 nm for functional performances falls under the category of nanotechnology (Corbett et al. 2000).

1.2   MARKET FOR PRECISION MANUFACTURING

Precision engineering has been largely focused on military applications in the 1940s and it was not till the 1970s that the concepts for precision engineering were adopted for commercial applications such as micro-optic arrays, display panels, structured films, and mirrors through the use of replicative molds (Evans 2012). Today, the impeccable accuracy of products can be found across a wide range of industries that demand performance, aesthetics, and safety requirements, largely in the semi-conductor, medical, optics, and sensor industries. The market continues to attract businesses and consumers with a global market valuation for precision parts projected to soar to US$350 billion by 2027 from a 2019 valuation of US$167 billion.

Semiconductors are the largest driving force in precision engineering that has become prevalent in daily life as semiconductors are found in most smart devices such as computers, electric vehicles, and aircraft. However, little has been shared about the massive hurdles faced in the manufacturing process. The most relatable product is the smartphone where consumers await yearly hardware upgrades of "faster chips" and "longer battery life" to existing models. The further miniaturizing of semiconductor chips corresponds to an increase in physical space for other components such as battery storage. Without diving too deeply into the circuitry of the semiconductors, the fundamental concept of miniaturizing demands a stringent level of precision to effectively integrate the nearly 40 billion components within the small, allotted space.

The second biggest market for precision products is in the medical industry where improvements to the geometrical accuracy of products could determine the survival of patients in the form of detection devices (e.g., collimators in CT scanners, MRI table components), implantable devices (e.g., pacemakers, coronary stents, cochlear implants, etc.), and surgical instruments (e.g., surgical drill bits, saw guides for bone surgery, laparoscopic staples, etc.).

The third market is in optical and sensory equipment, which are the backbone of the industry that validate manufacturing accuracy in precision metrology. Optics are particularly important in areas where the physical measurement of components is impractical due to restricted areas and the miniaturization of products, especially in integrated circuits of electronic products. Moreover, the development of sensory equipment is a vital support to the next movement in automated manufacturing (i.e., manufacturing 4.0) where the positioning, maneuverability, traceability, and repeatability of robotics are critical metrics for industrial internet of things (IoT). Therefore, it is evident that the market for precision products will remain highly attractive for the next generation of technology and the question now entails selecting the most efficient method to produce these precision products. An overview description of the various conventional processes and additive manufacturing technologies will be covered in subsequent sections.

1.3   MACHINE TOOLS AND PROCESSES

The precision engineering market is well diversified with machine tool suppliers situated all over the globe but largely dominated by companies from Germany, Japan, and the United States. These precision manufacturing machines have been designed to meet the demands from various industries and can be classified as contact and non-contact material removal processes, where the former mechanically removes material by typically driving a tool with one or more cutting edges into the work material, and the latter removes material by vaporizing work material through heat addition. The development of these technologies has been the main support toward the evolution of precision engineering over the past century. Thus, it is meaningful to acknowledge the features and characteristics of these subtractive processes to gain a better appreciation of how each may complement additive manufacturing technology.

1.3.1   MECHANICAL MACHINING

Mechanical material removal processes include turning, milling, and drilling operations where the machine tools are typically equipped with aerostatic bearing spindles for rotary motion, slides for linear motion, and highly rigid and sharp cutting tools (typically made of diamonds). The simplest representation of this subtractive manufacturing process can be found in the turning process, which uses a single cutting edge on the tool. The tool drives into the work material along the designated tool path and removes material via chip formation. This chip formation action occurs across all three mechanical material removal processes despite their differences (e.g., number of cutting edges, direction of cutting, etc.).

The workpiece typically rotates at a constant revolution speed during turning, while the cutting tool traverses along the length of the workpiece (in longitudinal turning)

or the end face of the workpiece (in face turning) as illustrated in Figures 1.1(a) and
1.1(b). The single-point cutting edge can also be programmed to produce freeform
structures along the designated tool path, otherwise known as grooving, as shown in
Figure 1.1(c).

Milling and drilling differ from turning where the tool vertically rotates while
the workpiece traverses laterally on a table for positioning. Additionally, milling and
drilling tools have multiple cutting edges attached to the shank, as shown in Figure 1.2.
Drilling tools do not vary too much and normally have a chisel cutting edge to exert
the pressure into the workpiece and induce the necessary friction to steady the drilling
process (Figure 1.3). Milling tools consist of multiple types designed to cut out specific geometries on the workpiece, such as face milling cutters that have cutting edges
protruding from the side of the cutting head to solely perform horizontal cutting and
end milling cutters that have cutting edges on the end and sides of the cutter to enable
axial and lateral machining (i.e., cutting along and perpendicular to the tool rotational axis).

Although milling and drilling may seem different from turning, the material
removal mechanisms can be traced back to the fundamentals derived from the single
point cutting process of turning (i.e., cutting with a single edge). Hence, fundamental
research on the mechanically based subtractive manufacturing process primarily
focuses on turning, while research in production engineering directly test the feasibility of the milling and drilling tools. The cutting process is described by the shear
deformation of material to be removed in the form of a cutting chip. A complex event
of shear strain occurs in the primary deformation zone alongside frictional forces
acting on the cutting tool rake and flank faces in the secondary and tertiary deformation zones, respectively (Figure 1.4).

With advances in rigidity control of machine tools, these machining processes
can achieve ultra-precision levels of manufacturing. Basic ultra-precision lathes, also
known as single-point diamond turning, typically arrive in a three-axis system configuration with up to 0.01 nm (linear) and 0.0000001° (rotary) programmable resolutions.
These machines operate like conventional systems but at smaller scales with greater
levels of precision. Manufacturers can supply machine tools to machine workpiece
diameters in the range of 150–1,000 mm with surface finishing of <1.25–1.5 nm

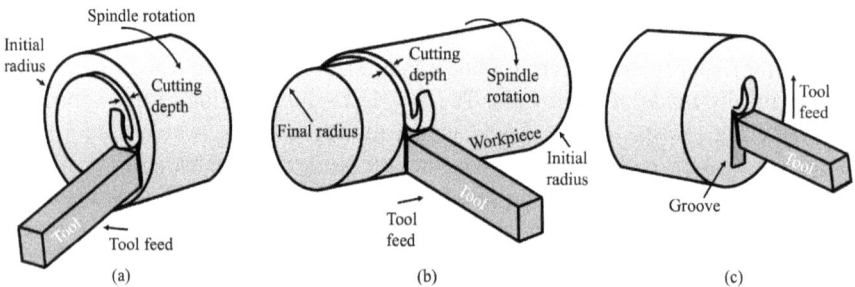

FIGURE 1.1 Material removal operations with single-point cutting tools: (a) face turning;
(b) longitudinal turning; (c) grooving.

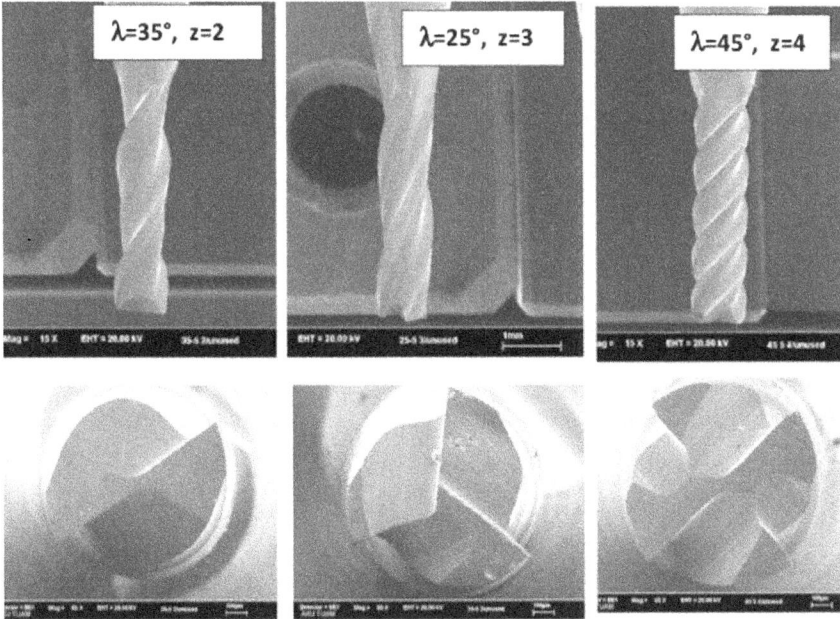

FIGURE 1.2 Micro-milling tools. Reprinted with permission from Springer Nature Figure 1.1(Aslantas and Alatrushi 2021).

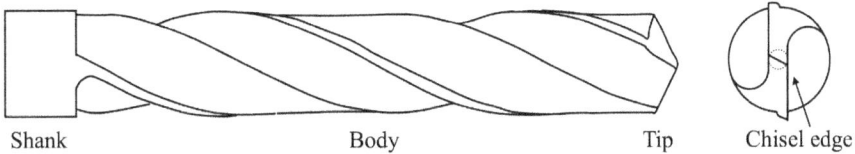

| Shank | Body | Tip | Chisel edge |

FIGURE 1.3 Drill bit nomenclature.

Sa and form accuracy <0.1–0.125 μm P-V. Micro-milling solutions available in the market offer machinable tolerances as low as 2 μm using miniaturized tools that are <1 mm (Chen et al. 2021) and can go down to 10 μm in diameter.

1.3.2 ELECTRIC DISCHARGE MACHINING

Electric discharge machining (EDM) and laser machining are common thermal-based machining technologies adopted in precision manufacturing. EDM removes material via a thermoelectric process that erodes material through sparks by discharging current through a dielectric fluid that flows between a tool and the workpiece (Figure 1.5). In other words, the flow of the current through the dielectric fluid results in spark generation that heats the workpiece and removes material by erosion and evaporation. With the input current supplied to the system at a high frequency, work

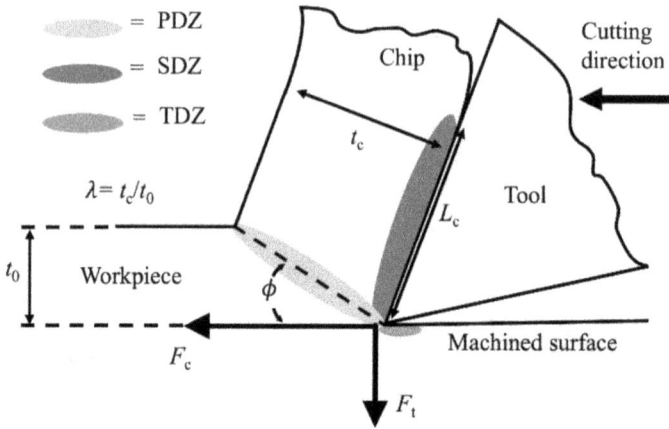

FIGURE 1.4 Deformation zone considerations in single point cutting. Reprinted with permission from MDPI (Lee, Shen, and Wang 2020).

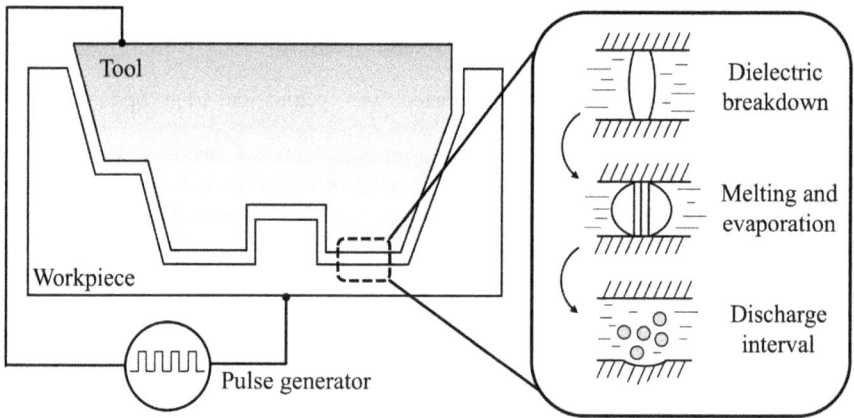

FIGURE 1.5 Electric discharge machining (EDM) material removal mechanism.

material can be spark eroded during current flow and the debris will be washed away by the flowing dielectric fluid when the current supply is intermittently disrupted. The nature of this material removal technique classifies EDM as a low volume production process due to the relatively low material removal rates.

The process necessitates both the workpiece and tool to be conductive for the flow of current, which limits the applicability of this precision manufacturing technique. Like the mechanical machining process, EDM utilizes different types of tools and material removal modes to perform machining. Common tool materials range from elemental metals such as copper and steel to tungsten alloys and graphite. The choice of material depends on the employed material removal process for EDM. Synonymous with the mechanical machining process, EDM also has variations such as die-sinking EDM, wire-cutting EDM, and hole-drilling EDM.

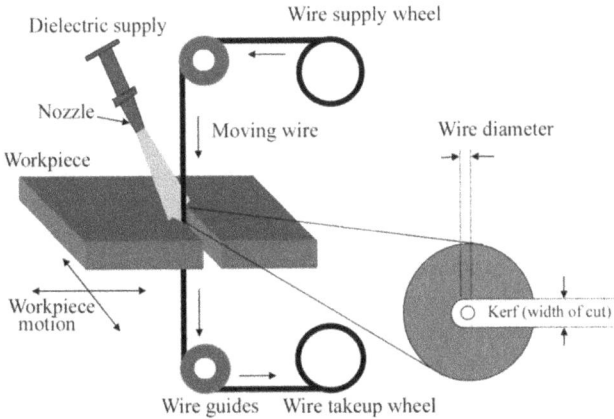

FIGURE 1.6 Illustration of wire-cutting EDM. Reprinted with permission from Springer Nature (Sreenivasa Rao and Venkaiah 2017).

Die-sinking EDM and hole-drilling EDM are effectively the same type of machining processes with the only difference being the geometry of the tool. The tool is designed to replicate the shape of the machined surface and removes material when the electrode tool is brought close to the workpiece for the electric current to form sparks within the gap filled with dielectric fluid (Figure 1.5). The geometry of the electrodes in die-sinking can be designed to imprint the final desired machined component, as commonly adopted in the production of molds (Rajeswari and Shunmugam 2020) and turbine blade cooling holes (Liang et al. 2018; Aas 2004). While this scenario is ideal in manufacturing the components without introducing mechanically induced residual stresses and vibrations, several challenges arise from the intricate details of the tool mold, such as the need to fabricate them with precision and the tendency for wear to occur more rapidly at sharp edges (Crookall and Moncrieff 1973). Therefore, electrode tools may also take the form of simple geometries that can be given a prescribed tool path to produce the final geometry, similar to the process of single-point grooving (Vidya, Wattal, and Rao 2021).

Vastly different to die-sinking, in wire-cutting EDM an electrode wire is fed from a spool and held in tension to slice through the workpiece (Figure 1.6). The benefit of this process is the superior precision of the machined geometry as the constant feed of new wire through the machining region effectively cancels the possibility of tool wear. Thus, the machined geometry would precisely correspond to the geometry of the wire, which can be as thin as 15 μm to achieve accuracies of 0.4 μm (Zou et al. 2021)

1.3.3 LASER MACHINING

Laser, a common term that is an acronym for "light amplification by stimulated emission of radiation", is the channeling of light energy into a focused beam to remove material by melting and vaporization (Olsen and Alting 1995). Just as light emits over different wavelengths, lasers also arrive in different wavelengths with the shorter wavelengths in the ultraviolet (UV) range having higher cutting efficiencies. The wavelength is also correlated with the spot size of the beam, which translates to achievable precision in machining. Again, shorter wavelengths typically enable smaller spot sizes to be used in ultra-precision fiber laser cutters. There are CO_2 lasers that produce light by introducing electricity into a gas-filled chamber and crystal lasers – i.e., neodymium-doped yttrium aluminum garnet (nd:YAG) or neodymium-doped yttrium ortho-vanadate (nd:YVO).

CO_2 lasers that emit high-intensity infrared light (10.6 μm wavelength) were the conventional favorites in metal cutting due to the high beam quality and output power (Thawari et al. 2005). These lasers have a typical power range of 1–40,000 W (Ghosal, Majumder, and Chattopadhyay 2018). However, advances in precision laser cutting adopted the use of Nd:YAG lasers that can emit laser wavelengths that are a tenth of the CO_2 lasers at 1.06 μm with ultra-short laser pulses in nano and femtoseconds (Pfeifer et al. 2010). These lasers with a slightly lower power range of 1–3,000 W are typically used for material processing and measurements. Ultra-precision fiber laser cutters in the market further enhance the capabilities of these lasers by offering ultra-short pulse lasers with picosecond ablation durations and linear positioning resolutions of 0.1 μm.

1.3.4 STATUS OF PRECISION SURFACE MANUFACTURING

These technologies are excellent options to meet the demand for precision manufacturing, but the extent of post-fabrication surface quality largely differs between processes. Surface quality is often characterized by the arithmetic mean roughness, Ra, which is defined as the average height (Z) over a measured length, L_s. It can be calculated by:

$$Ra = \frac{1}{L_s} \int_0^{L_s} |Z(x)| \, dx \qquad (1.1)$$

There are other common methods for characterizing roughness, such as the equivalent areal roughness, Sa, which measures the average height over an area in comparison to a sample line in Ra. Several researchers also prefer to use the root mean square (RMS) measurements, Rq or Sq, which correspond to the standard deviation of the height distribution and is calculated by:

$$Rq = \sqrt{\frac{1}{L_s} \int_0^{L_s} Z(x)^2 \, dx} \qquad (1.2)$$

For simplicity, this book will mainly characterize the surface roughness based on the arithmetic mean roughness and refer to the RMS value where appropriate. Table 1.2 lists a comparison of the various achievable surface quality between these precision machining technologies.

TABLE 1.2
Precision Machining Processes And The Achievable Surface Roughness

Process	Work Material	Surface Roughness	Reference
Diamond turning	Si	0.010 μm Ra	(M. Wang, Wang, and Lu 2012)
	Al6061	0.017 μm Ra	(H. Wang et al. 2010)
		0.007 μm Ra	(S. Wang et al. 2020)
	Ti-6Al-4V	0.017 μm Ra	(Yip and To 2019)
	CuBe	0.005 μm Ra	(Sharma et al. 2020)
	Cu	0.009 μm Ra	(Lucca, Klopfstein, and Riemer 2020)
	KDP crystal	0.003 μm Ra	(Zhang and Zong 2020)
Micro-milling	Ti-6Al-4V	0.32 μm Ra	(X. Lu et al. 2018)
	Cu	0.15 μm Ra	(X. Lu et al. 2019)
	Cu	0.11 μm Sa	(Sorgato, Bertolini, and Bruschi 2020)
	Al	0.72 μm Ra	(Vázquez et al. 2010)
	Cu	1.33 μm Ra	
	Zr-based bulk metallic glass	0.31 μm Ra	(Ray, Puri, and Hanumaiah 2020)
	Sapphire	0.004 μm Ra	(Katahira et al. 2017)
	3J33 maraging steel	0.3 μm Ra	(Yao et al. 2019)
Electric discharge machining	Ti-6Al-4V	1.05 μm Ra	(Ahmed, Ishfaq, et al. 2019)
		1.94 μm Ra	(Venkatarao and Anup Kumar 2019)
	Nimonic C263 alloy	10.89 μm Ra	(Shastri and Mohanty 2020)
	SiC	5.12 μm Ra	(Qidong and Wei 2018)
	Boron carbide	1.30 μm Ra	(Puertas and Luis 2004)
	Si-SiC	1.05 μm Ra	(Clijsters et al. 2010)
	Al6061-SiC	1.92 μm Ra	(Raza et al. 2018)
	DSS-2205	0.1 μm Ra	(Ablyaz et al. 2020)
Laser machining	CVD diamond	0.15 μm Sa	(Li et al. 2020)
	SS304	2.18 μm Ra	(Kotadiya and Pandya 2016)
	Nitinol	0.2 μm Ra	(Huang, Zheng, and Lim 2004)
	Inconel 718	2.67 μm Ra	(Ahmed, Rafaqat, et al. 2019)
	Ti-6Al-4V	0.97 μm Ra	(Ahmed, Ahmad, et al. 2019)

Advancements in precision engineering were largely dominated by subtractive manufacturing processes but the recent developments in additive manufacturing (AM) technology have widened the possibilities for manufacturing. Introduced in the 1980s, AM has revolutionized the manufacturing industry with its unconventional method of producing near-net-shape products. This involves the layer-by-layer construction of freeform three-dimensional solids from powder or liquid base materials. Today, a wide range of materials are processed using AM technology, which include metals, ceramics, glass, plastics, and even food.

At a glance, AM benefits the manufacturing industry in providing cost savings by reducing material wastage (Jiang, Xu, and Stringer 2019; Delgado Camacho et al. 2018; Laureijs et al. 2017). The volume of raw materials in the supply chain for additive manufacturing can be substantially lowered close to the final volume of the product compared to traditional material removal processes that begin with bulk stock materials. Yet, the cost per weight of raw material required by AM machines can be higher by a multiple of ten (Bauer and Malone 2015), which may be attributed to the considerable amount of energy required to prepare these raw materials. Waste material from traditional machining processes can also be recycled to promote sustainable manufacturing, which may level the overall cost comparison between AM and traditional machining (Paris et al. 2016). Ongoing debates suggest that there are no generic answers to the cost efficiency of adopting AM in the manufacturing line.

However, AM technology definitely has an exclusive advantage that has revolutionized product design and fabrication (Rosso et al. 2021; Simons 2018). This is the capability to disregard the assembly of parts to form the final product where products are produced as a single part, which cannot be achieved via conventional methods. The layer-by-layer construct also enables the fabrication of unique internal features that would conventionally increase the number of required parts for assembly. Despite the advantages of AM and its rapid development in achieving excellent form accuracies of desired precision geometries, the technology is still unable to meet the surface quality demanded in precision and ultra-precision engineering. This book covers the additional steps needed to produce the desired surface quality of AM products, which is known as post-processing for additive manufacturing. These post-processes typically utilize conventional ultra-precision manufacturing technologies to address the geometrical and surface requirements.

The question then arises on the need to further discuss the attributes and developments of each precision manufacturing process to be employed for AM post-processing. The key difference here would be the characteristics of materials produced by AM. In conventional material processing, research efforts are poured into studying new materials that have superior properties with the goal of promoting productivity and extending product lifespan. The simplest example is the addition of alloying elements to maintain the mechanical integrity of aircraft jet engine parts made of metals operating at high temperatures. While the introduction of these additives influences the mechanical, thermal, and electrical conductivity performances that benefit the function of the material, the processing of these materials during the manufacturing stage may face challenges to achieve the desired product quality. In the same way, AM technology presents a new range of materials that are microstructurally different while maintaining the same elemental composition, especially in additively manufactured metals. More detail on the characteristics of AM metals will be covered in Chapter 3. These differences inevitably lead to new considerations in the post-processes, which will form the crux of this book. Moreover, the understanding of post-processes enhances awareness of further considerations in designing the building procedure.

1.4 ADDITIVE MANUFACTURING TECHNOLOGY

Before proceeding into the post-processing of AM production, it is good to revisit the concepts and advances of AM technologies that can produce a wide variety of materials. This book focuses on metal production and, therefore, the more established metal fabrication technologies will be discussed in the following sections.

Metal AM processes are generally categorized based on the state of the initial raw material, which is commonly divided into powder and deposition-based processes as illustrated in Figure 1.7. Although sheet lamination (e.g., ultrasonic consolidation) is also recognized as a category of metal additive manufacturing involving the stacking and lamination of thin sheets of metal, it requires subsequent major processing, such as machining or laser cutting to produce the final shape of the product instead of producing a near-net-shape product. This section will focus on powder-bed fusion and direct energy deposition processes, which are more popular in the field of additive manufacturing.

Each of these processes has its own set of advantages and disadvantages in meeting the standards of precision engineering surface fabrication. This is on top of the basic requirements in AM, such as porosity or density, that have a strong influence on the final mechanical properties. The porosity of AM parts has been the central focus of research since its inception, but the advances in each technology have enabled a shift in the focus to the applicability of AM. According to the American Society for Testing Materials (ASTM) standard B311-17, the relative density of the part fabricated by powder-based processes can be calculated by (ASTM 2014):

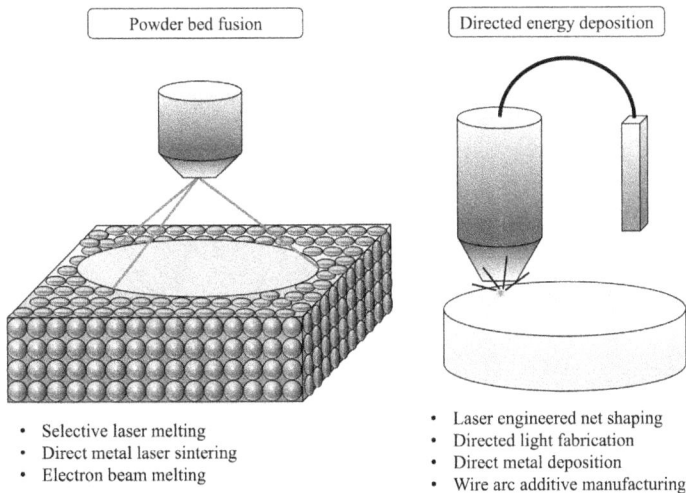

Powder bed fusion	Directed energy deposition

- Selective laser melting
- Direct metal laser sintering
- Electron beam melting

- Laser engineered net shaping
- Directed light fabrication
- Direct metal deposition
- Wire arc additive manufacturing

FIGURE 1.7 Categorization of metal AM processes by state of raw material.

$$\rho = m_{air} \frac{\rho_{liquid}}{m_{air} - m_{liquid}} \qquad (1.3)$$

where m is the mass and ρ is the density. However, the surface quality of these AM technologies have only recently gained traction in the research field in view of the feedback on the technological gaps provided by various industries.

1.4.1 POWDER BED FUSION (PBF) PROCESSES

There are three main types of metal AM processes that build parts from a bed of powder: (i) selective laser melting (SLM), (ii) direct metal laser sintering (DMLS), and (iii) electron beam melting (EBM). Conceptually, these processes are similar in utilizing a heat source (e.g., laser or electron beam) to selectively irradiate the bed of loose powder to form the final product, as illustrated by the SLM process in Figure 1.8. For each layer of the construction process, a layer of powder is deposited onto the build platform and flattened by administering a roller or scraper across the powder bed. The heat energy system subsequently irradiates the selected regions for the construction of the part by melting the powder particles and allow it to solidify with the bottom layer of material. This process repeats itself with laying the subsequent layer of powder, followed by scrapping, and irradiation until the final part is completed. The final part remains submerged in the bed of powder particles, before removal for post-processing.

Market leaders for SLM machines include SLM Solutions, 3D Systems, Renishaw, EOS, Laseradd Technology, and Sculpteo. These companies offer solutions to

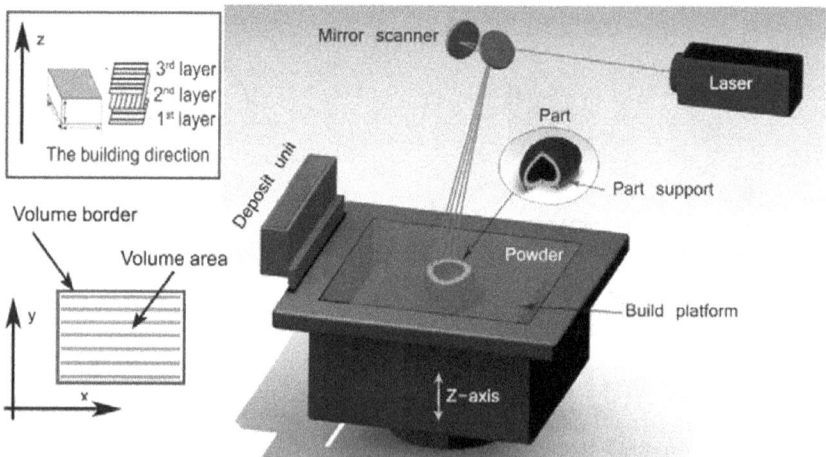

FIGURE 1.8 Principle of selective laser melting (SLM). Reprinted with permission from Elsevier (Fischer et al. 2016).

manufacture products from various metals, such as steel and aluminum, titanium, and nickel alloys, largely for the aerospace and medical industries. DMLS technology has the highest growth rate – led by North America followed by the Asia-Pacific region – with key players including 3D systems Inc., EOS GmbH Electro Optical Systems, Farsoon Technologies, Prodways Group, and Formlabs Inc. The leaders in EBM-based additive manufacturing technology largely targeted at processing titanium alloys and chromium-cobalt alloys are Arcam EBM, JEOL, Freemelt, and Sailong Metal.

There are several common process parameters used for these PBF processes, such as the scan speed, hatch space, powder size, and layer thickness. Despite having a very similar fabrication concept, a key difference between SLM and EBM is in the source of the irradiation energy. In SLM, thermal energy is delivered through a high power-density laser to melt the metal powder according to the designed cross-sectional geometry and gradually build the final product with each layer of resolidified material. Each layer of the building process is controlled by the displacement of the build platform, which incrementally lowers between steps to allow the deposition of a fresh plane of metal powders supplied from the deposition unit. Therefore, SLM considers additional parameters such as the laser power, spot size, and pulse duration and frequency.

Laser positioning in commercial systems is usually controlled using closed-looped galvanometer motors to deliver quick and precise motions (Rasoanarivo, Dumur, and Rodriguez-Ayerbe 2021; Aylward 2003). Hence, scanning speeds are often available over a wide range of 200–1,500 mm/s (Lu et al. 2017). The precision positioning of the laser is also critical in adhering to the scan strategy, which includes various aspects, such as the scan direction (Wang et al. 2018) and overlapping of melt tracks or hatch spacing (Chen et al. 2019). Contrary to the commonly depicted single laser that irradiates the powder bed, commercial SLM systems can be equipped with up to four lasers that operate simultaneously to increase fabrication efficiency (see Figure 1.8).

The key parameters in DMLS (see Figure 1.9) also comprise the laser power, scanning speed, and hatch spacing. The key difference between DMLS and SLM is that the former does not melt the powder but rather supplies sufficient heat for intermolecular fusion between particles (Owsiński and Niesłony 2018). Because of this, DMLS products are usually more porous compared to SLM parts. Selective laser sintering (SLS) is also a common term in AM technologies and is effectively identical to DMLS except that the latter is exclusively used for metals and no other materials, such as plastics glass, and ceramics. As DMLS does not involve the complete melting of powder, the supplied energy density is typically lower than those in SLM processes and is achieved by either reducing laser power (<400 W) or increasing scan speeds (>1,000 mm/s). As the hatch spacing and layer thickness rely more on the particle size and distribution, these parameters are often relatively similar in DMLS and SLM.

EBM differs from SLM and DMLS in that an electron beam is the main source of thermal energy to fuse the metallic powders during fabrication. Electrons are emitted from a tungsten filament (heated to 2,500–3,000 °C) or lanthanum

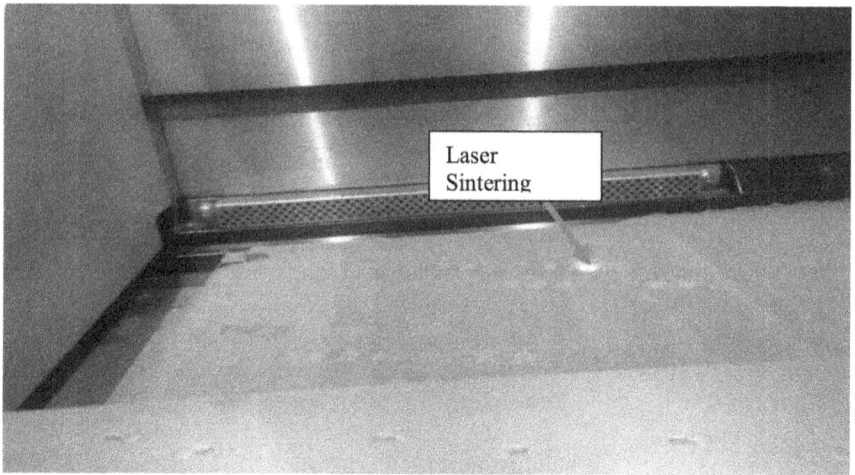

FIGURE 1.9 Visual of the sintering process in the DMLS chamber. Reprinted with permission from IOP Publishing (Shah and Dey 2019).

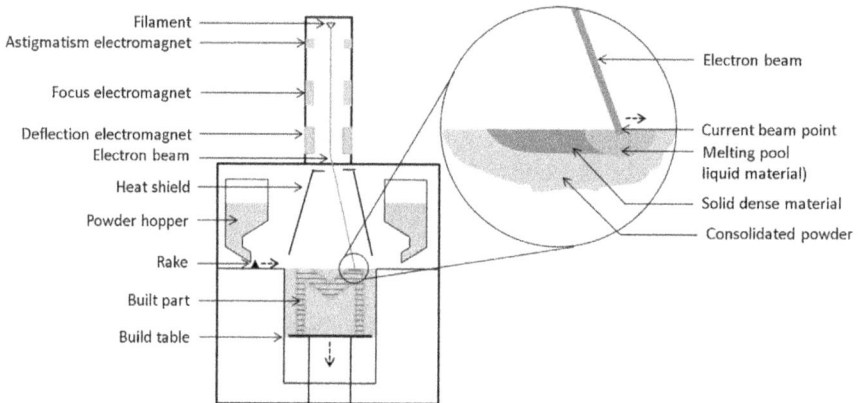

FIGURE 1.10 Schematic of the electron beam melting operating principle. Reprinted with permission from Elsevier (Béraud et al. 2017).

hexaboride (LaB$_6$) cathode (heated to 1,500–1,700 °C) and accelerated through a high voltage electron gun (60 kV) to the worktable, as shown in Figure 1.10. The direction of the electron beam is controlled by electromagnetic lenses and deflectors within the gun without the use of any mechanically driven components. This indicates that the beam direction can be repositioned extremely quickly up to 10,000 m/s within the build area (Körner 2016) and beam positioning accuracies of up to 25 μm.

The electron beam voltage refers to the available energy for electron acceleration and penetration into substrate surfaces while the electron current quantifies the number of emitted electrons, usually measured in the range of 3–17 mA. In other scenarios, the dose or current density (A/cm^2) is used as a better measure for comparison among different machines that have different working areas in the chamber.

Between the different energy carriers (i.e., electron and laser beams), the absorption efficiencies also largely differ and affect the conversion of energy for heating and evaporation (Körner 2016). Absorption is an important aspect of the melting process as it determines the complete or partial melting of the powders during the build process. Lasers are strongly affected by the reflectivity of the material where higher reflectivity, reduces the absorption efficiency, and can differ by 2%–60% among metals. The electron beam is mainly influenced by the density of the material and thus can maintain relatively higher absorption efficiencies ranging from 65% to 85%, which boosts the energy efficiency of the manufacturing process with low power requirements.

As electrons are the primary mode of transferring energy, the work material must be electrically conductive, which means that EBM is exclusively for metal manufacturing. Vested interest has been particularly shown in the EBM of titanium alloys for medical implants due to the relatively higher build rates in comparison to laser-based powder bed fusion processes (Karunakaran et al. 2012). However, EBM is inferior to SLM in the level of precision.

1.4.2 DIRECTED ENERGY DEPOSITION (DED)

The family of DED comprises various process names including laser engineered net shaping (LENS), directed light fabrication (DLF), direct metal deposition (DMD), 3D laser cladding, electron beam additive manufacturing (EBAM), and wire arc additive manufacturing (WAAM) or DED-arc. Unlike PBF processes where raw material is laid across the worktable for melting, DED processes deposit material onto existing surfaces of the fabricated part in wire or powder form before being melted by a focused heat energy source (e.g., laser, electron beam, and plasma arc). Additionally, DED processes control material and energy delivery with the use of position-controlled nozzle heads and worktables, giving great flexibility in terms of positioning compared to PBF processes where the heat source is typically fixed and the worktable mainly controls positioning. The nature of this additive process makes DED useful for repair work or remanufacturing.

Multiple laser-based metal deposition systems adopt the same concept for additive fabrication where each comprises a laser system fitted with a powder-feeding system to melt raw material as it is deposited (i.e., a laser cladding process). LENS (Figure 1.11) is one of the most well-known DED processes, which was conceptualized in 1997 by Sandia National Laboratories and Pratt & Whitney. In the same year, directed light fabrication (DLF) was developed at Los Alamos National Laboratory (Lewis et al. 1994) with a concept similar to LENS in the use of a laser to fuse gas-delivered metal powder particles. Direct metal deposition (DMD) was developed at the University of Michigan specializing in the use of a robotic arm to position the coaxial nozzle during the fabrication of complex geometries.

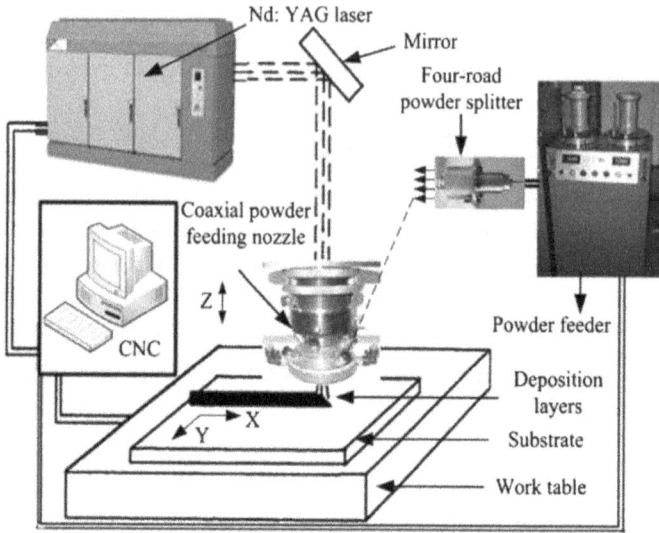

FIGURE 1.11   Schematic laser engineered net shaping (LENS) machine configuration. Reprinted with permission from Elsevier (Z. L. Lu et al. 2010).

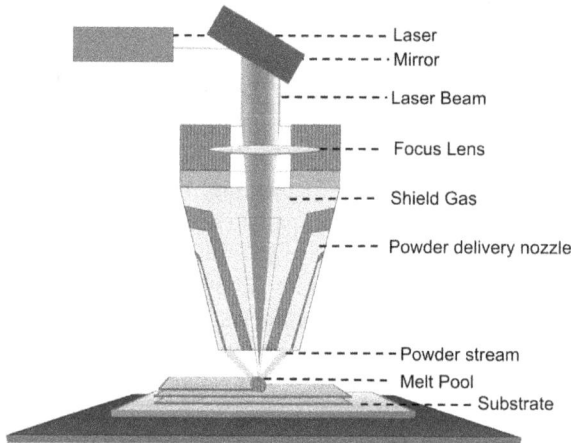

FIGURE 1.12   Schematic laser engineered net shaping (LENS) material deposition concept. Reprinted with permission from Springer Nature (Guddati et al. 2019).

Within the main nozzle head, the laser beam is passed through a focusing lens and directed through the center of the nozzle while metal powders are streamed through the sides to be melted and solidified together with the substrate (Figure 1.12). This is otherwise known as the coaxial supply of powder. On the other hand, there are also lateral powder supply systems that feed powder at an angle off-axis from the laser beam. The method of powder delivery in LENS opens the possibility of fabricating multi-material components by simply configuring the proportion of powders deposited (España, Balla, and Bandyopadhyay 2010). In most cases, the nozzle is also fitted with a stream of shielding gases to direct the powder flow and create an inert environment for highly reactive metals that tend to oxidize.

Like SLM, the laser power and scanning speed are also critical parameters that affect the build quality, with the most influential parameter being the laser power as with other laser-based systems. While the laser power ranges similarly in both PBF and DED processes up to 1,000 W, the considerations for power selection are different. As the molten powders in LENS flow freely with lesser boundary constraints, the excessive heat energy added to the system can lead to high thermal gradients between the highest and lowest temperatures in the melt pool. As the material cools and shrinks, the large thermal gradient then results in the development of residual stresses (Ngoveni et al. 2019). Thus, higher laser power can increase the input heat energy and lead to higher residual stress between layers (Pratt et al. 2008).

Powder delivery, typically measured in g/min or cm^3/min, is unique to LENS where the feeding rate can significantly affect the size of the melt pool, porosity and the consequential surface finishing (Amano and Rohatgi 2011; Mahamood and Akinlabi 2018). In some cases, the porosity of the fabricated parts are desirable, such as stress shielding in the medical industry (Bandyopadhyay et al. 2010). These powders' size typically range 38–150 μm (Long, Nand, and Ray 2021) and are taken from the feeders and carried to the nozzle by a gas.

Generally, raw materials in powder form are preferred in DED processes but there are also wire-fed raw materials used in wire arc additive manufacturing (WAAM). This manufacturing technique uses the concept of arc welding where an electric arc is used to heat and melt adjoining metals. During additive manufacturing, the feed of molten raw material is deposited onto the work surface to form layers of weld beads that form the final product.

Available industrial welding technology includes metal inert gas (MIG) or gas metal arc welding (GMAW), tungsten inert gas (TIG), and plasma arc welding. In each technology, an electric current is sent through the electrode that breaks through the insulated air medium in the form of an arc discharge. The discharge of electric current converts into light and heat energy that records temperatures of up to 20,000 °C. The main difference between MIG and TIG is in the way the arc is created. MIG feeds the raw material through a gun (electrode) at a constant rate to melt and form the weld bead, while TIG employs a non-sacrificial tungsten electrode and the raw material is added to the arc from a separate feedstock. Between the two concepts, TIG is generally preferred due to a higher level of precision but at the cost of lower material deposition rates. Plasma welding also uses a tungsten electrode but is capable of concentrating the arc by ionizing an inert gas to serve as the conductor for arc discharging to the workpiece. The concentrated arc results in a denser energy output from the electrode gun to further improve precision during melting and the additional directional capabilities. In all three types of welding technologies, shielding gases (e.g., argon, nitrogen, and helium) are necessarily supplied to the melt zone to prevent contamination and oxidation during melting.

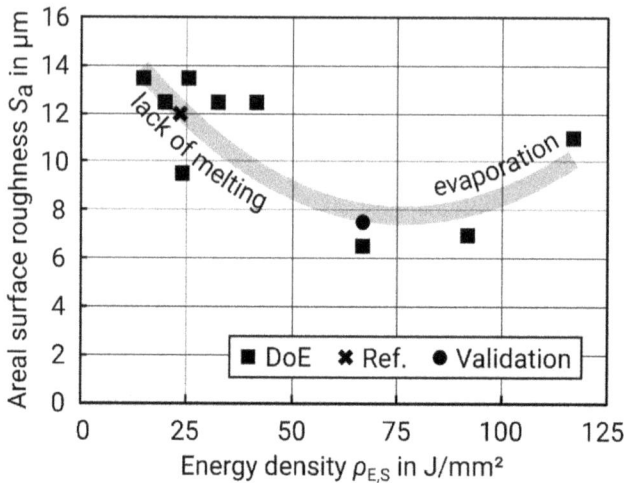

FIGURE 1.13   Relationship between energy density and surface quality. Reprinted with permission from MDPI (Wüst, Edelmann, and Hellmann 2020).

One advantage of WAAM over other powder-based additive manufacturing processes is the higher material deposition rates, ranging 1–10 kg/hr (Cunningham et al. 2018) compared to other laser and electron beam-based processes that are limited to less than 0.6 kg/hr (Liberini et al. 2017). The downside of arc-based manufacturing processes is the limited freedom of tool paths to produce large-sized components (Singh and Khanna 2021).

1.5   CONCLUDING REMARKS

Having gone through the details of each process, where do we stand today in terms of the achievable surface finishing of as-produced metal AM parts? Researchers have found a common relationship between the input energy density and the surface roughness (Spierings, Herres, and Levy 2011; Casalino et al. 2015; Yang et al. 2019; Han et al. 2018). In general, the surface roughness decreases with the increase in energy density to an optimal value before increasing again with a further increase in energy density (Figure 1.13).

The phenomenon occurs as the increase in energy density ensures the melting of the raw materials that then fill porous holes between powder particles. However, as the energy density increases further, the raw material is removed by evaporation and holes are left on the part's surface, which would increase the surface roughness. As the surface roughness is highly dependent on the energy density, which is affected by the combination of several key factors, each type of metal AM process and work material can result in a wide range of surface quality, as illustrated in Figure 1.14. Several predictive surface quality models (Reeves and Cobb 1997; Campbell, Martorelli, and Lee 2002; Ahn, Kim, and Lee 2009; Luis Pérez, Vivancos Calvet, and Sebastián Pérez 2001; Galati et al. 2021) have been proposed but will not be discussed in further detail in this book.

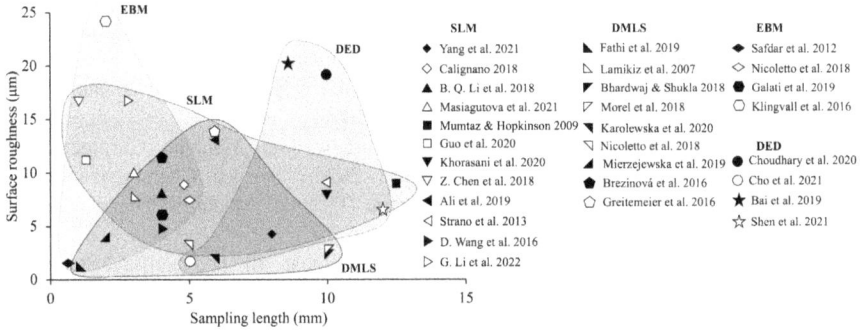

FIGURE 1.14 Surface quality achievable in metal additive manufacturing (Yang et al. 2021; Calignano 2018; B. Q. Li et al. 2018; Masiagutova et al. 2021; Mumtaz and Hopkinson 2009; Guo et al. 2020; Khorasani et al. 2020; Z. Chen et al. 2018; Ali et al. 2019; Strano et al. 2013; D. Wang et al. 2016; G. Li et al. 2022; Fathi et al. 2019; Lamikiz et al. 2007; Bhardwaj and Shukla 2018; Morel et al. 2018; Karolewska, Ligaj, and Boroński 2020; Nicoletto et al. 2018; Mierzejewska, Hudák, and Sidun 2019; Brezinová et al. 2016; Greitemeier et al. 2016; Safdar et al. 2012; Galati, Minetola, and Rizza 2019; Klingvall Ek et al. 2016; Choudhary et al. 2020; Cho, Shin, and Shim 2021; Bai, Chaudhari, and Wang 2020; Shen et al. 2021).

A comparison between the surface roughness achievable in as-built metal surfaces across the different types of additive manufacturing processes and the conventional machining processes presents the technological gap in additive technology toward achieving high precision quality. The road toward that state may be achievable with further advancements in the miniaturizing of current processes such as particle sizes and energy source control. Yet, it is still uncertain if that advancement will be possible. Hence, post-processes are made readily available to complete the additive manufacturing of precision engineering products, in the form of conventional precision machining methods and it is, therefore, an important discussion to account for in additive manufacturing technologies.

Conventional manufacturing begins with large bulk materials that are cut into the rough shape of the final product using rough cutting parameters and heavy-duty machinery before being processed using precision/ultra-precision machining technologies. The standard procedure is typically adopted for the preservation of machine tool precision positioning systems, the preservation of cutting tools, and the reduction in vibrations induced during heavy mechanical loading. In the same way, additive manufacturing also serves to produce the near-net-shape product that will require precision finishing using these precision/ultra-precision machining technologies. The benefits and limitations of adopting the latter strategy for precision manufacturing are found in the advantages and disadvantages of additive manufacturing technology. This book does not discriminate traditional methods for precision manufacturing but discusses the attributes of precision post-processing of additively manufactured high-quality surfaces.

REFERENCES

Aas, Kristian L. 2004. "Performance of Two Graphite Electrode Qualities in EDM of Seal Slots in a Jet Engine Turbine Vane." *Journal of Materials Processing Technology* 149 (1): 152–156. doi:10.1016/j.jmatprotec.2004.02.005.

Ablyaz, Timur R, Evgeny S Shlykov, Karim R Muratov, Amit Mahajan, Gurpreet Singh, Sandeep Devgan, and Sarabjeet S Sidhu. 2020. "Surface Characterization and Tribological Performance Analysis of Electric Discharge Machined Duplex Stainless Steel." *Micromachines*. doi:10.3390/mi11100926.

Ahmed, Naveed, Kashif Ishfaq, Khaja Moiduddin, Rafaqat Ali, and Naif Al-Shammary. 2019. "Machinability of Titanium Alloy through Electric Discharge Machining." *Materials and Manufacturing Processes* 34 (1). Taylor & Francis: 93–102. doi:10.1080/10426914.2018.1532092.

Ahmed, Naveed, Madiha Rafaqat, Salman Pervaiz, Usama Umer, Hesham Alkhalefa, Muhammad Ali Shar, and Syed Hammad Mian. 2019. "Controlling the Material Removal and Roughness of Inconel 718 in Laser Machining." *Materials and Manufacturing Processes* 34 (10). Taylor & Francis: 1169–1181. doi:10.1080/10426914.2019.1615082.

Ahmed, Naveed, Shafiq Ahmad, Saqib Anwar, Amjad Hussain, Madiha Rafaqat, and Mazen Zaindin. 2019. "Machinability of Titanium Alloy through Laser Machining: Material Removal and Surface Roughness Analysis." *The International Journal of Advanced Manufacturing Technology* 105 (7): 3303–3323. doi:10.1007/s00170-019-04564-7.

Ahn, Daekeon, Hochan Kim, and Seokhee Lee. 2009. "Surface Roughness Prediction Using Measured Data and Interpolation in Layered Manufacturing." *Journal of Materials Processing Technology* 209 (2): 664–671. doi:10.1016/j.jmatprotec.2008.02.050.

Ali, Usman, Reza Esmaeilizadeh, Farid Ahmed, Dyuti Sarker, Waqas Muhammad, Ali Keshavarzkermani, Yahya Mahmoodkhani, Ehsan Marzbanrad, and Ehsan Toyserkani. 2019. "Identification and Characterization of Spatter Particles and Their Effect on Surface Roughness, Density and Mechanical Response of 17-4 PH Stainless Steel Laser Powder-Bed Fusion Parts." *Materials Science and Engineering: A* 756: 98–107. doi:10.1016/j.msea.2019.04.026.

Amano, Ryoichi Samuel, and Pradeep Kumar Rohatgi. 2011. "Laser Engineered Net Shaping Process for SAE 4140 Low Alloy Steel." *Materials Science and Engineering: A* 528 (22): 6680–6693. doi:10.1016/j.msea.2011.05.036.

Aslantas, Kubilay, and Luqman K H Alatrushi. 2021. "Experimental Study on the Effect of Cutting Tool Geometry in Micro-Milling of Inconel 718." *Arabian Journal for Science and Engineering* 46 (3): 2327–2342. doi:10.1007/s13369-020-05034-z.

ASTM, B311-13. 2014. "Standard Test Method for Density of Powder Metallurgy (PM) Materials Containing Less than Two Percent Porosity." In *ASTM Volume 02.05: Metallic and Inorganic Coatings; Metal Powders and Metal Powder Products*. ASTM: New York, NY, USA. doi:10.1520/B0311-17.

Aylward, Redmond P. 2003. "Advanced Galvanometer-based Optical Scanner Design." *Sensor Review* 23 (3). MCB UP Ltd: 216–222. doi:10.1108/02602280310481968.

Bai, Yuchao, Akshay Chaudhari, and Hao Wang. 2020. "Investigation on the Microstructure and Machinability of ASTM A131 Steel Manufactured by Directed Energy Deposition." *Journal of Materials Processing Technology* 276: 116410. doi:10.1016/j.jmatprotec.2019.116410.

Bandyopadhyay, Amit, Felix Espana, Vamsi Krishna Balla, Susmita Bose, Yusuke Ohgami, and Neal M Davies. 2010. "Influence of Porosity on Mechanical Properties and in Vivo Response of Ti6Al4V Implants." *Acta Biomaterialia* 6 (4): 1640–1648. doi:10.1016/j.actbio.2009.11.011.

Bauer, Joe, and Patrick Malone. 2015. "Cost Estimating Challenges in Additive Manufacturing."
    In *International Cost Estimating and Analysis Association Professional Development
    and Training Workshop*. Vol. 4.

Béraud, Nicolas, Frédéric Vignat, François Villeneuve, and Rémy Dendievel. 2017.
    "Improving Dimensional Accuracy in EBM Using Beam Characterization
    and Trajectory Optimization." *Additive Manufacturing* 14: 1–6. doi:10.1016/
    j.addma.2016.12.002.

Bhardwaj, Tarun, and Mukul Shukla. 2018. "Direct Metal Laser Sintering of Maraging
    Steel: Effect of Building Orientation on Surface Roughness and Microhardness." *Materials
    Today: Proceedings* 5 (9, Part 3): 20485–20491. doi:10.1016/j.matpr.2018.06.425.

Brezinová, Janette, Radovan Hudák, Anna Guzanová, Dagmar Draganovská, Gabriela
    Ižaríková, and Juraj Koncz. 2016. "Direct Metal Laser Sintering of Ti6Al4V for
    Biomedical Applications: Microstructure, Corrosion Properties, and Mechanical
    Treatment of Implants." *Metals*. doi:10.3390/met6070171.

Brinksmeier, Ekkard. 2014. "Ultra Precision Machining." In *CIRP Encyclopedia of Production
    Engineering*, 1277–1280. doi:10.1007/978-3-642-20617-7_6403.

Calignano, Flaviana. 2018. "Investigation of the Accuracy and Roughness in the Laser Powder
    Bed Fusion Process." *Virtual and Physical Prototyping* 13 (2). Taylor & Francis: 97–
    104. doi:10.1080/17452759.2018.1426368.

Campbell, R Ian, Massimo Martorelli, and Han Sang Lee. 2002. "Surface Roughness
    Visualisation for Rapid Prototyping Models." *Computer-Aided Design* 34 (10): 717–
    725. doi:10.1016/S0010-4485(01)00201-9.

Casalino, Giusppe, Sabina Luisa Campanelli, Nicola Contuzzi, and Antonio Domenico
    Ludovico. 2015. "Experimental Investigation and Statistical Optimisation of the
    Selective Laser Melting Process of a Maraging Steel." *Optics & Laser Technology*
    65: 151–158. doi:10.1016/j.optlastec.2014.07.021.

Chen, Changpeng, Jie Yin, Haihong Zhu, Zhongxu Xiao, Luo Zhang, and Xiaoyan Zeng. 2019.
    "Effect of Overlap Rate and Pattern on Residual Stress in Selective Laser Melting."
    International Journal of Machine Tools and Manufacture 145: 103433. doi:10.1016/
    j.ijmachtools.2019.103433.

Chen, Ni, Hao Nan Li, Jinming Wu, Zhenjun Li, Liang Li, Gongyu Liu, and Ning He.
    2021. "Advances in Micro Milling: From Tool Fabrication to Process Outcomes."
    International Journal of Machine Tools and Manufacture 160: 103670. doi:10.1016/
    j.ijmachtools.2020.103670.

Chen, Zhuoer, Xinhua Wu, Dacian Tomus, and Chris H J Davies. 2018. "Surface Roughness of
    Selective Laser Melted Ti-6Al-4V Alloy Components." *Additive Manufacturing* 21: 91–
    103. doi:10.1016/j.addma.2018.02.009.

Cho, Seung Yeong, Gwang Yong Shin, and Do Sik Shim. 2021. "Effect of Laser Remelting
    on the Surface Characteristics of 316L Stainless Steel Fabricated via Directed
    Energy Deposition." *Journal of Materials Research and Technology* 15: 5814–5832.
    doi:10.1016/j.jmrt.2021.11.054.

Choudhary, Amit, Abhijit Sadhu, Sagar Sarkar, Ashish Kumar Nath, and Gopinath Muvvala.
    2020. "Laser Surface Polishing of NiCrSiBC – 60WC Ceramic-Metal Matrix Composite
    Deposited by Laser Directed Energy Deposition Process." *Surface and Coatings
    Technology* 404: 126480. doi:10.1016/j.surfcoat.2020.126480.

Clijsters, Stjin, Kun Liu, Dominiek Reynaerts, and Bert Lauwers. 2010. "EDM Technology
    and Strategy Development for the Manufacturing of Complex Parts in SiSiC."
    Journal of Materials Processing Technology 210 (4): 631–641. doi:10.1016/
    j.jmatprotec.2009.11.012.

Corbett, John, R A McKeown, Graham N Peggs, and Roger Whatmore. 2000. "Nanotechnology: International Developments and Emerging Products." *CIRP Annals* 49 (2): 523–545. doi:10.1016/S0007-8506(07)63454-4.

Crookall, J R, and A J R Moncrieff. 1973. "A Theory and Evaluation of Tool-Electrode Shape Degeneration in Electro-Discharge Machining." *Proceedings of the Institution of Mechanical Engineers* 187 (1). IMECHE: 51–61. doi:10.1243/PIME_PROC_1973_187_082_02.

Cunningham, C R, Joseph Michael Flynn, Alborz Shokrani, V Dhokia, and Stephen T Newman. 2018. "Invited Review Article: Strategies and Processes for High Quality Wire Arc Additive Manufacturing." *Additive Manufacturing* 22: 672–686. doi:10.1016/j.addma.2018.06.020.

Delgado Camacho, Daniel, Patricia Clayton, William J O'Brien, Carolyn Seepersad, Maria Juenger, Raissa Ferron, and Salvatore Salamone. 2018. "Applications of Additive Manufacturing in the Construction Industry – A Forward-Looking Review." *Automation in Construction* 89: 110–119. doi:10.1016/j.autcon.2017.12.031.

España, Félix A, Vamsi Krishna Balla, and Amit Bandyopadhyay. 2010. "Laser Surface Modification of AISI 410 Stainless Steel with Brass for Enhanced Thermal Properties." *Surface and Coatings Technology* 204 (15): 2510–2517. doi:10.1016/j.surfcoat.2010.01.029.

Evans, Chris J. 2012. "Precision Engineering: An Evolutionary Perspective." *Philosophical Transactions of the Royal Society A: Mathematical, Physical and Engineering Sciences* 370 (1973). Royal Society: 3835–3851. doi:10.1098/rsta.2011.0050.

Fathi, Parisa, Mehran Rafieazad, Xili Duan, Mohsen Mohammadi, and Ali M Nasiri. 2019. "On Microstructure and Corrosion Behaviour of AlSi10Mg Alloy with Low Surface Roughness Fabricated by Direct Metal Laser Sintering." *Corrosion Science* 157: 126–145. doi:10.1016/j.corsci.2019.05.032.

Fischer, Marie, David Joguet, Guillaume Robin, Laurent Peltier, and Pascal Laheurte. 2016. "In Situ Elaboration of a Binary Ti–26Nb Alloy by Selective Laser Melting of Elemental Titanium and Niobium Mixed Powders." *Materials Science and Engineering: C* 62: 852–859. doi:10.1016/j.msec.2016.02.033.

Galati, Manuela, Paolo Minetola, and Giovanni Rizza. 2019. "Surface Roughness Characterisation and Analysis of the Electron Beam Melting (EBM) Process." *Materials*. doi:10.3390/ma12132211.

Galati, Manuela, Giovanni Rizza, Silvio Defanti, and Lucia Denti. 2021. "Surface Roughness Prediction Model for Electron Beam Melting (EBM) Processing Ti6Al4V." *Precision Engineering* 69: 19–28. doi:10.1016/j.precisioneng.2021.01.002.

Ghosal, Puja, Manik Chandra Majumder, and Anangamohan Chattopadhyay. 2018. "Study on Direct Laser Metal Deposition." *Materials Today: Proceedings* 5 (5, Part 2): 12509–12518. doi:10.1016/j.matpr.2018.02.232.

Greitemeier, Daniel, C Dalle Donne, F Syassen, J Eufinger, and T Melz. 2016. "Effect of Surface Roughness on Fatigue Performance of Additive Manufactured Ti–6Al–4V." *Materials Science and Technology* 32 (7). Taylor & Francis: 629–634. doi:10.1179/1743284715Y.0000000053.

Guddati, Subhash, A Sandeep Kranthi Kiran, Montray Leavy, and Seeram Ramakrishna. 2019. "Recent Advancements in Additive Manufacturing Technologies for Porous Material Applications." *The International Journal of Advanced Manufacturing Technology* 105 (1): 193–215. doi:10.1007/s00170-019-04116-z.

Guo, Chuan, Sheng Li, Shi Shi, Xinggang Li, Xiaogang Hu, Qiang Zhu, and R Mark Ward. 2020. "Effect of Processing Parameters on Surface Roughness, Porosity and Cracking of as-Built IN738LC Parts Fabricated by Laser Powder Bed Fusion." *Journal of Materials Processing Technology* 285: 116788. doi:10.1016/j.jmatprotec.2020.116788.

Han, Xuesong, Haihong Zhu, Xiaojia Nie, Guoqing Wang, and Xiaoyan Zeng. 2018. "Investigation on Selective Laser Melting AlSi10Mg Cellular Lattice Strut: Molten Pool Morphology, Surface Roughness and Dimensional Accuracy." *Materials*. doi:10.3390/ma11030392.

Huang, Han, Hong Yu Zheng, and Gnian Cher Lim. 2004. "Femtosecond Laser Machining Characteristics of Nitinol." *Applied Surface Science* 228 (1): 201–206. doi:10.1016/j.apsusc.2004.01.018.

Jiang, Jingchao, Xun Xu, and Jonathan Stringer. 2019. "Optimization of Process Planning for Reducing Material Waste in Extrusion Based Additive Manufacturing." *Robotics and Computer-Integrated Manufacturing* 59: 317–325. doi:10.1016/j.rcim.2019.05.007.

Karolewska, Karolina, Bogdan Ligaj, and Dariusz Boroński. 2020. "Strain Analysis of Ti6Al4V Titanium Alloy Samples Using Digital Image Correlation." *Materials*. doi:10.3390/ma13153398.

Karunakaran, K P, Alain Bernard, S Suryakumar, Lucas Dembinski, and Georges Taillandier. 2012. "Rapid Manufacturing of Metallic Objects." *Rapid Prototyping Journal* 18 (4). Emerald Group Publishing Limited: 264–280. doi:10.1108/13552541211231644.

Katahira, Kazutoshi, Yuhei Matsumoto, Jun Komotori, and Kazuo Yamazaki. 2017. "Experimental Investigation of Machinability and Surface Quality of Sapphire Machined with Polycrystalline Diamond Micro-Milling Tool." *The International Journal of Advanced Manufacturing Technology* 93 (9): 4389–4398. doi:10.1007/s00170-017-0881-1.

Khorasani, Amir Mahyar, Ian Gibson, AmirHossein Ghasemi, and Alireza Ghaderi. 2020. "Modelling of Laser Powder Bed Fusion Process and Analysing the Effective Parameters on Surface Characteristics of Ti-6Al-4V." *International Journal of Mechanical Sciences* 168: 105299. doi:10.1016/j.ijmecsci.2019.105299.

Klingvall Ek, Rebecca, Lars-Erik Rännar, Mikael Bäckstöm, and Peter Carlsson. 2016. "The Effect of EBM Process Parameters upon Surface Roughness." *Rapid Prototyping Journal* 22 (3). Emerald Group Publishing Limited: 495–503. doi:10.1108/RPJ-10-2013-0102.

Körner, Carolin. 2016. "Additive Manufacturing of Metallic Components by Selective Electron Beam Melting — a Review." *International Materials Reviews* 61 (5). Taylor & Francis: 361–377. doi:10.1080/09506608.2016.1176289.

Kotadiya, D J, and D H Pandya. 2016. "Parametric Analysis of Laser Machining with Response Surface Method on SS-304." *Procedia Technology* 23: 376–382. doi:10.1016/j.protcy.2016.03.040.

Lamikiz, Aitzol, José Antonio Sánchez, Luis Noberto López de Lacalle, and Josí Lu Arana. 2007. "Laser Polishing of Parts Built up by Selective Laser Sintering." *International Journal of Machine Tools and Manufacture* 47 (12). Elsevier: 2040–2050. doi:10.1016/j.ijmachtools.2007.01.013.

Laureijs, Rianne E, Jaime Bonnín Roca, Sneha Prabha Narra, Colt Montgomery, Jack L Beuth, and Erica R H Fuchs. 2017. "Metal Additive Manufacturing: Cost Competitive beyond Low Volumes." *Journal of Manufacturing Science and Engineering* 139 (8). doi:10.1115/1.4035420.

Leach, Richard, and Simone Carmignato. 2020. *Precision Metal Additive Manufacturing.* Boca Raton: CRC Press. doi:10.1201/9780429436543.

Lee, Yan Jin, Yung-Kang Shen, and Hao Wang. 2020. "Suppression of Polycrystalline Diamond Tool Wear with Mechanochemical Effects in Micromachining of Ferrous Metal." *Journal of Manufacturing and Materials Processing* 4 (3): 81. doi:10.3390/jmmp4030081.

Lewis, Gary K, Ron Nemec, John Milewski, Dan J Thoma, Dave Cremers, and Mike Barbe. 1994. "Directed Light Fabrication." *International Congress on Applications of Lasers & Electro-Optics* 1994 (1). Laser Institute of America: 17–26. doi:10.2351/1.5058786.

Li, Bao Qiang, Zhonghua Li, Peikang Bai, Bin Liu, and Zezhou Kuai. 2018. "Research on Surface Roughness of AlSi10Mg Parts Fabricated by Laser Powder Bed Fusion." *Metals*. doi:10.3390/met8070524.

Li, Gan, Xinwei Li, Chuan Guo, Yang Zhou, Qiyang Tan, Wenying Qu, Xinggang Li, Xiaogang Hu, Ming-Xing Zhang, and Qiang Zhu. 2022. "Investigation into the Effect of Energy Density on Densification, Surface Roughness and Loss of Alloying Elements of 7075 Aluminium Alloy Processed by Laser Powder Bed Fusion." *Optics & Laser Technology* 147: 107621. doi:10.1016/j.optlastec.2021.107621.

Li, Zhenjun, Ni Chen, Liang Li, Yang Wu, and Ning He. 2020. "Influence of the Grain Size of CVD Diamond on the Thermal Conductivity, Material Removal Depth and Surface Roughness in Nanosecond Laser Machining." *Ceramics International* 46 (12): 20510–20520. doi:10.1016/j.ceramint.2020.05.157.

Liang, Wei, Yong Li, Hao Tong, and Quancun Kong. 2018. "A Block Divided EDM Process for Diffuser Shaped Film Cooling Holes." *Procedia CIRP* 68: 415–419. doi:10.1016/j.procir.2017.12.105.

Liberini, Mariacira, Antonello Astarita, Gianni Campatelli, Antonio Scippa, Filippo Montevecchi, Giuseppe Venturini, Massimo Durante, Luca Boccarusso, Fabrizio Memola Capece Minutolo, and A Squillace. 2017. "Selection of Optimal Process Parameters for Wire Arc Additive Manufacturing." *Procedia CIRP* 62: 470–474. doi:10.1016/j.procir.2016.06.124.

Long, Jingjunjiao, Ashveen Nand, and Sudip Ray. 2021. "Application of Spectroscopy in Additive Manufacturing." *Materials*. doi:10.3390/ma14010203.

Lu, Xiaohong, Xiaochen Hu, Zhenyuan Jia, Mingyang Liu, Song Gao, Chenglin Qu, and Steven Y Liang. 2018. "Model for the Prediction of 3D Surface Topography and Surface Roughness in Micro-Milling Inconel 718." *The International Journal of Advanced Manufacturing Technology* 94 (5): 2043–2056. doi:10.1007/s00170-017-1001-y.

Lu, Xiaohong, Liang Xue, Feixiang Ruan, Kun Yang, and Steven Y Liang. 2019. "Prediction Model of the Surface Roughness of Micro-Milling Single Crystal Copper." *Journal of Mechanical Science and Technology* 33 (11): 5369–5374. doi:10.1007/s12206-019-1030-6.

Lu, Yanjin, Yiliang Gan, Junjie Lin, Sai Guo, Songquan Wu, and Jinxin Lin. 2017. "Effect of Laser Speeds on the Mechanical Property and Corrosion Resistance of CoCrW Alloy Fabricated by SLM." *Rapid Prototyping Journal* 23 (1). Emerald Publishing Limited: 28–33. doi:10.1108/RPJ-07-2015-0085.

Lu, Zhong Liang, Di Chen Li, Bing Heng Lu, Ang Feng Zhang, Gang Xian Zhu, and Gang Pi. 2010. "The Prediction of the Building Precision in the Laser Engineered Net Shaping Process Using Advanced Networks." *Optics and Lasers in Engineering* 48 (5): 519–525. doi:10.1016/j.optlaseng.2010.01.002.

Lucca, Don A, Matthew J Klopfstein, and Oltmann Riemer. 2020. "Ultra-Precision Machining: Cutting with Diamond Tools." *Journal of Manufacturing Science and Engineering* 142 (11). doi:10.1115/1.4048194.

Luis Pérez, C J, Joan Vivancos Calvet, and Miguel A Sebastián Pérez. 2001. "Geometric Roughness Analysis in Solid Free-Form Manufacturing Processes." *Journal of Materials Processing Technology* 119 (1): 52–57. doi:10.1016/S0924-0136(01)00897-4.

Mahamood, R M, and E T Akinlabi. 2018. "Effect of Powder Flow Rate on Surface Finish in Laser Additive Manufacturing Process." *IOP Conference Series: Materials Science and Engineering* 391. IOP Publishing: 12005. doi:10.1088/1757-899x/391/1/012005.

Masiagutova, E, F Cabanettes, A Sova, Mehmet Cici, Guillaume Bidron, and P Bertrand. 2021. "Side Surface Topography Generation during Laser Powder Bed Fusion of AlSi10Mg." *Additive Manufacturing* 47: 102230. doi:10.1016/j.addma.2021.102230.

Mierzejewska, Żaneta A, Radovan Hudák, and Jarosław Sidun. 2019. "Mechanical Properties and Microstructure of DMLS Ti6Al4V Alloy Dedicated to Biomedical Applications." *Materials*. doi:10.3390/ma12010176.

Morel, Claire, Vitor Valentin Cioca, Sylvain Lavernhe, A Jardini, and Erik Gustavo del Conte. 2018. "Part Surface Roughness on Laser Sintering and Milling of Maraging Steel 300." In *14th International Conference on High Speed Manufacturing*. San-Sebastian, Spain.

Mumtaz, Kamran, and Neil Hopkinson. 2009. "Top Surface and Side Roughness of Inconel 625 Parts Processed Using Selective Laser Melting." *Rapid Prototyping Journal* 15 (2). Emerald Group Publishing Limited: 96–103. doi:10.1108/13552540910943397.

Ngoveni, A S, Abimola Patricia I Popoola, Nana K Arthur, and Sisa Lesley Pityana. 2019. "Residual Stress Modelling and Experimental Analyses of Ti6Al4V ELI Additive Manufactured by Laser Engineered Net Shaping." *Procedia Manufacturing* 35: 1001–1006. doi:10.1016/j.promfg.2019.06.048.

Nicoletto, Gianni, Radomila Konečná, Martin Frkáň, and Enrica Riva. 2018. "Surface Roughness and Directional Fatigue Behavior of As-Built EBM and DMLS Ti6Al4V." *International Journal of Fatigue* 116: 140–148. doi:10.1016/j.ijfatigue.2018.06.011.

Olsen, Flemming Ove, and Leo Alting. 1995. "Pulsed Laser Materials Processing, ND-YAG versus CO2 Lasers." *CIRP Annals* 44 (1): 141–145. doi:10.1016/S0007-8506(07)62293-8.

Owsiński, Robert, and Adam Niesłony. 2018. "Fatigue Properties in Additive Manufacturing Methods Applying Ti6Al4V." *AIP Conference Proceedings* 2029 (1). American Institute of Physics: 20049. doi:10.1063/1.5066511.

Pagano, Claudia. 2014a. "Precision." In *CIRP Encyclopedia of Production Engineering*, 968–972.

Pagano, Claudia. 2014b. "Positioning." In *CIRP Encyclopedia of Production Engineering*, 962–968.

Paris, Henri, Hossein Mokhtarian, Eric Coatanéa, Matthieu Museau, and Inigo Flores Ituarte. 2016. "Comparative Environmental Impacts of Additive and Subtractive Manufacturing Technologies." *CIRP Annals* 65 (1): 29–32. doi:10.1016/j.cirp.2016.04.036.

Pfeifer, Ronny, Dirk Herzog, Michael Hustedt, and Stephan Barcikowski. 2010. "Pulsed Nd:YAG Laser Cutting of NiTi Shape Memory Alloys—Influence of Process Parameters." *Journal of Materials Processing Technology* 210 (14): 1918–1925. doi:10.1016/j.jmatprotec.2010.07.004.

Pratt, P, S D Felicelli, L Wang, and C R Hubbard. 2008. "Residual Stress Measurement of Laser-Engineered Net Shaping AISI 410 Thin Plates Using Neutron Diffraction." *Metallurgical and Materials Transactions A* 39 (13): 3155–3163. doi:10.1007/s11661-008-9660-9.

Puertas, Inaki, and C J Luis. 2004. "A Study of Optimization of Machining Parameters for Electrical Discharge Machining of Boron Carbide." *Materials and Manufacturing Processes* 19 (6). Taylor & Francis: 1041–1070. doi:10.1081/AMP-200035200.

Qidong, Geng, and Wang Wei. 2018. "Experimental Study on Planetary Electric Discharge Machining of Substrate of Large-Aperture Silicon Carbide (SiC) Mirrors." *The International Journal of Advanced Manufacturing Technology* 94 (1): 711–717. doi:10.1007/s00170-017-0933-6.

Rajeswari, R, and M S Shunmugam. 2020. "Finishing Performance of Die-Sinking EDM with Ultrasonic Vibration and Powder Addition through Pulse Train Studies." *Machining Science and Technology* 24 (2). Taylor & Francis: 245–273. doi:10.1080/10910344.2019.1636276.

Rasoanarivo, Fetra, Didier Dumur, and Pedro Rodriguez-Ayerbe. 2021. "Improving SLM Additive Manufacturing Operation Precision with H-Infinity Controller Structure." *CIRP Journal of Manufacturing Science and Technology* 33: 82–90. doi:10.1016/j.cirpj.2020.09.007.

Ray, Debajyoti, Asit Baran Puri, and Naga Hanumaiah. 2020. "Experimental Analysis on the Quality Aspects of Micro-Channels in Mechanical Micro Milling of Zr-Based Bulk Metallic Glass." *Measurement* 158: 107622. doi:10.1016/j.measurement.2020.107622.

Raza, Muhammad Huzaifa, Ahmad Wasim, Muhammad Asad Ali, Salman Hussain, and Mirza Jahanzaib. 2018. "Investigating the Effects of Different Electrodes on Al6061-SiC-7.5 Wt% during Electric Discharge Machining." *The International Journal of Advanced Manufacturing Technology* 99 (9): 3017–3034. doi:10.1007/s00170-018-2694-2.

Reeves, Philip E, and Richard C Cobb. 1997. "Reducing the Surface Deviation of Stereolithography Using In-process Techniques." *Rapid Prototyping Journal* 3 (1). MCB UP Ltd: 20–31. doi:10.1108/13552549710169255.

Rosso, Stefano, Federico Uriati, Luca Grigolato, Roberto Meneghello, Gianmaria Concheri, and Gianpaolo Savio. 2021. "An Optimization Workflow in Design for Additive Manufacturing." *Applied Sciences*. doi:10.3390/app11062572.

Safdar, Adnan, H Z He, Liu-Ying Wei, Anders Snis, and Luis E Chavez de Paz. 2012. "Effect of Process Parameters Settings and Thickness on Surface Roughness of EBM Produced Ti-6Al-4V." *Rapid Prototyping Journal* 18 (5). Emerald Group Publishing Limited: 401–408. doi:10.1108/13552541211250391.

Shah, R K, and Partha Pratim Dey. 2019. "Process Parameter Optimization of Dmls Process to Produce AlSi10Mg Components." *Journal of Physics: Conference Series* 1240 (1). IOP Publishing: 12011. doi:10.1088/1742-6596/1240/1/012011.

Sharma, Anuj, Suhas S Joshi, D Datta, and R Balasubramaniam. 2020. "Investigation of Tool and Workpiece Interaction on Surface Quality While Diamond Turning of Copper Beryllium Alloy." *Journal of Manufacturing Science and Engineering* 142 (2). doi:10.1115/1.4045721.

Shastri, Renu K, and Chinmaya P Mohanty. 2020. "Machinability Investigation on Nimonic C263 Alloy in Electric Discharge Machine." *Materials Today: Proceedings* 26: 529–533. doi:10.1016/j.matpr.2019.12.133.

Shen, Hong, Conghao Liao, Jing Zhou, and Kai Zhao. 2021. "Two-Step Laser Based Surface Treatments of Laser Metal Deposition Manufactured Ti6Al4V Components." *Journal of Manufacturing Processes* 64: 239–252. doi:10.1016/j.jmapro.2021.01.028.

Simons, Magnus. 2018. "Additive Manufacturing—a Revolution in Progress? Insights from a Multiple Case Study." *The International Journal of Advanced Manufacturing Technology* 96 (1): 735–749. doi:10.1007/s00170-018-1601-1.

Singh, Sudhanshu Ranjan, and Pradeep Khanna. 2021. "Wire Arc Additive Manufacturing (WAAM): A New Process to Shape Engineering Materials." *Materials Today: Proceedings* 44: 118–128. doi:10.1016/j.matpr.2020.08.030.

Sorgato, Marco, Rachele Bertolini, and Stefania Bruschi. 2020. "On the Correlation between Surface Quality and Tool Wear in Micro–Milling of Pure Copper." *Journal of Manufacturing Processes* 50: 547–560. doi:10.1016/j.jmapro.2020.01.015.

Spierings, Adriaan B, Nikolaus Herres, and Gideon Levy. 2011. "Influence of the Particle Size Distribution on Surface Quality and Mechanical Properties in AM Steel Parts." Edited by Dave Bourell and Brent Stucker. *Rapid Prototyping Journal* 17 (3). Emerald Group Publishing Limited: 195–202. doi:10.1108/13552541111124770.

Sreenivasa Rao, Maddina, and N Venkaiah. 2017. "A Modified Cuckoo Search Algorithm to Optimize Wire-EDM Process While Machining Inconel-690." *Journal of the Brazilian*

Society of Mechanical Sciences and Engineering 39 (5): 1647–1661. doi:10.1007/s40430-016-0568-9.

Strano, Giovanni, Liang Hao, Richard M Everson, and Kenneth E Evans. 2013. "Surface Roughness Analysis, Modelling and Prediction in Selective Laser Melting." *Journal of Materials Processing Technology* 213 (4). Elsevier: 589–597. doi:10.1016/j.jmatprotec.2012.11.011.

Thawari, G, J K Sarin Sundar, Govindan Sundararajan, and Shrikant V Joshi. 2005. "Influence of Process Parameters during Pulsed Nd:YAG Laser Cutting of Nickel-Base Superalloys." *Journal of Materials Processing Technology* 170 (1): 229–239. doi:10.1016/j.jmatprotec.2005.05.021.

Vázquez, Elisa, Ciro A Rodríguez, Alex Elías-Zúñiga, and Joaquim Ciurana. 2010. "An Experimental Analysis of Process Parameters to Manufacture Metallic Micro-Channels by Micro-Milling." *The International Journal of Advanced Manufacturing Technology* 51 (9): 945–955. doi:10.1007/s00170-010-2685-4.

Venkatarao, K, and T Anup Kumar. 2019. "An Experimental Parametric Analysis on Performance Characteristics in Wire Electric Discharge Machining of Inconel 718." *Proceedings of the Institution of Mechanical Engineers, Part C: Journal of Mechanical Engineering Science* 233 (14). IMECHE: 4836–4849. doi:10.1177/0954406219840677.

Vidya, Shrikant, Reeta Wattal, and P Venkateswara Rao. 2021. "Investigation of Machining Performance in Die-Sinking Electrical Discharge Machining of Pentagonal Micro-Cavities Using Cylindrical Electrode." *Journal of the Brazilian Society of Mechanical Sciences and Engineering* 43 (6): 288. doi:10.1007/s40430-021-03012-6.

Wang, Di, Yang Liu, Yongqiang Yang, and Dongming Xiao. 2016. "Theoretical and Experimental Study on Surface Roughness of 316L Stainless Steel Metal Parts Obtained through Selective Laser Melting." *Rapid Prototyping Journal* 22 (4). Emerald Group Publishing Limited: 706–716. doi:10.1108/RPJ-06-2015-0078.

Wang, Hao, Sandy To, Chang Yuen Chan, C F Cheung, and W B Lee. 2010. "A Theoretical and Experimental Investigation of the Tool-Tip Vibration and Its Influence upon Surface Generation in Single-Point Diamond Turning." *International Journal of Machine Tools and Manufacture* 50: 241–252. doi:10.1016/j.ijmachtools.2009.12.003.

Wang, Di, Shibiao Wu, Yongqiang Yang, Wenhao Dou, Shishi Deng, Zhi Wang, and Sheng Li. 2018. "The Effect of a Scanning Strategy on the Residual Stress of 316L Steel Parts Fabricated by Selective Laser Melting (SLM)." *Materials*. doi:10.3390/ma11101821.

Wang, Minghai, Wei Wang, and Zesheng Lu. 2012. "Anisotropy of Machined Surfaces Involved in the Ultra-Precision Turning of Single-Crystal Silicon—a Simulation and Experimental Study." *The International Journal of Advanced Manufacturing Technology* 60 (5): 473–485. doi:10.1007/s00170-011-3633-7.

Wang, Sujuan, Senbin Xia, Hailong Wang, Ziqiang Yin, and Zhanwen Sun. 2020. "Prediction of Surface Roughness in Diamond Turning of Al6061 with Precipitation Effect." *Journal of Manufacturing Processes* 60: 292–298. doi:10.1016/j.jmapro.2020.10.070.

Wüst, Philipp, André Edelmann, and Ralf Hellmann. 2020. "Areal Surface Roughness Optimization of Maraging Steel Parts Produced by Hybrid Additive Manufacturing." *Materials* 13 (2). doi:10.3390/ma13020418.

Yang, Tao, Tingting Liu, Wenhe Liao, Eric MacDonald, Huiliang Wei, Xiangyuan Chen, and Liyi Jiang. 2019. "The Influence of Process Parameters on Vertical Surface Roughness of the AlSi10Mg Parts Fabricated by Selective Laser Melting." *Journal of Materials Processing Technology* 266: 26–36. doi:10.1016/j.jmatprotec.2018.10.015.

Yang, Tao, Tingting Liu, Wenhe Liao, Huiliang Wei, Changdong Zhang, Xiangyuan Chen, and Kai Zhang. 2021. "Effect of Processing Parameters on Overhanging Surface Roughness during Laser Powder Bed Fusion of AlSi10Mg." *Journal of Manufacturing Processes* 61: 440–453. doi:10.1016/j.jmapro.2020.11.030.

Yao, Yang, Hongtao Zhu, Chuanzhen Huang, Jun Wang, Pu Zhang, and Peng Yao. 2019. "On the Relations between the Specific Cutting Energy and Surface Generation in Micro-Milling of Maraging Steel." *The International Journal of Advanced Manufacturing Technology* 104 (1): 585–598. doi:10.1007/s00170-019-03911-y.

Yip, Wai Sze, and Suet To. 2019. "Theoretical and Experimental Investigations of Tool Tip Vibration in Single Point Diamond Turning of Titanium Alloys." *Micromachines* 10 (4). Multidisciplinary Digital Publishing Institute: 231. doi:10.3390/mi10040231.

Zhang, Shuo, and Wenjun Zong. 2020. "A Novel Surface Roughness Model for Potassium Dihydrogen Phosphate (KDP) Crystal in Oblique Diamond Turning." *International Journal of Mechanical Sciences* 173: 105462. doi:10.1016/j.ijmecsci.2020.105462.

Zou, Zhixiang, Zhongning Guo, Qinming Huang, Taiman Yue, Jiangwen Liu, and Xiaolei Chen. 2021. "Precision EDM of Micron-Scale Diameter Hole Array Using in-Process Wire Electro-Discharge Grinding High-Aspect-Ratio Microelectrodes." *Micromachines*. doi:10.3390/mi12010017.

2 Unique Properties of AM Metals

2.1 MATERIAL CHARACTERIZATION AND TESTING TECHNIQUES

The purpose of material characterization and testing techniques is to understand the chemical composition, microstructure, mechanical performance, and the interrelationships among each property. Characterization mainly comprises analysis and test methods on material chemical composition, microstructure, physical properties, mechanical properties, and so forth. In other words, it is used to reveal the basic properties of the different materials that will serve as key indicators of the material performance and subsequently aid design and fabrication considerations. Alternatively, the material composition and microstructure may also be correspondingly modified to account for the limitations of manufacturing processes. By establishing the material characteristics, an optimized plan can be developed to process the material and obtain the most economical and reasonable design during production of high-quality, lightweight, long-lasting, and reliable products.

Before diving into the characterization of additively manufactured metals, this section will first provide an outlay of the different test methods that are commonly used in material characterization. For the characterization of additive manufacturing, the tests are categorized as:

- Chemical composition
- Microstructure and phase composition
- Physical properties
- Mechanical properties

2.1.1 CHEMICAL COMPOSITION ANALYSIS

Additively manufactured metal parts generally undergo high temperatures during fabrication. On the one hand, the high temperature in the melt pool formed during the additive manufacturing process will cause the evaporation of some elements (e.g., Al and Mg). On the other hand, some elements (e.g., O, C, H, and N) will be

DOI: 10.1201/9781003272601-2

adsorbed into the molten pool and may undergo chemical reactions. It is known that the change in chemical composition may highly affect the final performance of the additively manufactured parts. The segregation of chemical elements and formation of compounds may occur in the final parts, which needs a detailed analysis of their composition. Therefore, chemical composition analysis is necessary in metal additive manufacturing.

2.1.1.1 Spectroscopic Analysis

(1) Atomic Absorption Spectrometry (AAS)

AAS is an instrument to determine the proportion of elemental contents within samples. Concentrations are determined based on the amount of light absorbed in the sample that matches the ground state atom resonance of a specific element of interest in its vapor phase. The increase in absorbed light will indicate a higher concentration of the element under study. The technique is instrumental in determining the content composition of the known elements within a sample.

(2) Inductively Coupled Plasma Atomic Emission Spectrometry (ICP-AES)

ICP is used as the excitation source to analyze specific elements according to the characteristic spectral line emitted when atoms of the target element return to the ground state from excitation. It can perform simultaneous analysis of multiple elements and is suitable for detecting approximately 70 elements. It has high precision with good stability (relative deviation <1%) but a low detection limit of approximately $10^{-1}\sim10^{-5}$ µg/cm^{-3} and poor detection sensitivity for non-metallic elements.

(3) X-Ray Fluorescence Spectrometry (XFS)

XFS is a non-destructive analytical method used to perform composition analysis on solid nanomaterial samples. There are two basic types of XFS systems: (i) wavelength dispersive type, and (ii) energy dispersive type. Measurements are qualitative and can detect all elements with atomic numbers greater than 3. The analytical sensitivity is high, and the detection limit reaches $10^{-5}\sim10^{-9}$ g/g (or g/cm^3). Additionally, the thin film thickness can also be measured ranging from several nanometers to tens of micrometers.

2.1.1.2 Mass Spectrometry Analysis

(1) Inductively Coupled Plasma Mass Spectrometry (ICP-MS)

ICP-MS is an elemental mass spectroscopy method that uses ICP as an ion source to determine the mass spectrum of elements within the sample. It has a low detection limit (ppb-ppt level for most elements), wide linear range (up to 7 orders of magnitude), fast analysis speed (results for 70 elements per minute), low interference in the spectrum to differentiate elements by 1 atomic weight, and capabilities of performing isotopic analysis.

(2)   Time of Flight Secondary Ion Mass Spectrometry (TOF-SIMS)

TOF-SIMS excites the surface of the sample with primary ions to reflect small amounts of secondary ions, which disperse according to the mass of the target elements. It is a high-resolution measurement technique that determines the mass of an ion by measuring the time taken for the secondary ions to reach the detector.

2.1.1.3   Energy Spectrum Analysis

(1)   X-Ray Photoelectron Spectroscopy (XPS)

XPS uses X-ray irradiation to excite the atoms and molecules on the sample surface so that electrons will be reflected and the energy distribution of the photoelectrons can be measured to obtain the chemical composition. Advances in microelectronic technology have also enabled XPS to target small areas with improved spatial resolution, and large areas for single-scan measurements. It is a powerful surface analysis instrument that not only detects surface element composition, but also the chemical state of elements within several atomic layers of surface energy band structure for depth distribution analysis of heterogenous layers, and micro-region chemical state imaging. It is one of the most common methods employed to study the electronic and atomic structure of the surface and interfaces of materials. It can detect all elements except hydrogen and helium.

(2)   Auger Electron Spectroscopy (AES)

AES employs the Auger effect using an electron beam (or X-ray) with designated energy to excite the sample and obtain information about the chemical composition and structure of the material surface by detecting the energy and intensity of the Auger electrons reflected from the surface. Qualitative and quantitative analysis of the surface composition of the samples in a micro-region is typically achieved in this process.

An Auger spectrometer combined with a low-energy electron diffractometer can analyze the surface composition and crystal structures of the sample, otherwise known as surface probing. In addition, the combination of electron microscopy (using an electron beam) and energy spectrum analysis (using X-rays) is an efficient method to perform energy spectrum analysis (EDAX) within a micro-area. Further characterization of micro-region composition can also be obtained when used in tandem with advanced electron microscopes (SEM, TEM).

2.1.2   MICROSTRUCTURE AND PHASE COMPOSITION

Due to the track-wise and layer-wise manufacturing method and high cooling rates, unique microstructures are formed during metal additive manufacturing, which include fine cellular structures, orientation-dependent epitaxial growth and process-dependent texture. These microstructures are different from that observed in the traditionally manufactured counterparts. Moreover, due to the complex thermal behavior (i.e., in-situ tempering, local and global heat flow, etc.), multiple phases may directly form in the as-built parts, which generally will not occur in the casting ones. It is widely known that microstructure determines macroscopic performance. Therefore, microstructure analysis must be performed on the additively manufactured metals for an in-depth understanding of the effect of process parameters and the evolution of part performance, as well as to provide a reference for the subsequent microstructure adjustment.

2.1.2.1   Microstructure Analysis
(1)   Optical Microscopy (OM)

OM is an optical instrument that employs fundamental physics of light to magnify and reflect or transmit optical images to microscopes that can be further digitalized with cameras. Samples are generally opaque where light is emitted onto the object from above, while light reflected off the sample enters the lens of the microscope for observation. The method is most effective for solid samples and is used in most fields of engineering. Translucent and transparent material surfaces may also be visualized to a certain extent. These upright microscopes are also called metallographic microscopes when used for observation of material microstructure.

(2)   Scanning Electron Microscopy (SEM)

SEM image surface features range from several nanometers to millimeters in size, with a large observation field and an approximate resolution of 6 nm. Field emission scanning electron microscopes (FE-SEM) can reach superior spatial resolutions of up to 0.5 nm. These capabilities allow the observation of intricate features and morphology of conductive material, such as the presence of defects (i.e., cracks, pits, etc.) and nano-sized particles. With the appropriate software, the SEM can also be fitted with elemental composition detection and phase structure analysis on the region of interest.

(3)   Transmission Electron Microscopy (TEM)

TEM operates like SEM but with a far superior spatial resolution capability and is particularly used for the analysis of smaller regions at higher magnification. Transmission electron microscopy is more suitable for the morphology analysis of nano-powder samples, but the particle size should be less than 300 nm, otherwise the electron beam cannot pass through. For the analysis of bulk samples, TEM generally requires thinning the samples.

TEM can be used to observe the size, shape, distribution, distribution range of particles, and calculate particle size using a statistical average method. Generally, an electron microscope observes the size of product particles instead of grain size. In contrast, high-resolution electron microscopy (HRTEM) can directly observe the microcrystalline structure, especially providing an effective means for the analysis of interface atomic structure. It can also observe the solid appearance of tiny particles. According to the crystal morphology and the corresponding diffraction pattern, high-resolution images can be employed to study the direction of crystal growth.

(4)   Scanning Tunneling Microscopy (STM)

STM uses a sharp conductive tip that is close to the sample surface and a voltage is applied to both the tip and the sample such that electron tunneling will occur within the vacuum space between the two. A tunnel current is then generated and used to determine the surface profile and electronic structure of the sample (i.e., local density of states). It has an atomic-level resolution (0.1 nm parallel and 0.01 nm perpendicular to the surface) to achieve near-atomic imaging of the sample such that the crystal surface can be observed alongside a three-dimensional image of the surface. However, stringent ambient conditions are required, such as ultra-high vacuum and extremely low operating temperatures. Additionally, it is only suitable for the analysis of thin films and conductive materials.

2.1.2.2   Phase Analysis

(1)   X-Ray Diffraction Analysis (XRD)

XRD employs X-rays to perform phase analysis based on the diffraction effect and can be used on mono- and polycrystalline samples. Qualitative and quantitative crystal phase, grain size, mesoporous structure, multilayer film analysis, crystalline structure, and bonding state are common output results from XRD. Despite the many types of results that can be obtained, XRD has low sensitivity and is only capable of detecting elements that are >1% content in the sample. There is also a minimum size of the sample (larger than 0.1 g) for accurate measurement. Lastly, it is ineffective for amorphous materials.

(2)   Raman Analysis

Incident light energy directed onto a sample will interact with the surface atoms and cause scattering of photons. As some energy is absorbed by the substrate atoms, there will be a change in the scattered light direction and frequency, which are used in determining elemental signatures during Raman spectroscopy. Light captured with no changes in frequency is typically defined as Rayleigh scattering, while light that changes in both direction and frequency is defined as Raman scattering. These changes are attributed to molecular vibrations, atomic-level interactions (e.g., electron transfer, phonon, plasma generation, etc.), and interaction with defects and impurities. Using the concept of energy transfer between photons and molecules, Raman spectroscopy can be used to analyze the molecular structure, physical and chemical properties, and qualitatively identify materials, and can reveal information about vacancies, interstitial atoms, dislocations, grain boundaries, and phase boundaries in materials.

(3)   Infrared (IR) Analysis

Infrared spectroscopy using the Fourier transform is typically used to detect organic functional groups. This is useful for checking chemical environment conditions and changes, such as the bonding of metal ions and non-metal ions, and the coordination of metal ions.

*(4)   Micro-Area Electron Diffraction Analysis (Selected Area Electron
       Diffraction, SAED)*

Like X-rays, electron diffraction also follows Bragg's equation to determine the
phase and orientation relationships, as well as structural defects in materials. As the
beam employed in SAED is extremely small, analysis is typically performed for
micro-regions.

2.1.3   PHYSICAL PROPERTIES

2.1.3.1   Thermal Properties

(1)   Thermal Capacity

Differential scanning calorimetry (DSC) is a versatile method used to measure the
heat flow that occurs when a sample is heated, cooled, or held at a constant temperature. This is the measurement of specific heat capacity of the amount of heat
energy needed to change the temperature of the material. Conventional DSCs typically measure temperatures up to 700 °C while more advanced equipment is needed
for accurate measurements of the heat capacity above this temperature.

(2)   Thermal Expansion

The mandrel method is a classic technique used to measure the thermal expansion
of a material, which is the change in volume or length of a material with temperature change, and uses mechanical measurement. Thermal expansion occurs with the
change in distance between atoms within a crystal lattice, which sums up to result in
volumetric differences as a function of temperature or applied heat. The expansivity is
usually defined in terms of linear expansion or volumetric expansion. In the mandrel
technique, one end of the sample is fixed on a holder while the other end is in contact
with the mandrel. Heat is applied to the setup and the differential thermal expansion
of the part is transmitted to an ejector pin where the displacement is measured. There
are many variations of the mandrel setup where samples may be positioned vertically
or horizontally. The measurement may be conducted by direct mechanical displacement, and even electronic or optical observation. However, in most cases, it is challenging to achieve uniform heating conditions in the furnace for non-biased thermal
expansion measurements. Correction algorithms are typically employed to handle
these experimental flaws.

Optical measurements come in the form of telescope reading or more advanced
laser measurement. In telescopic readings, a set of binoculars are used to observe
the expansion and determine the linear expansion coefficient through calculation.
Measurement temperatures of these systems can reach up to 2,000 °C. However,
automated measurement systems based on this concept are difficult to achieve.
Therefore, laser measurements are preferred, which scan the sample and continuously measure the expansion throughout the heating process. It has high accuracy
and automated control and recording. Hence, it is evident that the temperature range,
measurement accuracy, and size of the sample are important considerations to determine the measurement method.

(3) Thermal Conduction

Thermal conductivity is another important material property that serves many engineering purposes and refers to the transfer of thermal energy by conduction per unit thickness of the material and temperature change. This property, however, does not account for other forms of heat transfer such as convection and radiation. The inclusion of these modes of heat transfer takes the form of many terms such as apparent thermal conductivity, explicit thermal conductivity, or thermal transmissivity. The property can be affected by other material characteristics such as porosity, multi-layers, multi-structural components, and anisotropies. Thermal conduction is also differentiated into different types as classified by several test standards from the American Society for Testing and Materials (ASTM) – ASTM D5470, ASTM E1461, and ASTM E1530.

ASTM D5470: A test method for the heat transfer characteristics of thermally conductive electrical insulating materials, using the steady-state heat flow method. The principle is to apply heat flow and pressure to the test sample of defined thickness using a hot plate and a cold plate, while measuring the temperature difference between the hot and cold plates to determine the conductivity. Samples using this test standard are typically larger for observation of temperature difference.

ASTM E1461: A test method to determine the thermal diffusivity of solids using the flash method (laser flash method). Small and thin specimens are irradiated using a laser over a short period of time with high-energy pulses and the energy of the pulses is absorbed by the front surface of the sample, which results in a rise in temperature on the back surface. An infrared detector is used to measure the temperature of the rear surface, which is then used to calculate the thermal diffusivity based on the thickness of the specimen. With the thermal diffusivity, the thermal conductivity can then be calculated with additional information of the specific heat and density of the test sample.

ASTM E1530: A test method for evaluating the heat transfer resistance of materials using the heat-shielded heat flow meter technique. The test concept is like the heat flow thermal conductivity measurement with a difference in the heater being located around the measurement area to give an average temperature of the sample.

2.1.3.2 Electrical Properties

Electrical conductivity refers to the resistance to electrical current flow as measured using either the ammeter-voltmeter method, bridge method, potentiometer method, or the DC four-terminal method. Different levels of electrical resistance will classify the material as an insulator, semiconductor, or conductor. Examples of electrical testing equipment include the HP4284A precision LCR tester, HP4191A RF impedance analyzer, HP4329A high impedance meter, and KEITHLTY6517A electrometer. These devices can also be used to characterize the dielectric temperature spectrum, dielectric spectrum, volume resistivity, electric strength, hysteresis loop, dielectric properties under bias electric field, and pyroelectric coefficient, among others.

2.1.3.3 Magnetic Properties

Magnetism is a measure of the susceptibility of the material to be magnetized and be equipped with magnetic properties based on the magnetic moment per unit volume ($M=\chi \bullet H$) where χ is the magnetic susceptibility of the medium and H is the magnetic field strength (A/m). The different classes of materials susceptible to magnetization are categorized into: (i) diamagnetic materials with negative χ values, (ii) paramagnetic that is minutely magnetizable, (iii) ferrimagnetic materials with slightly larger χ values, (iv) ferromagnetic with large χ values typically the most magnetizable, and (v) antiferromagnetic materials that is magnetizable under the appropriate temperature conditions. Magnetic properties are measured using (i) the impact method to produce the magnetization curve and hysteresis loop and (ii) induction thermomagnetic instruments to measure the C curve.

2.1.4 MECHANICAL PROPERTIES

2.1.4.1 Hardness

Hardness is a mechanical property that refers to the ability of a material to resist deformation or rupture. There are over ten types of hardness testing methods that can be divided into two categories of macroscopic compression/tension and localized compression (e.g., indentation). In each of these categories, the type of load can also vary in terms of loading rate, dynamic loading, and static loading. Ultrasonic hardness, Shore hardness, and hammer-type Brinell hardness testing belong to the dynamic loading category while Brinell hardness, Rockwell hardness, and Vickers hardness are classified as static loading. Engraving tests are also used to manually evaluate the hardness of materials with the file knife method or scratching to identify the Mohs hardness. Each type of test method will have its range of hardness values whereas some of the more popular test methods have tables of comparison to refer between the different hardness scales. In other cases, such as engraving, the hardness may not be comparable due to the different type of deformation that occurs within the material.

2.1.4.2 Tensile Strength

The strength of a material is mostly obtained through tensile testing, that is, a test method used to determine several properties of a material under an axial tensile load. These properties include the elastic limit, elongation, elastic modulus, proportional limit, area reduction, tensile strength, yield point, yield strength, ultimate tensile strength, and so forth. As the setup is relatively simple, tensile tests on metals follow the ASTM E-8 standard to ensure comparable results between universal tensile strength testing machines.

2.1.4.3 Compression Strength

Compression testing employs the action of an axial static pressure acting on the entirety of a material. The maximum compressive load at which the specimen fails divided by the cross-sectional area of the specimen is defined as the compressive strength limit or compressive strength. Compression testing is mainly applicable to brittle materials, such as cast iron, bearing alloys, and building materials. For plastic materials, the compressive strength limit cannot be accurately measured, but the elastic modulus, proportional limit and yield strength can still be obtained. Similar to the tensile test, a compression curve can be plotted during the course of the compression test, which follows the internationally recognized ASTM E-9 standard for metal testing.

2.1.4.4 Torsional Strength

The ability of a material to resist an applied torque is characterized by its torsional strength, which can be measured through a torsion test. Additional properties obtained from the test include the torsional stiffness and plastic deformation capacity. The test follows the ASTM A938 standard with a torque applied to both ends of a sample on a torsional test machine and is most often used for shafts and springs that are subjected to these moments and torque during service.

2.1.4.5 Bending

Aside from the basic uniaxial tensile and compression tests to characterize material strength, bending tests are slightly more complex at evaluating material properties subjected to more appropriate loading conditions that resemble actual engineering applications. During bending, the material is subjected to tension on the upper surface while the lower surface is subjected to compression, which will both be used to determine the maximum normal stresses of the material and its geometry. Three-point bending and four-point bending tests are the most common test machine configurations and the sample is typically round or rectangular with the length of the beams approximately 10 times larger than the cross-sectional diameter. These tests are typically used to evaluate the ductility of tough materials (e.g., cast iron, high-carbon steel, tool steel, etc.) that are used as structural components. They are also used to evaluate the bending fracture strength of brittle materials that are only able to sustain a minute degree of deformation before failure. For consistency, bending tests follow the ASTM E290 standard for comprehensive evaluations.

2.1.4.6 Toughness

Impact toughness refers to the ability of a material to absorb plastic deformation work and fracture work for a given impact load. The practical significance of the impact toughness index is to reveal the brittle tendency of the material and to reflect the resistance of the metal material to external impact loads. It is generally expressed by the impact toughness value (ak) and impact energy (Ak) with units of J/cm^2 and J, respectively. Impact toughness testing follow the international standard ASTM E23.

2.1.4.7 Fatigue Strength

Aside from material failure based on instantaneous loading, materials may fail over prolonged use when repeatedly subjected to cyclical plastic deformation or micro-plastic deformation. Defects (e.g., cracks) begin to form in the material and serve as weak points that consequently result in the failure of the material below its typical loading limits. Fatigue strength refers to the maximum allowable stress of a material under repeatedly applied loads before failure, and is also called fatigue limit. The fatigue life of a part is related to the stress and strain levels of the part, and their relationship can be represented by a stress-life curve (σ-N curve) and a strain-life curve (δ-N curve). The stress-life curve and the strain-life curve are collectively referred to as the S-N curve. Testing equipment follows ASTM E466 standards with the use of a fatigue testing machine and a set of Vernier calipers.

2.1.4.8   Friction and Wear

Friction and wear are common phenomena in daily life where contact between two parts exist. They can be found in rockets, airplanes, automobiles, machine tools, and human bodies. Typical examples are shafts, bearings, gears, and joints. Any relative motion between two components will be subjected to a certain degree of friction and wear, which causes the loss of the surface material. The type of surface damage is characterized by the type of wear, such as abrasive, adhesive, contact fatigue, fretting, cavitation, and so forth. Wear tests are typically performed by repeatedly scribing a material until failure and follow the ASTM G99 standard. Measures of wear can be determined by surface characterization or mass loss of the sample.

2.1.4.9   Creep

Unlike fatigue testing where the material is subjected to cyclical loading conditions, creep strength is the evaluation of a material under constant stress but over a prolonged period of time. Materials tend to gradually fail even when subjected to stresses below the elastic limit as long as the stress is applied over a prolonged period. Creep deformation gradually occurs as dislocations slip and atoms diffuse into favorable configurations under the applied stress. The creep limit or endurance strength limit is the result obtained from creep tests performed in accordance with the ASTM E139 standard.

Creep limits are also typically evaluated under high temperatures, which simulate most operating conditions of engineering materials. For these tests, creep limits are determined based on the maximum stress that a material can withstand for specified temperatures and loading periods. Another form of evaluating the material creep limit is by measuring the creep elongation over the test period. The endurance strength limit of metals is determined through high-temperature tensile endurance testing. The test method is similar to that of determining the creep limit with a difference in that the specimen is loaded until failure.

2.2   CHARACTERIZATION AND TESTING OF AM METALS

Having established that metal additive manufacturing is a promising process to revolutionize manufacturing (see Chapter 1), it is important to now discuss the various methods that have been employed to analyze these materials. With the differences in heating and cooling of the metal during fabrication, the microstructure and material properties can be expected to be different from their traditionally manufactured counterparts. Material characterization of these metals will serve as a reference for further advances in scientific research and industrial applications. For AM materials, the most common investigations include the surface topography, internal defect morphology, chemical composition, microstructure, phase analysis, and mechanical properties. Hence, this section will briefly describe factors contributing to each material aspect and cover the various methods that have been employed to characterize additively manufactured metals.

2.2.1 SURFACE MORPHOLOGY

During the metal additive manufacturing process, metal powders are melted to form molten pools of material during construction. As the heat source moves across the building platform to construct the part, periodic formation and solidification of the molten pool lead to the formation of melt tracks that tend to overlap each other when forming each layer. The molten pools, melt tracks, and overlapping layers result in a complex interaction with each other, which consequently affect the final surface morphology with the combination of remnant surface features arising from each occurrence. While analyzing the surface quality of as-built metal parts is useful for understanding the limitations of AM for direct engineering applications, the investigation is also useful for understanding the manufacturing mechanisms such as the molten pool behavior during the SLM process that describes the flow of material during heating and melting. Therefore, the characterization of surface morphology on additively manufactured metals is a very attractive topic to researchers.

A large depth of field is a key factor for obtaining a clear surface image on as-built AM metal surfaces due to the high peak-to-valley features. Most optical microscopes should be able to observe the surface of an as-built metal but newer models that feature ultra-depth-of-fields such as the Keyence VHX-6000 that allows for the collection of greater details using a combination of white light interferometry (WLI) and confocal microscopy (Bai, Zhao, Wang, et al. 2022). These advanced microscopes are capable of not only capturing top-view 2D images but also providing the corresponding 3D-surface topographies and calculation of surface roughness measurements. SEMs are also commonly used to observe the surface morphologies (see Figure 2.1) and are typically used in combination with a laser confocal microscope to profile the 3D surface (Hirt et al. 2017).

As post-processing is most often needed to lower the roughness of as-built metal parts (Maleki et al. 2021), considerations for surface profilers should also include the range of roughness achievable on the metal surfaces after post-processing. Fortunately, most of the modern surface analysis equipment (i.e., the SEM, WLI, laser confocal microscope, etc.) can accommodate the wide range of surface roughness stretching from the as-built morphology to the finished surfaces.

2.2.2 INTERNAL DEFECTS

Internal defects are important contributors to the mechanical properties of the material and are often characterized using 2D cross-sectional views for time and cost

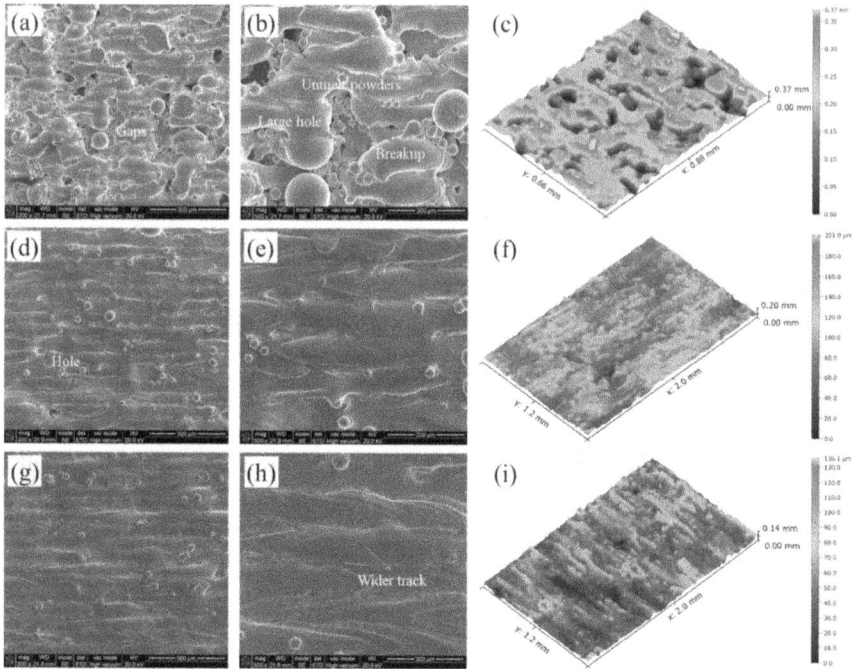

FIGURE 2.1 SEM and 3D-surface morphologies on the top surface under different laser power: (a, b, c) 100 W; (d, e, f) 160 W; (g, h, i) 200 W (scanning speed 500 mm/s, hatch spacing 85 μm, layer thickness 35 μm). Reprinted with permission from Elsevier (Bai, Zhao, Wang, et al. 2022).

efficiency. Optical microscopes and SEMs are common equipment used to reveal the defects found on the cross-sections that are typically grinded and polished before observation. Recent technological advances have also included in-situ X-ray imaging to observe the dynamic formation of internal defects (e.g., holes) within the molten pools during the AM construction process. A schematic of a laser powder bed fusion configured with an in-situ high-speed and high-energy X-ray imaging system is shown in Figures 2.2(a) and 2.2(b). With X-ray radiography, defects such as powder consolidation (Figures 2.2(c–f)), spattering (Figure 2.2(g)), and porosity (Figure 2.2(h)) can be clearly visualized during the AM process (Leung et al. 2018).

The distribution of internal defects can also be visualized in 3D using computed tomography (CT) scanning, which is common in the AM field so much that the potential for CT scanning being a quality control measure for AM was proposed (Karme et al. 2015). Advanced variations of CT technology such as micro-CT were also shown to be effective in determining internal pore distribution, which greatly assisted process optimization and proof of concept for quality assurance of AM parts (Du Plessis et al. 2019). These internal defects may form in different shapes and sizes (Figure 2.3), which can be easily observed through CT scanning (Wang et al. 2020). The corresponding volume of defects can then be derived using integrated software to quantify the quality of the AM part.

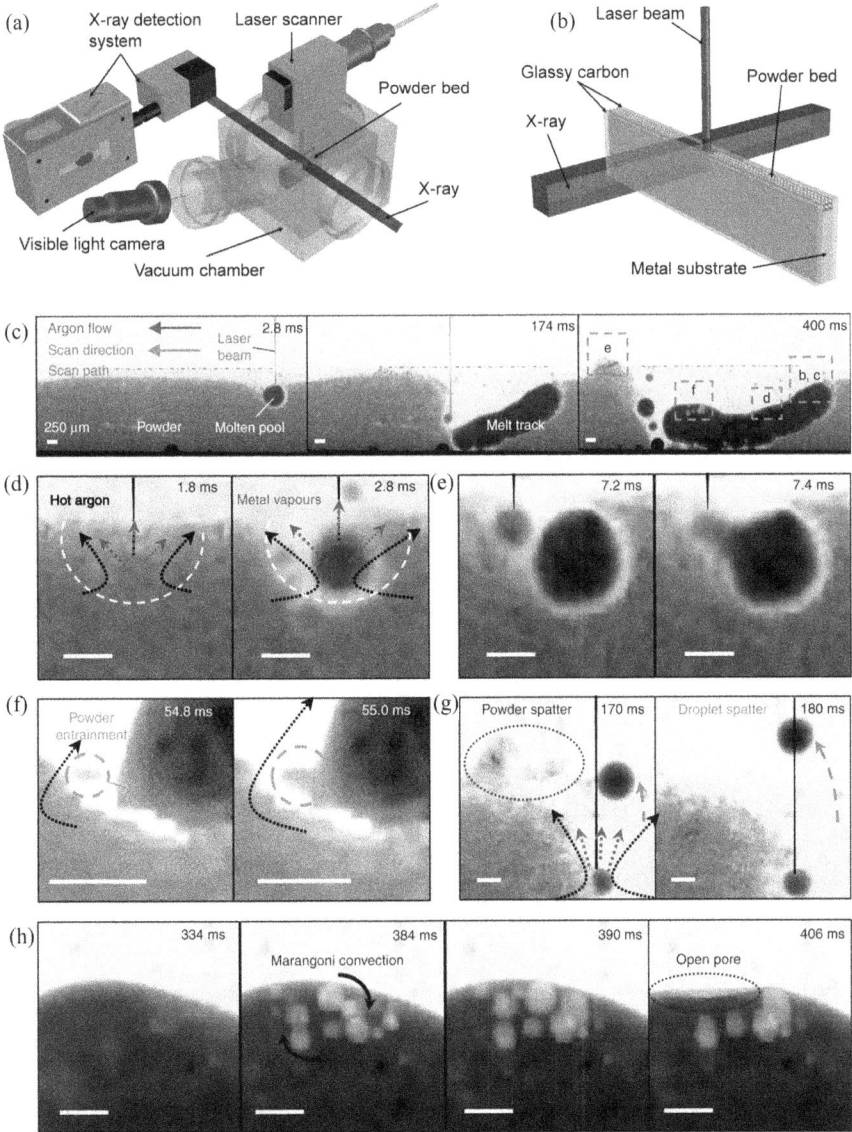

FIGURE 2.2 Schematic of the experimental setup for in-situ laser powder bed fusion high-speed high-energy X-ray imaging: (a) Schematic of the whole system. The experiments are carried out in a vacuum chamber with refilled argon gas. The powder bed (the sample) is mounted on a motion stage (not shown in the image) and placed on the X-ray beam path. The laser beam is applied from the top of the powder bed and controlled by a galvanometer scanning system. An off-axis visible light camera is placed 45° to the powder bed surface for sample alignment. (b) Schematic of powder bed assembly. A layer of metal powder is applied on top of the metal substrate (build plate) made of the same material as the powder. To ensure X-ray transparency along the X-ray beam path, glassy carbon is used for the wall of the powder bed.

FIGURE 2.2   (Continued)

Reprinted with permission from Elsevier (Guo et al. 2018); Time-series radiographs acquired during LAM of an Invar 36 single layer melt track (MT1) under P = 209 W, v = 13 mm s^{-1} and LED = 16.1 J mm^{-1}: (c) The melt track morphology at three key stages of LAM. (d) The formation of a molten pool and a denuded zone (dotted line). The laser beam causes metal vaporization, generating a recoil pressure at the interaction zone (dark dotted arrows) while indirectly heating up the surrounding argon gas (light dotted arrows). The molten pool/track grows while enlarging the denuded zone by (e) molten pool wetting and (f) vapor-driven powder entrainment (dashed circles) which can lead to the formation of (g) powder spatter (dotted ellipse) and droplet spatter (its trajectory path is indicated by the dashed arrows). After the laser switches off at t = 334 ms, (h) pores nucleate, coalesce and collapse, resulting in an open pore (dotted ellipse). All scale bars = 250 μm. Reprinted with permission from Springer Nature (Leung et al. 2018).

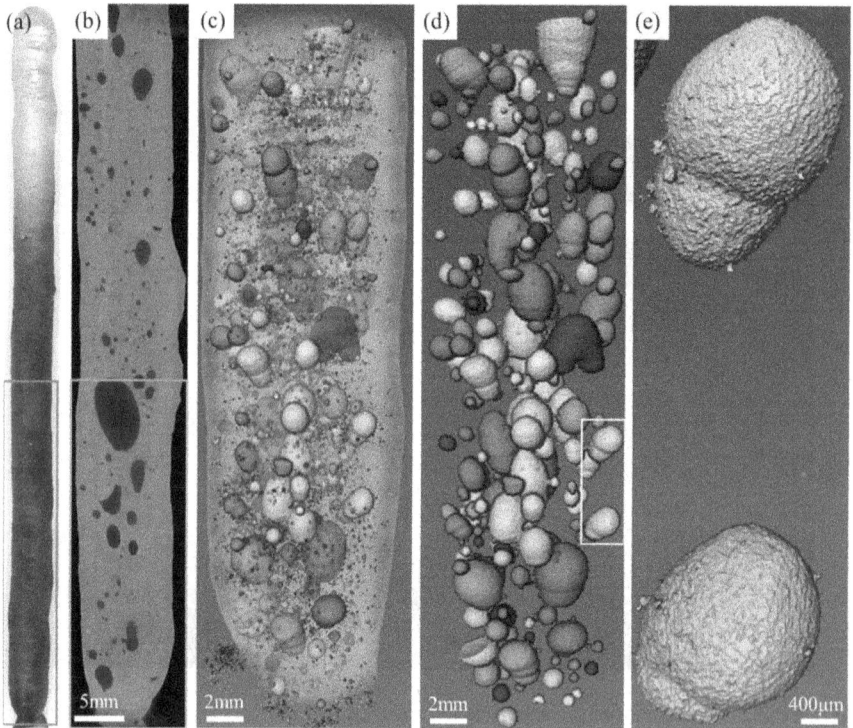

FIGURE 2.3   (a) Sample fabricated with I_p = 200 A and I_B = 60 A; (b) An original scan tomogram of the boxed area in (a); (c) The reconstructed image of the boxed area in (b). For better visualization; (d) is the image of (c) without the SSPs; (e) The image of the boxed area in (d). Reprinted with permission from Elsevier (Wang et al. 2020).

As one of the most undesirable defects is porosity (i.e., the presence of holes or gaps in the material), density measurements using the Archimedes drainage method (Spierings, Schneider, and Eggenberger 2011) is also commonly employed for AM parts. Although it cannot provide critical information such as the size and distribution of the internal defects, it is a quick method to quantify the presence of internal defects by comparing with the standard density or relative density. The relative density is measured using the Archimedes principle that compares the weight of the sample in air and water, and is calculated as:

$$R = \frac{\rho_{ml}}{\rho_{m0}} \times 100\% = \frac{m_{air} \times \rho_{H_2O}}{\rho_{m0} \times \left(m_{can} - m_{H_2O}\right)} \times 100\% \qquad (2.1)$$

where, ρ_{ml} is the density of the AM part, ρ_{m0} is the standard density of the conventional counterpart, m_{air} is the weight of the sample, ρ_{H_2O} is the density of the water, m_{can} is the weight of the sample in air after wax sealing, and m_{H_2O} is the weight of the sample in water. This method has been successfully adopted to characterize the build quality of AM parts (Bai et al. 2017; Bai et al. 2018; Zhang et al. 2020).

2.2.3 CHEMICAL COMPOSITION AND ELEMENT DISTRIBUTION

While different melting and cooling rates employed in additive manufacturing are expected to result in changes in material properties, the combination of these changes with the different mix of elements within the powders and multi-materials further complicates the material evolution process during AM. This gives rise to the importance in confirming the element distribution and classification of material after additive manufacturing.

In the AM field, the most common method to evaluate chemical composition is the energy dispersive spectroscopy (EDS), which is the measurement of X-rays emitted off the sample surface that has been irradiated with electrons during SEM/TEM analysis. The EDS software is typically configured to determine the elemental composition of the scanned surface by either detailing the concentration of elements at a designated point or location. With this information, element maps of the selected material elements can also be plotted out across the scanned areas by color coding the different elements detected within the scan. The information provides the element distribution of the sample, which will also be useful in reasoning for the various properties of AM parts. For instance, differences in corrosion can be explained to be associated with the cellular structures of the granular precipitates that were observed during EDS (Xie, Xue, and Ren 2020). Interfacial transition regions between multi-materials can also be detected by EDS (Figures 2.4(a) and 2.4(b)), which follow with element diffusion as observed in a 316LSS-C52400 copper combination (Bai, Zhang, et al. 2020). EDS can also be used in tandem with TEM to analyze materials at a smaller length scale (e.g., a 2 μm × 2 μm observation window), which enables a higher resolution of observing the way different elements aggregate within the microstructure of the material (Figures 2.4(c–h)) (Godec et al. 2021).

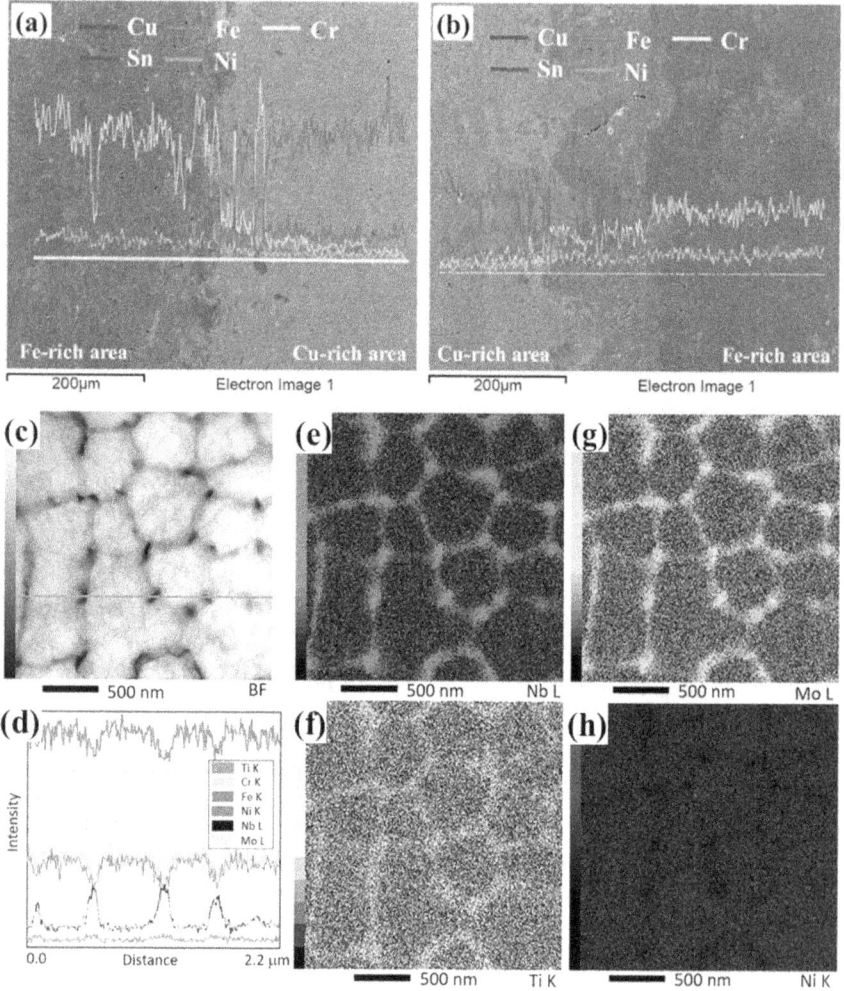

FIGURE 2.4   (a) EDS line scan at the C52400/316L interface, (b) EDS line scan at the 316L/ C52400 interface. Reprinted with permission from Elsevier (Bai, Zhang, et al. 2020); TEM/ EDS measurements of chemical analyses across cell boundaries, (c) STEM-BF image with marked line of EDS analyses, (d) STEM-EDS concentration profile of Ti, Cr, Fe, Ni, Nb, Mo, and STEM-EDS elemental mapping of (e) Nb, (f) Ti, (g) Mo, (h) Ni. Reprinted with permission from Elsevier (Godec et al. 2021).

2.2.4   MICROSTRUCTURE CHARACTERIZATION

As briefly mentioned earlier, the microstructure is as important as identifying the distribution of elements, which work together to determine the macroscopic mechanical properties. Microstructure metallurgy can be studied using a range of equipment depending on the amount of information required. Basic optical imaging of grain boundaries is the simplest form of analysis but requires etching of the sample surface

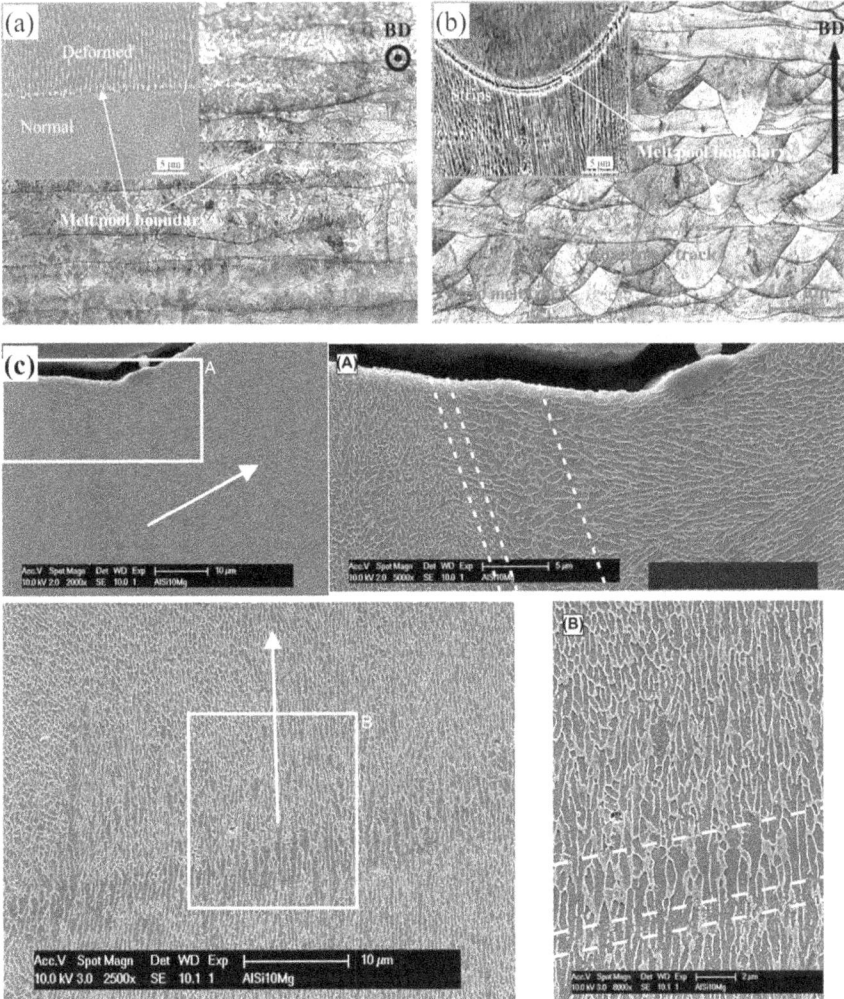

FIGURE 2.5 OM images of as-built additively manufactured maraging steel: (a) top surface, (b) side surface. Reprinted with permission from Elsevier (Bai et al. 2021); (c) microstructure of a AlSi10Mg part in front and side view by SEM Reprinted with permission from Elsevier (Thijs et al. 2013).

prior to analysis and is limited in terms of magnification and resolution. Yet, it has been proven capable of observing melt pool boundaries, as shown in Figures 2.5(a) and 2.5(b) (Bai et al. 2021), and large grained materials (Donoghue et al. 2016). SEM is also capable of observing grain boundaries with higher resolution but, like OM, requires that the sample surface is polished first. However, high cooling rates are involved in most AM processes, which results in the refinement of microstructure (i.e., microscopic grains) (Kong et al. 2021). These are not easily observable by OM, and advanced SEM equipment is required to image the grain morphology at this scale, such as the fine cellular–dendritic solidification structure (< 1 μm in size) and white fibrous Si particles shown in the analysis of AlSi10Mg (see Figure 2.5(c)) (Thijs et al. 2013).

Textures of AM metals may be very strong due to way the material is constructed (i.e., bottom-up building), which causes a thermal gradient along the building direction and a local thermal gradient towards the center of the melt pool. This results in severe element segregation and complex thermodynamic behavior, which can create difficulties in grain identification. OM and SEM technology is unable to capture grain information due to the strong textures and require advanced equipment to evaluate the material. Electron backscatter diffraction (EBSD) analysis can provide rich material information of the grain texture, misorientation between grains, grain size and shape distribution analysis, grain boundary, sub-grain, and twin boundary property analysis. It is a popular microstructure characterization method used for AM metal parts despite requiring more stringent sample preparation techniques such as ion milling for very fine polishing of the target surface. It should be noted that parameter selection for EBSD analysis such as the acceleration voltage and scanning step is critical to obtaining accurate results. As shown in Figures 2.6(a–f), EBSD was used to obtain the crystal features (e.g., inverse pole figures (IPF), grain boundaries, morphology, orientation, and grain size) where unique differences could be observed between AM scanning strategies and building directions (Zhao et al. 2021). AM 316L stainless steel was used as an example for the large variation in microstructure where the $0°$-XY plane consisted of large quadrilateral-like columnar grains arranged along the melt track, while the $90°$-XY and $67.5°$-XY planes contained square-like grains surrounded by many small grains. On the $0°$-XZ plane, the scanning direction of each constructed layer was consistent with elongated columnar crystals along the building direction. On the other hand, the $67.5°$-XZ plane showed very complex crystallographic orientations due to the disturbance of continuous grain growth along adjacent layers, which was induced by the different scanning direction of each layer.

Another advanced technique for microstructure characterization is TEM, which is the most expensive method among the listed techniques due to the high cost of sample preparation and machine operation (typically requiring strong technical expertise). TEM can display extremely fine sub-nano features with high resolution (<1 nm) to reveal dislocations, nano-precipitates, and even the crystal lattice of atoms. Using 316L stainless steel again as an example, TEM demonstrates the capabilities in observing different crystallographic textures and dislocation densities (higher density in sample manufactured with higher laser power) (Figures 2.6(g–j)), dislocation twinning (Figure 2.6(g–i)), stacking faults (Figure 2.6(k)), and even nano-twinning (Figure 2.6(l)) that allows for strength and ductility (Sun et al. 2018).

2.2.5 PHASE COMPOSITION AND TRANSFORMATION

X-ray diffraction (XRD) is the most common method used to obtain phase composition. Not only can it easily and quickly determine the types of phases in the material but it can also calculate the proportion of each phase. Within the XRD plots, the broadness of the peak intensities representing the different elements (e.g., the Si(111) and Al(111) peaks in AlSi10Mg in Figures 2.7(a) and 2.7(b))

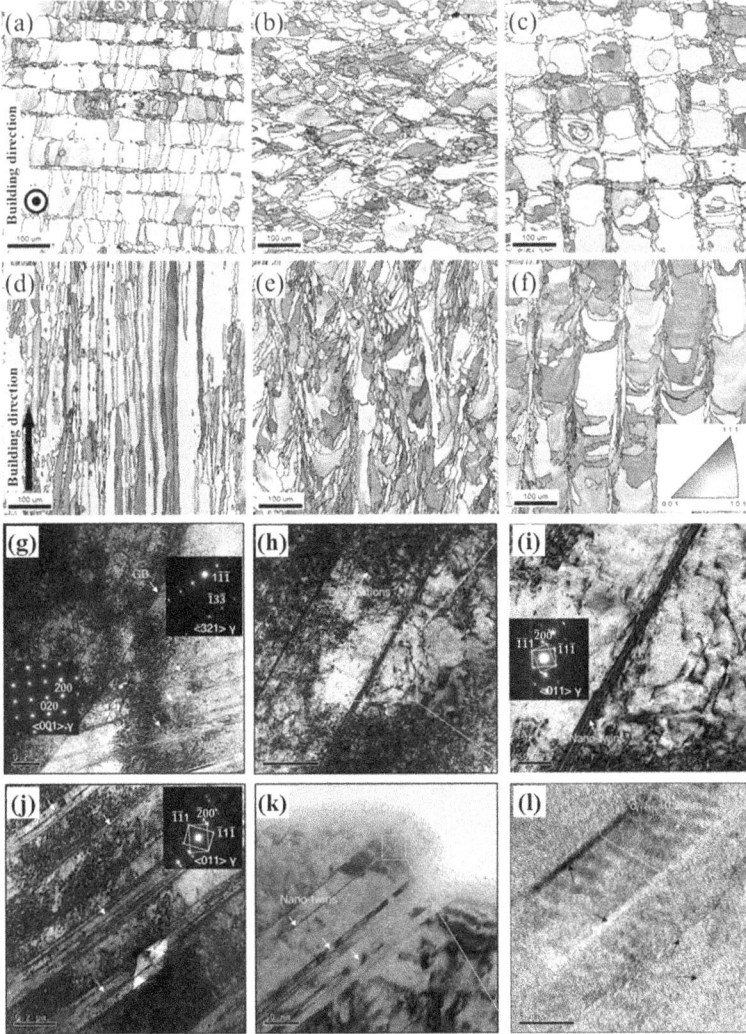

FIGURE 2.6 EBSD-IPF images with grain boundaries: (a, d) XY and XZ planes with 0° rotation, (b, e) XY and XZ planes with 67.5° rotation and (c, f) XY and XZ planes with 90° rotation. Reprinted with permission from Elsevier (Zhao et al. 2021); (g) TEM image showing two adjacent grains with <001> and <321> orientations in the deformed 380 W sample. The grain boundary (GB) is indicated. The right three arrows point to the single-twin lamellae. (h) TEM image showing dislocations and stacking faults between two single-twin lamellae in the deformed 380 W sample. (i) Enlarged TEM image of the enclosed region in b showing a nano-twin and surrounding stacking faults (SFs) along the <011> zone axis. (j) TEM image showing a large number of multiple twins indicated by arrows in the deformed 950 W sample along the <011> zone axis. (k) TEM image showing the multiple nano-twins in the deformed 950 W sample. (l) HRTEM image of the enclosed region in (k) showing the nano-twin lamellae with twin boundaries (TBs) and SFs. Scale bars for (g), (h), (j) are 200 nm, for (i), (k) 50 nm, and for f 5 nm. Reprinted with permission from Springer Nature (Sun et al. 2018).

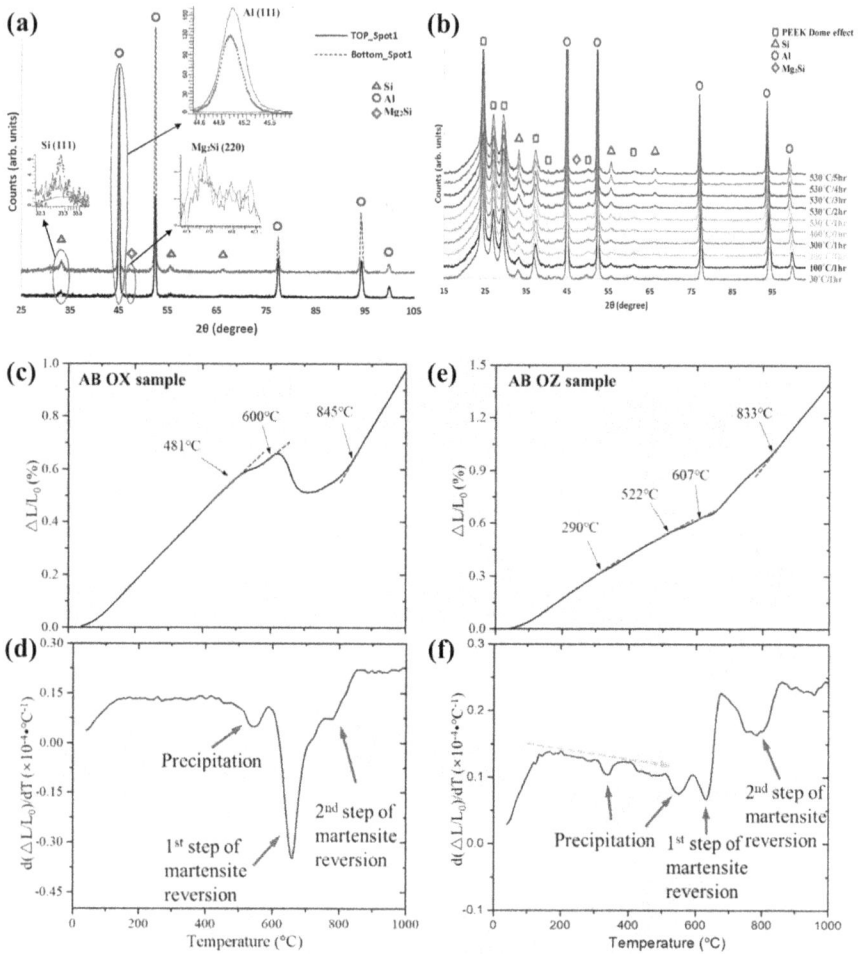

FIGURE 2.7 (a) XRD phase pattern comparison of the as-built AlSi10Mg sample for the top and bottom surfaces and (b) XRD phase pattern of the AlSi10Mg sample under different solution heat treatment conditions. Reprinted with permission from Elsevier (Maamoun et al. 2018); Thermal expansion curves of the as-built maraging steel sample along (c) OX direction (horizontal) and (e) OZ direction (vertical), and the corresponding CTE as a function of temperature along (d) OX direction and (f) OZ direction. Reprinted with permission from Elsevier (Bai, Zhao, Zhang, et al. 2022).

can be indicative of the crystal size, which may differ between the top and bottom surface of the AM metal (Maamoun et al. 2018). The change in concentration of precipitates (e.g., Mg_2Si in AlSi10Mg) with the change in temperatures can also be measured using XRD. Unfortunately, the measurement is only limited to constant temperatures and is unable to perform in-situ measurements with the change in sample temperature.

To understand dynamic phase transformation behavior, thermal expansion curves (Figures 2.7(c) and 2.7(e)) are used to record the volumetric change caused by the phase transformation (Bai, Zhao, Zhang, et al. 2022). In an example of maraging steel, two phase transitions can be found (Figures 2.7(d) and 2.7(f)), which show the change in coefficient of thermal expansion during heating. The first change in temperature (481–580 °C) represents the precipitation of fine particles, while the second change in temperature (600–845 °C) corresponds to the transformation from martensite to austenite.

2.2.6 Mechanical Properties

Mechanical properties are important indicators for evaluating the structural integrity of these materials to be used in machinery, transportation, aerospace, and construction, among others. Due to the bottom-up building process in AM, anisotropy in mechanical properties can be significantly found in most AM metal parts, which include hardness, tensile strength, compressive strength, bending property, fracture toughness, fatigue property, and friction and wear properties. Importantly, the anisotropy can result in regions and directions with properties that are lower than expected and could compromise the structural integrity of the material to be used during service. This, however, remains a topic for further discussion.

In an example of Ti6Al4V manufactured by SLM, the hardness on the XY plane was lower than those on the YZ and ZX planes by approximately 20% (Chen et al. 2017). Compressive yield strengths also differed by approximately 10% when loading Ti6Al4V in a universal testing machine and a split Hopkinson pressure bar to evaluate the quasi-static and dynamic compression properties as illustrated in Figures 2.8(a–c) (Zheng et al. 2022). The material strength was also found to be proportional to the strain rate during quasi-static and dynamic tensile testing (Figures 2.8(d–f)). On the other hand, heat treatment lowered the strength and enhanced ductility (Yang et al. 2019). Anisotropy will also be found in the fracture toughness and fatigue crack growth behavior (Seifi et al. 2017).

In another example on the anisotropy of AM maraging steel, higher tensile strengths were reported when loaded perpendicular to the build direction while higher elongation (indicator of ductility) was observed during tension along 45° to the build direction (Bai, Lee, et al. 2020). Friction and wear behavior of M50 steel were also found to be correlated to the coral-like microstructure that promoted the formation of graphene for surface lubrication (Liu et al. 2018).

The layer-by-layer method of manufacturing will also give heterogeneity (in the form of variant melt pool shapes and dissymmetry) to the material due to differences between the boundary and center of the melt track as observed (in Figures 2.9(a) and 2.9(b)) by the combination of "interpool" and "intrapool" fracture modes in AlSi10Mg (Laurençon et al. 2019). Different microstructures are also expected due to the high cooling rates, such as the formation of fine cellular structures in 316L stainless steel that contain a high density of dislocations (Wang et al. 2018). For this case, the cellular structures allowed for more uniform deformation due to the pinning and twinning of dislocations that result in higher yield strength of the material while still maintaining ductility with high elongation (Figures 2.9(c) and 2.9(d)).

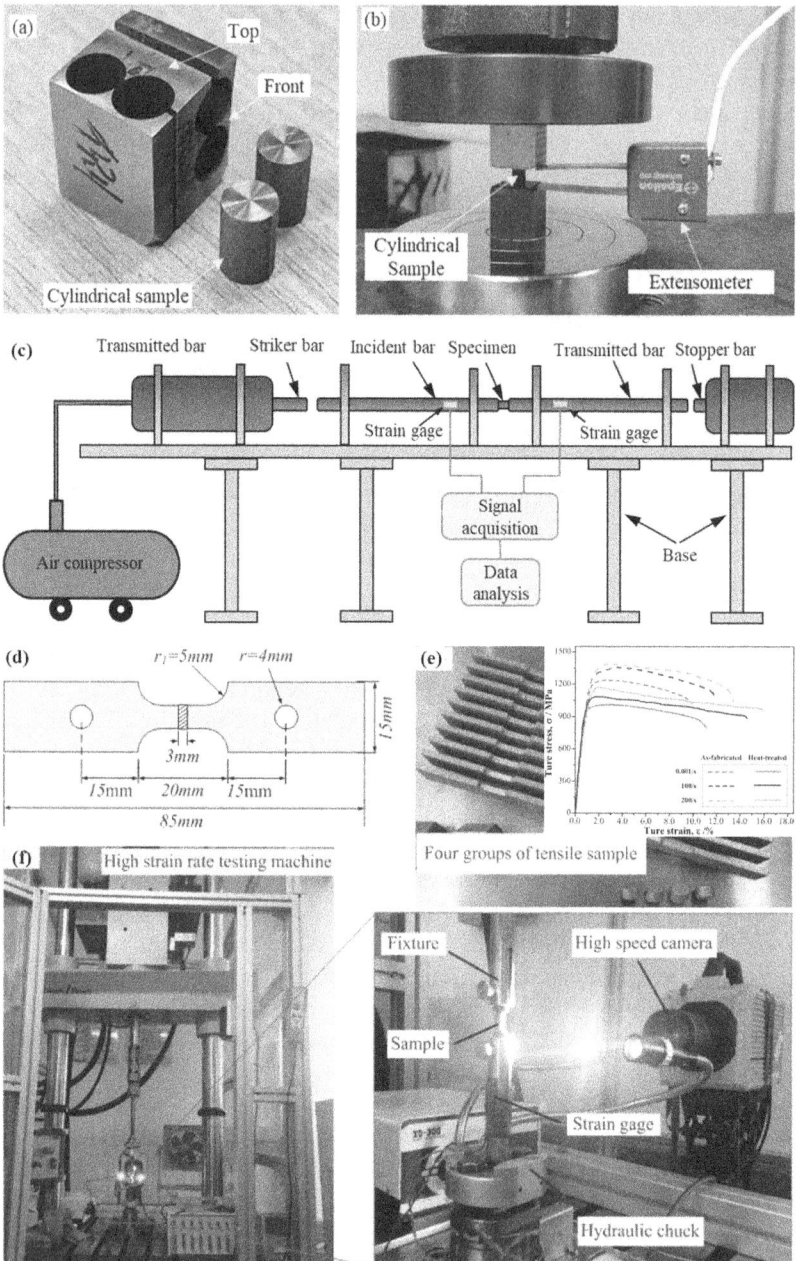

FIGURE 2.8 (a) Wire-cut cylindrical samples ($\phi 5 \times 8$ mm), (b) the setup for quasi-static compressive experiment and (c) schematic showing the principle of the split Hopkinson pressure bar test. Reprinted with permission from Elsevier (Zheng et al. 2022); (d, e) tensile specimens and testing results of Ti-6Al-4V alloy manufactured by selective laser melting and (f) high strain rate tensile testing system. Reprinted with permission from Elsevier (Yang et al. 2019).

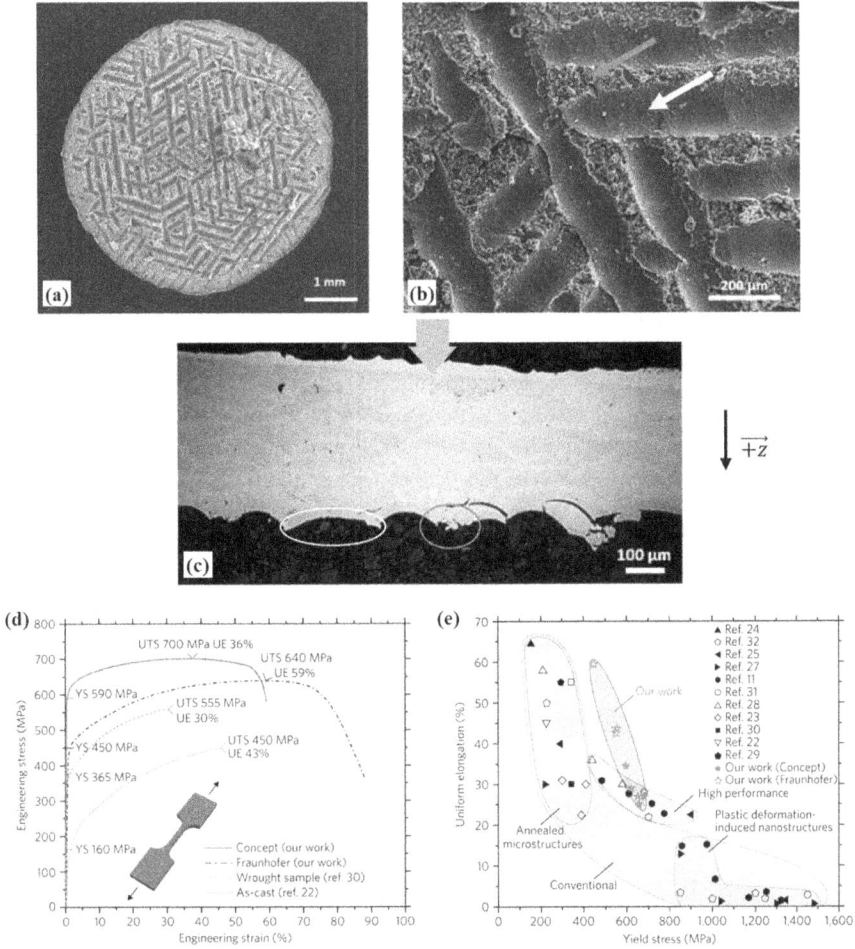

FIGURE 2.9 SEM fracture images of the spall crater (a, b) and OM image of a cross section (c): arrows and circles show both intrapool (dark arrow) and interpool (light arrow) fractures. Reprinted with permission from Elsevier (Laurençon et al. 2019); (d) Representative tensile engineering stress–strain curves for two L-PBF 316L SS (line curve with UTS 700 MPa: Concept sample; dot line curve with UTS 640 MPa: Fraunhofer sample), compared to those of as-cast and as-wrought materials. And (e) a summary of yield stress versus uniform elongation for various 316L SS, including our work, high-performance materials (strengthened by nanotwin bundles and bimodal grain materials), conventional coarse-grained materials (annealed microstructures), and materials strengthened through traditional plastic deformation. Reprinted with permission from Springer Nature (Wang et al. 2018).

2.3 CONCLUDING REMARKS

Due to the bottom-up and layer-by-layer manufacturing method and high cooling rates in metal additive manufacturing (with reference to PBF and DED), some unique characteristics that are different from the conventionally fabricated counterparts are observed, including melt track morphology, step stairs, powder adhesion, internal defects, strong texture, anisotropy in mechanical properties, microscopic composition segregation, fine cellular structure, high strength, in-situ heat treatment outcomes, and so forth. A deep understanding of the above characteristics can not only establish a foundation for building the process-microstructure-property relationship and optimizing the performance of the additively manufactured components, but also provide a reference for the corresponding post-processing procedures (i.e., machining, polishing, heat treatment, etc.). Fortunately, a large number of well-established characterization methods can be used to complete the characterization, which is summarized in Figure 2.10. For example, OM, SEM, and TEM are still effective ways to observe the microstructure, and XRD is useful for determining phase composition. In the future, new characterization and testing methods whould be developed based on the metal additive manufacturing process features. To date, there have been a few interesting methods. Examples include the high-speed high-energy X-ray used to monitor the printing process by revealing the melt pool behavior and formation mechanism of the keyhole, and computed tomography (CT) scanning used to observe the distribution of internal defects.

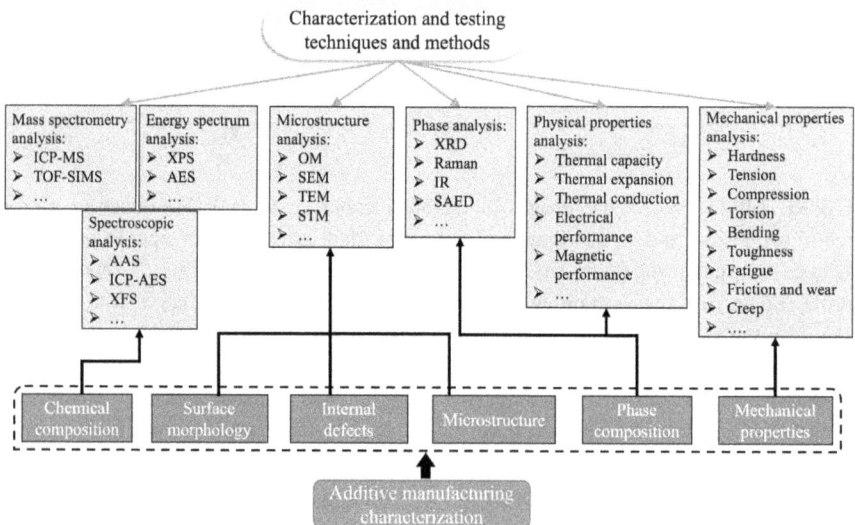

FIGURE 2.10 An overview diagram of materials characterization and testing techniques for additively manufactured metals.

REFERENCES

Bai, Yuchao, Yan Jin Lee, Chaojiang Li, and Hao Wang. 2020. "Densification Behavior and Influence of Building Direction on High Anisotropy in Selective Laser Melting of High-Strength 18Ni-Co-Mo-Ti Maraging Steel." *Metallurgical and Materials Transactions A* 51 (11). Springer US: 5861–5879. doi:10.1007/s11661-020-05978-9.

Bai, Yuchao, Yongqiang Yang, Di Wang, and Mingkang Zhang. 2017. "Influence Mechanism of Parameters Process and Mechanical Properties Evolution Mechanism of Maraging Steel 300 by Selective Laser Melting." *Materials Science and Engineering A.* doi:10.1016/j.msea.2017.06.033.

Bai, Yuchao, Yongqiang Yang, Zefeng Xiao, Mingkang Zhang, and Di Wang. 2018. "Process Optimization and Mechanical Property Evolution of AlSiMg0.75 by Selective Laser Melting." *Materials and Design* 140. Elsevier Ltd: 257–266. doi:10.1016/j.matdes.2017.11.045.

Bai, Yuchao, Jiayi Zhang, Cuiling Zhao, Chaojiang Li, and Hao Wang. 2020. "Dual Interfacial Characterization and Property in Multi-Material Selective Laser Melting of 316L Stainless Steel and C52400 Copper Alloy." *Materials Characterization* 167: 110489. doi:10.1016/j.matchar.2020.110489.

Bai, Yuchao, Cuiling Zhao, Di Wang, and Hao Wang. 2022. "Evolution Mechanism of Surface Morphology and Internal Hole Defect of 18Ni300 Maraging Steel Fabricated by Selective Laser Melting." *Journal of Materials Processing Technology* 299 (March 2021). Elsevier B.V.: 117328. doi:10.1016/j.jmatprotec.2021.117328.

Bai, Yuchao, Cuiling Zhao, Jin Yang, Ruochen Hong, Can Weng, and Hao Wang. 2021. "Microstructure and Machinability of Selective Laser Melted High-Strength Maraging Steel with Heat Treatment." *Journal of Materials Processing Technology* 288: 116906. doi:10.1016/j.jmatprotec.2020.116906.

Bai, Yuchao, Cuiling Zhao, Jiayi Zhang, and Hao Wang. 2022. "Abnormal Thermal Expansion Behaviour and Phase Transition of Laser Powder Bed Fusion Maraging Steel with Different Thermal Histories during Continuous Heating." *Additive Manufacturing* 53 (December 2021). Elsevier B.V.: 102712. doi:10.1016/j.addma.2022.102712.

Chen, Liang Yu, Jason Chih Huang, Che Hsin Lin, Cheng Tang Pan, Sih Yue Chen, Tsung Lin Yang, De Yao Lin, Hsuan Kai Lin, and Jason Shian Ching Jang. 2017. "Anisotropic Response of Ti-6Al-4V Alloy Fabricated by 3D Printing Selective Laser Melting." *Materials Science and Engineering A* 682 (October 2016). Elsevier: 389–395. doi:10.1016/j.msea.2016.11.061.

Donoghue, Jack, Alphons Anandaraj Antonysamy, Filomeno Martina, Paul A. Colegrove, Steward W. Williams, and Philip B. Prangnell. 2016. "The Effectiveness of Combining Rolling Deformation with Wire-Arc Additive Manufacture on β-Grain Refinement and Texture Modification in Ti-6Al-4V." *Materials Characterization* 114. The Authors: 103–114. doi:10.1016/j.matchar.2016.02.001.

du Plessis, Anton, Stephan G. le Roux, Jess Waller, Philip Sperling, Nils Achilles, Andre Beerlink, Jean François Métayer, et al. 2019. "Laboratory X-Ray Tomography for Metal Additive Manufacturing: Round Robin Test." *Additive Manufacturing* 30 (July). Elsevier: 100837. doi:10.1016/j.addma.2019.100837.

Godec, Matjaž, Simon Malej, Darjia Feizpour, Črtomir Donik, Matej Balažic, Damjan Klobčar, Laurent Pambaguian, Marjetka Conradi, and Aleksandra Kocijan. 2021. "Hybrid Additive Manufacturing of Inconel 718 for Future Space Applications." *Materials Characterization* 172. doi:10.1016/j.matchar.2020.110842.

Guo, Qilin, Cang Zhao, Luis I. Escano, Zachary Young, Lianghua Xiong, Kamel Fezzaa, Wes Everhart, Ben Brown, Tao Sun, and Lianyi Chen. 2018. "Transient Dynamics of Powder

Spattering in Laser Powder Bed Fusion Additive Manufacturing Process Revealed by In-Situ High-Speed High-Energy X-Ray Imaging." *Acta Materialia* 151. Elsevier Ltd: 169–180. doi:10.1016/j.actamat.2018.03.036.

Hirt, Luca, Alain Reiser, Ralph Spolenak, and Tomaso Zambelli. 2017. "Additive Manufacturing of Metal Structures at the Micrometer Scale." *Advanced Materials* 29 (17). doi:10.1002/adma.201604211.

Karme, Aleksis, Aki Kallonen, Ville Pekka Matilainen, Heidi Piili, and Antti Salminen. 2015. "Possibilities of CT Scanning as Analysis Method in Laser Additive Manufacturing." *Physics Procedia* 78 (August). Elsevier B.V.: 347–356. doi:10.1016/j.phpro.2015.11.049.

Kong, Decheng, Chaofang Dong, Shaolou Wei, Xiaoqing Ni, Liang Zhang, Ruixue Li, Li Wang, Cheng Man, and Xiaogang Li. 2021. "About Metastable Cellular Structure in Additively Manufactured Austenitic Stainless Steels." *Additive Manufacturing* 38 (December 2020). Elsevier B.V.: 101804. doi:10.1016/j.addma.2020.101804.

Laurençon, M., Thibaut de Rességuier, Didier Loison, Jacques Baillargeat, Julius Noel Domfang Ngnekou, and Yves Nadot. 2019. "Effects of Additive Manufacturing on the Dynamic Response of AlSi10Mg to Laser Shock Loading." *Materials Science and Engineering A* 748 (January). Elsevier B.V.: 407–417. doi:10.1016/j.msea.2019.02.001.

Leung, Chu Lun Alex, Sebastian Marussi, Robert C. Atwood, Michael Towrie, Philip J. Withers, and Peter D. Lee. 2018. "In Situ X-Ray Imaging of Defect and Molten Pool Dynamics in Laser Additive Manufacturing." *Nature Communications* 9 (1). Springer US: 1–9. doi:10.1038/s41467-018-03734-7.

Liu, Xiyao, Xiaoliang Shi, Yuchun Huang, Xiaobin Deng, Guanchen Lu, Zhao Yan, Hongyan Zhou, and Bing Xue. 2018. "The Sliding Wear and Friction Behavior of M50-Graphene Self-Lubricating Composites Prepared by Laser Additive Manufacturing at Elevated Temperature." *Journal of Materials Engineering and Performance* 27 (3). Springer US: 985–996. doi:10.1007/s11665-018-3187-z.

Maamoun, Ahmed H., Mohamed Elbestawi, Goulnara K. Dosbaeva, and Stephen C. Veldhuis. 2018. "Thermal Post-Processing of AlSi10Mg Parts Produced by Selective Laser Melting Using Recycled Powder." *Additive Manufacturing* 21 (January). Elsevier: 234–247. doi:10.1016/j.addma.2018.03.014.

Maleki, Erfan, Sara Bagherifard, Michele Bandini, and Mario Guagliano. 2021. "Surface Post-Treatments for Metal Additive Manufacturing: Progress, Challenges, and Opportunities." *Additive Manufacturing* 37 (May 2020). Elsevier B.V.: 101619. doi:10.1016/j.addma.2020.101619.

Seifi, Mohsen, Ayman Salem, Daniel Satko, Joshua Shaffer, and John J. Lewandowski. 2017. "Defect Distribution and Microstructure Heterogeneity Effects on Fracture Resistance and Fatigue Behavior of EBM Ti–6Al–4V." *International Journal of Fatigue* 94. Elsevier Ltd: 263–287. doi:10.1016/j.ijfatigue.2016.06.001.

Spierings, A. B., M. Schneider, and R. Eggenberger. 2011. "Comparison of Density Measurement Techniques for Additive Manufactured Metallic Parts." *Rapid Prototyping Journal* 17 (5): 380–386. doi:10.1108/13552541111156504.

Sun, Zhongji, Xipeng Tan, Shu Beng Tor, and Chee Kai Chua. 2018. "Simultaneously Enhanced Strength and Ductility for 3D-Printed Stainless Steel 316L by Selective Laser Melting." *NPG Asia Materials* 10 (4). Springer US: 127–136. doi:10.1038/s41427-018-0018-5.

Thijs, Lore, Karolien Kempen, Jean Pierre Kruth, and Jan Van Humbeeck. 2013. "Fine-Structured Aluminium Products with Controllable Texture by Selective Laser Melting of Pre-Alloyed AlSi10Mg Powder." *Acta Materialia* 61 (5): 1809–1819. doi:10.1016/j.actamat.2012.11.052.

Wang, Jiachen, Yinan Cui, Changmeng Liu, Zixiang Li, Qianru Wu, and Daining Fang. 2020. "Understanding Internal Defects in Mo Fabricated by Wire Arc Additive Manufacturing through 3D Computed Tomography." *Journal of Alloys and Compounds* 840. Elsevier B.V: 155753. doi:10.1016/j.jallcom.2020.155753.

Wang, Y. Morris, Thomas Voisin, Joseph T. McKeown, Jianchao Ye, Nicholas P. Calta, Zan Li, Zhi Zeng, et al. 2018. "Additively Manufactured Hierarchical Stainless Steels with High Strength and Ductility." *Nature Materials* 17 (1): 63–70. doi:10.1038/NMAT5021.

Xie, Bin, Jiaxiang Xue, and Xianghui Ren. 2020. "Wire Arc Deposition Additive Manufacturing and Experimental Study of 316l Stainless Steel by Cmt + p Process." *Metals* 10 (11): 1–19. doi:10.3390/met10111419.

Yang, Liu, Pang Zhicong, Li Ming, Wang Yonggang, Wang Di, Song Changhui, and Li Shuxin. 2019. "Investigation into the Dynamic Mechanical Properties of Selective Laser Melted Ti-6Al-4V Alloy at High Strain Rate Tensile Loading." *Materials Science and Engineering A* 745 (January). Elsevier B.V.: 440–449. doi:10.1016/j.msea.2019.01.010.

Zhang, Xiang Yu, Xing Chen Yan, Gang Fang, and Min Liu. 2020. "Biomechanical Influence of Structural Variation Strategies on Functionally Graded Scaffolds Constructed with Triply Periodic Minimal Surface." *Additive Manufacturing* 32 (September 2019). Elsevier: 101015. doi:10.1016/j.addma.2019.101015.

Zhao, Cuiling, Yuchao Bai, Yu Zhang, Xiaopeng Wang, Jun Min Xue, and Hao Wang. 2021. "Influence of Scanning Strategy and Building Direction on Microstructure and Corrosion Behaviour of Selective Laser Melted 316L Stainless Steel." *Materials and Design* 209. The Authors: 109999. doi:10.1016/j.matdes.2021.109999.

Zheng, Zhongpeng, Xin Jin, Yuchao Bai, Yun Yang, Chenbing Ni, Wen Feng Lu, and Hao Wang. 2022. "Microstructure and Anisotropic Mechanical Properties of Selective Laser Melted Ti6Al4V Alloy under Different Scanning Strategies." *Materials Science and Engineering: A* 831 (August 2021). Elsevier B.V.: 142236. doi:10.1016/j.msea.2021.142236.

3 Manufacturing Workflow

3.1 MANUFACTURING PROCESS CHAIN

Before going into detailed discussions on the post-processes, it is important to understand the sequence of procedures that lead to those stages. The additive manufacturing (AM) process chain comprises five main sequential procedures (Figure 3.1):

1. Three-dimensional (3D) model rendering
2. Machine setup
3. Building
4. Part removal
5. Post-processing

As it is with other manufacturing processes, the requirements and design of a part must be provided as a starting point, which include the geometrical tolerances, surface quality, and material selection. Several planning procedures will follow such as the digital model generation of the part using computer-aided drawing (CAD) software for the rendering of build orientation and support structures, raw material preparation processes, and AM parameter selection. The AM procedure is followed by the removal of the part from the powder bed and the corresponding support structures. Secondary post-processes will then perform the finishing touches to meet the specified requirements of the part. These typically involve conventional material removal processes to achieve the desired surface finishing and geometrical precision.

3.1.1 3D MODEL CREATION

Like traditional manufacturing, AM parts are first designed and drawn to meet the specific requirements of the product, which is normally performed using modern CAD compared to predecessor manual engineering drawings portrayed via various projection styles (e.g., isometric, oblique, perspective, section views, etc.). The most common multiview projection style provides engineers with sufficient information on

DOI: 10.1201/9781003272601-3

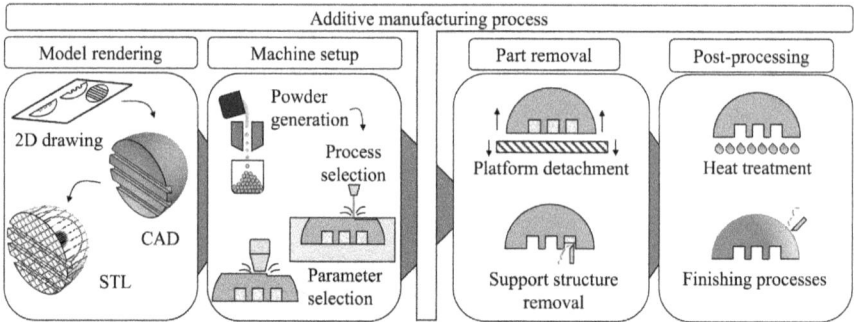

FIGURE 3.1   Workflow in additive manufacturing.

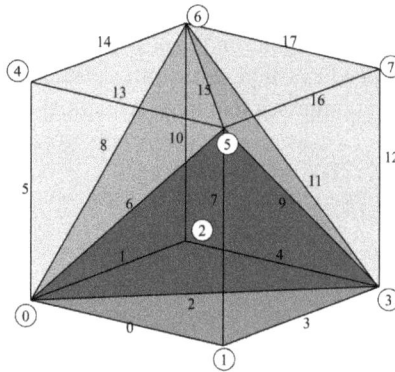

FIGURE 3.2   Concept of tessellation in a cubic space to form twelve triangle surfaces.

a two-dimensional (2D) source (i.e., printed drawings) to understand the final form of the product and plan for the most efficient manufacturing parameters (e.g., tool path, cutting parameters, tool specifications). AM has also promoted the replication of existing parts such as the geometries of unique human body parts, which may be obtained by performing CT or MRI scans. These scans are subsequently processed in a CAD model as the 3D part design. The rapid advancements in digitization for product design enable the simple import of data from the 3D rendered model into the AM machine for analysis, format conversion, and determining the construction process through optimization of the scanning path.

The data is mostly stored in the STL file format, which stands for standard tessellation language or standard triangle language, utilizing a series of triangles to map out the surface geometry of the designed model. Oddly, the abbreviation STL is also widely used to represent the stereolithographic printing process. The function of the STL file is to tile the 3D surface of the model with 2D triangles without overlapping or any missing gaps. A simple example is that of a cube with six square surfaces, which can be tessellated with triangle facets totaling twelve equally dimensioned triangle surfaces (Figure 3.2). In this example, the facets have the same dimensions, but

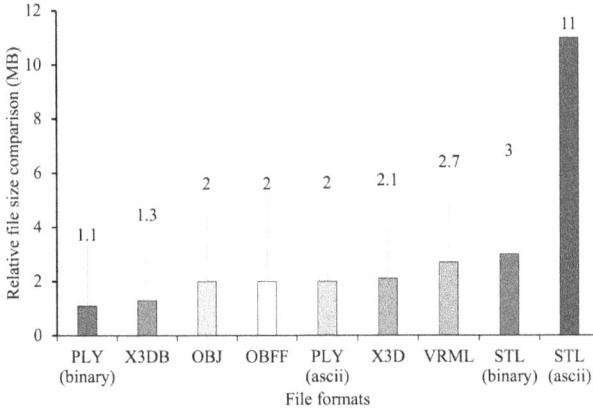

FIGURE 3.3   Storage disk space usage for different file formats.

these facets need not be designed with the same dimension for complex geometries with different surface dimensions and can be adjusted accordingly to fit the complexity of the designed surface with the final goal of rendering the 3D model surface with triangles without any overlaps or gaps.

Complex geometries will result in more triangles that are sized according to the intricate features and consequently larger file sizes. As the triangle facet is used to map the surface, the intersection of two surfaces (both surface-to-surface and surface-to-gaps) is dependent on the graphical software to accurately render these edges. The CAD design can also be difficult to modify using the STL file, which is a challenge when correcting design errors discovered during the later stage of the manufacturing process chain. Another less important limitation of the STL file format is the lack of final product information such as internal texture, color, and material information that can be provided using other file formats such as VMRL, OBJ, and PLY. The use of the latter file formats has been increasing in recent years due to the efficient use of disk space compared to STL that is currently ranked the highest in file storage. Figure 3.3 illustrates the various types of file formats used in additive manufacturing and the approximate comparison between file storage space usage.

3.1.2   RAW MATERIAL PREPARATION

Most AM processes for metal production begin with raw material in powder form with some exceptions of filament-based processes (i.e., fused deposit modeling), which are typically intended for prototyping purposes. Powder preparation is crucial for the final build quality, which involves powder spreading method, powder particle size and size distribution (Yang et al. 2020), and the thermal, optical, and chemical properties of individual powder particles (Meier et al. 2019).

The major processes for manufacturing AM powders are listed in Table 3.1, which typically produce spherical powders (Sun et al. 2016). Fundamentally, each process requires the metal to be melted before going through an atomization procedure and

TABLE 3.1

Additive Manufacturing Powder Preparation Processes

Fabrication process	Description	Advantages	Disadvantages
Gas atomization (GA)	Free fall of molten metal in a high-pressure gas stream through a tower to solidify before collection	• Low cost • High powder flow rate • Modifiable alloy compositions • Controllable metallurgical properties	• Large variation in quality between suppliers
Induction melted bar atomization (EIGA)	A metal bar is melted in an induction coil and the molten stream forms to be atomized in high-pressure gas	• Capable of producing Ti6Al4V powders • High powder flow rate • High production rates	• High-quality stock bar required • Limited alloy selection
Plasma atomized wire (PAW)	Using plasma torches to melt metal wires that flow through a gas stream for solidification and collection	• Extremely high flow rates • Suitable for high melting point metals	• Limited suppliers • Require stock material in wire-form • High costs
Plasma rotating electrode process (PREP)	Plasma torch melting of a bar rotating at a high revolution speed such that the molten metal is formed and solidifies radially due to the centrifugal force	• Very high flow rates • Near-perfect spherical particles • Suitable for high melting point metals	• Limited suppliers • Require high-quality stock material • High costs
Water atomization (WA)	Disintegration of molten metal droplets using high-pressure water to quickly cool and solidify	• Low cost • Scalable	• Poor metallurgical quality

finally solidification. Each process has its own merits and downsides (Table 3.1) that determine the ability to produce the desirable powder quality.

Powder specifications are typically determined based on several factors such as the particle size distribution (PSD), apparent density, packing density, and hall flow rate. The PSD is used to describe the percentage of particles that are of a particular size, which enables manufacturers to gauge the average size of the powders within a particular batch of raw material. SLM systems often employ powder sizes ranging 15–60 µm (Nguyen et al. 2017), while EBM systems typically range 50–150 µm (Vock et al. 2019).

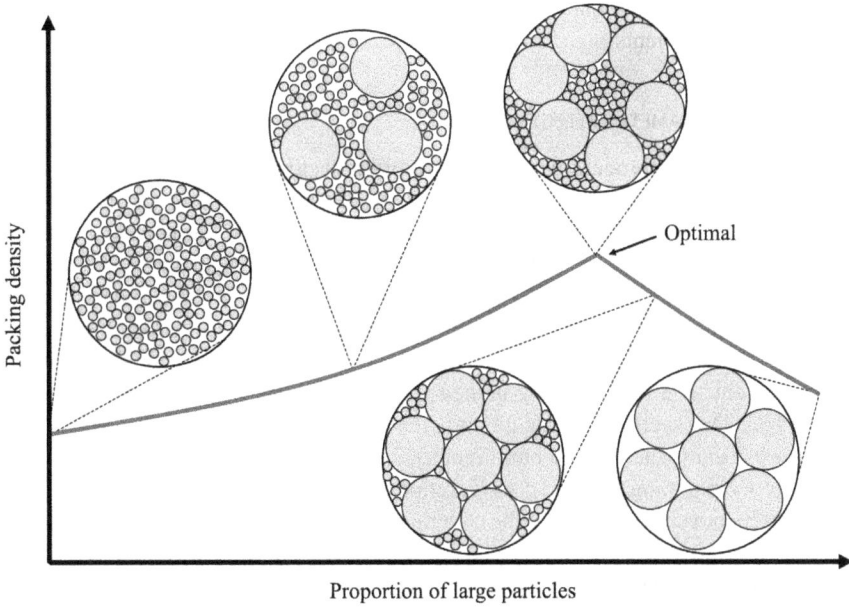

FIGURE 3.4 Combination and relationship between powder particle size and packing density.

The apparent density (AD) is a measure of loosely packed powder where the collected powder naturally flows under gravity. Packing density is the measure of rearranged powder particles for the minimization of the volume and the increase in density by compacting or tapping. Generally, higher homogeneity in the particle distribution and evenness leads to higher packing density, which supposedly translates to superior quality of the manufactured part. However, there are also instances where the mix of large and fine particles can lead to higher packing densities, as illustrated in Figure 3.4. The illustrated low packing density with smaller particles can be mitigated by proper powder spreading that prepares the powder bed for irradiation.

The hall flow rate is a measure of the flowability for the powder particles, which is critical in AM for unrestricted flow during powder spreading. Powder spreading is affected by the particle adhesion and friction effects (Wang et al. 2020). In some cases, finer particles could achieve higher densities such as that in steel, but may not always immediately affect the mechanical properties of the parts (Li et al. 2020).

While metal filament feed stocks employed in fused filament fabrication (FFF) appear in wire form to be spooled through an AM machine, the metal in these filaments comprise powder particles bounded together with organic binder polymers that are to be dissolved by catalytic depolymerization or thermally decomposed by sintering after fabrication of the component (Wagner et al. 2022). The mix of these constitutive materials is typically extruded to form the spools of filaments for storage and direct feeding to the AM machine. Appropriate combinations of the powder and filament preparation parameters are necessary for the efficient production of high-quality AM components.

3.1.3 AM PARAMETER SELECTION

A standard series of process parameters are often decided on the AM machine itself such as the scan speed, hatch space, layer thickness, and other process-specific parameters (e.g., laser power, spot size, and pulse duration and frequency). Figure 3.5 illustrates an overview of the various process parameters to consider in a PBF process. Layer thickness is the measure of height between each successive addition of powder onto the existing powder bed for the subsequent step of irradiation. Hatch spacing is defined as the distance between the center of the laser beam along one scanning line to the adjacent. The scan velocity is the speed at which the heat source traverses the powder bed. This section will not go into significant detail discussing the influence of each process parameter on the build quality, but rather give a general outline for the various process parameters and considerations to be made during the AM procedure.

The laser spot size determines the coverage of powder irradiated using an adequate energy density. In some cases, it is also necessary to be comparably small to precisely produce the intricate details of the fabricated part. Typical spot sizes are in the range of 50–80 µm (Attar et al. 2014; Wang et al. 2013) for standard SLM operations, while precision microfabrication spot sizes can be as small as 15 µm (Chen, Wang, and Zuo 2003). As laser beams emit with a Gaussian profile, the laser spot size is determined by the focal length (Equation 3.1). The focal length physically represents the distance between the laser and the irradiated object. While this does not immediately affect the

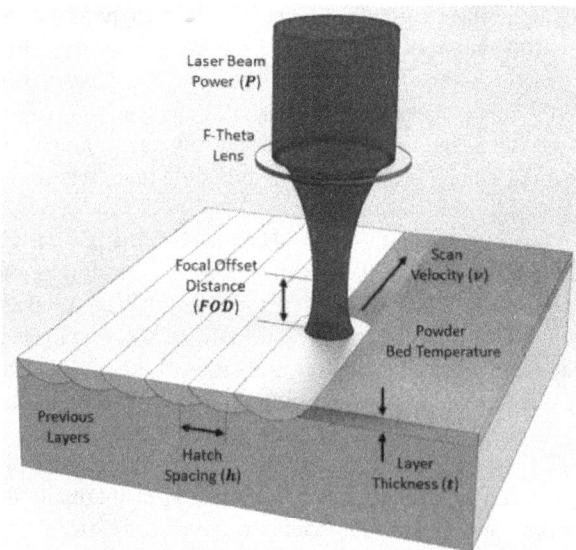

FIGURE 3.5 Illustration of the process parameters involved in a SLM process. Reprinted with permission from Elsevier (Shipley et al. 2018).

reachability of the laser to the powder bed in the SLM process, the change in focal position may cause disruptions to gas flow within the chamber (Bean et al. 2018).

$$d = \frac{4f\lambda M^2}{\pi D} \tag{3.1}$$

where d is the spot size, f is the focal length, λ is the laser wavelength, M is the beam quality, and D is the laser beam diameter.

Standard SLM solutions are equipped with 400 W lasers with the option for additional high-power lasers (700–1,000 W). High-power lasers are important to effectively process metals, particularly those with high reflectivity and thermal conductivities such as aluminum, which requires a minimum of 150 W with 50 mm/s scanning speed for adequate build quality (Buchbinder et al. 2011). Scanning speeds can also be increased with higher energy added to quicken the melting of the powder, which effectively shortens the overall build time. However, an excessively powerful laser (≥1 kW) would lead to the equivalent of laser machining where the work material ends up being vaporized. Thus, lasers in commercial solutions are rarely equipped with lasers with power above 1 kW. On the other hand, EBM machines are equipped with electron guns that administer a beam of high-energy electrons at high velocity to melt the target. Acceleration voltages are typically kept constant while the beam power is determined by the variable beam current in the range of 1–50 mA.

It is easy to understand that a thicker layer would reduce the production time due to the reduction in time spent on spreading new layers of powder onto the bed. However, each new layer of melted powder should also adhere to the previous layer to construct the solid part, which requires the applied thermal energy to reach the interface between layers and limits the maximum allowable layer thickness. Typical thicknesses used in conventional SLM systems are in the range of 60–150 μm, while precision AM machines are in the 20–60 μm range to meet the density demands of the intricate features and surface quality in microfabrication (Shi et al. 2020).

Undoubtedly, an appropriate combination of parameters is needed to achieve desirable build quality. For instance, the inappropriate combination of a small laser spot size and large powder particle size would be expected to diminish the build quality. Another example is where a high-frequency pulse and high-scanning speed would reduce the effective irradiation period. Thus, a simple metric to best predict the part quality is defined by the energy density as (Thijs et al. 2010):

$$E = \frac{P}{vht} \tag{3.2}$$

where P is the laser power (J/s), v is the scanning velocity (mm/s), h is the hatch spacing (mm), and t is the layer thickness.

Despite the energy density being one of the most used metrics to predict the build quality, there is no definite relationship between the energy density and porosity of the fabricated parts. Densification of the parts also depends on other factors such as the chemical composition of the material – e.g., lower energy density required for steels with higher carbon content (Nakamoto et al. 2009). Beyond using the energy density as a predictive gauge for porosity, the energy density was also shown to influence the microstructure of the metals such as the increase in refined grains with lower energy densities in Inconel 718 (Liu et al. 2020).

The final consideration for PBF-AM is the building orientation and scanning sequence, which would inevitably affect the microstructure (Yan et al. 2021), residual stresses (Ali, Ghadbeigi, and Mumtaz 2018), and surface roughness (Xie et al. 2020) of the constructed part. Building sequence can vary through a combination of scanning direction (Figure 3.6) and building angles (Figure 3.7). These figures show a basic flat surface in rudimentary studies on the influence of building sequence as a stepping

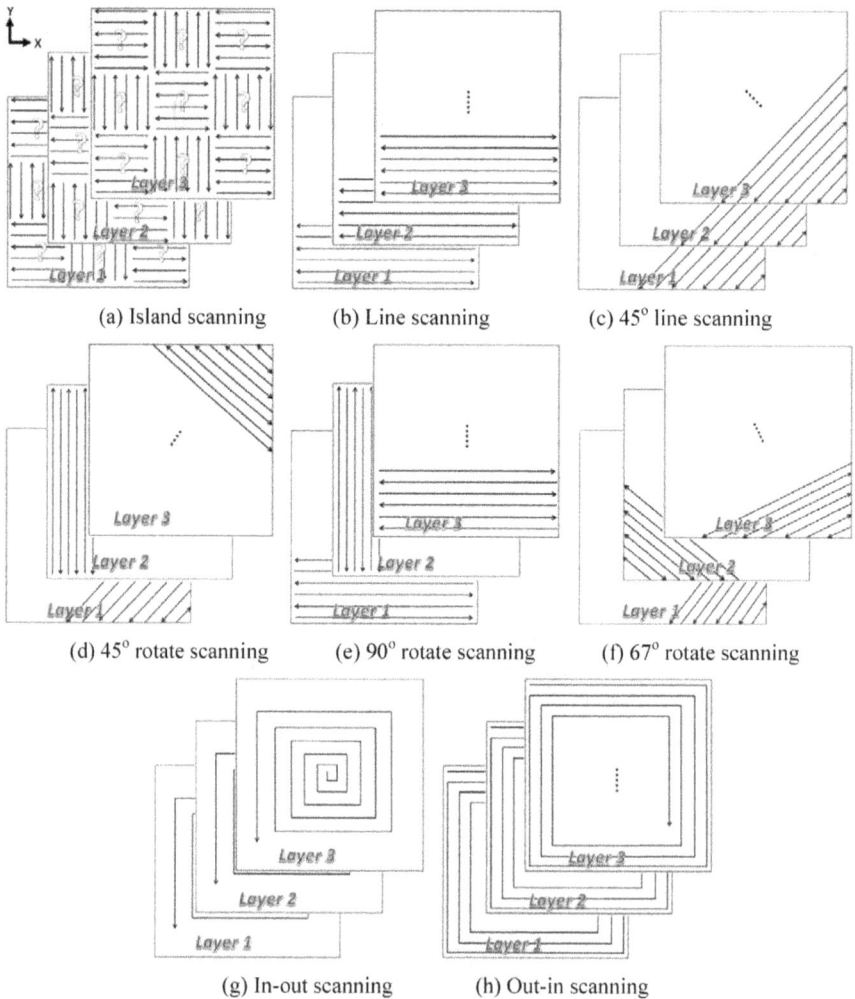

(a) Island scanning   (b) Line scanning   (c) 45° line scanning

(d) 45° rotate scanning   (e) 90° rotate scanning   (f) 67° rotate scanning

(g) In-out scanning   (h) Out-in scanning

FIGURE 3.6 Example of scanning strategies. Reprinted with permission from Elsevier (Cheng, Shrestha, and Chou 2016).

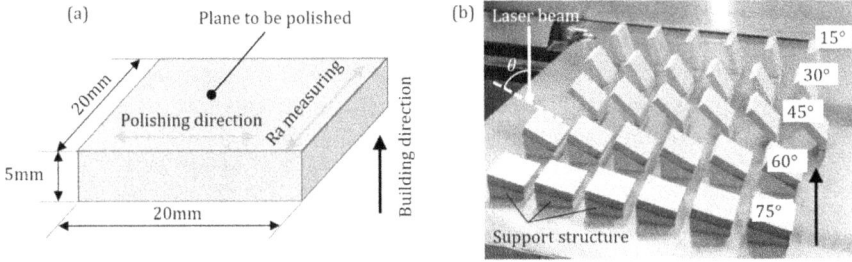

FIGURE 3.7 An example study on polishing of additively manufactured 316L steel: (a) illustration of the polishing surface relative to the build direction; (b) steel test samples constructed at different build angles by SLM. Reprinted with permission from Elsevier (Zhang, Chaudhari, and Wang 2019).

stone for more complex structures. The influence of build orientation on mechanical properties may be apparent for certain materials such as Ti6Al4V but was reported to be negligible for others such as AlSi10Mg (Maconachie et al. 2020).

DED processes also have similar process parameters such as laser power, scanning speed, and hatch spacing, which are sufficient to determine the energy density for DED as given by Equation 3.3. Due to the nature of the process, additional parameters are also available such as the powder feed rate typically ranging 3–15 g/min. In DED, the variation of powder feed rate was reported to be more influential on defect formation at interfaces of repair works (Kim and Shim 2021) and has a substantial impact on the final surface quality in terms of bead height (Lee et al. 2017).

$$E = \frac{P}{vD} \tag{3.3}$$

where P is the laser power (J/s), v is the scanning velocity (mm/s), and D is the laser spot size.

Upon completion of the AM process, the support structures and manufactured part will be removed from the building platform before further processing of the part, such as heat treatment and surface treatment. The importance of considering support structure removal and the influence of heat treatment will be discussed in the subsequent sub-sections before moving on into further detail on the various material removal processes available for implementation to achieve the desirable surface quality of the final part.

3.2 SUPPORT STRUCTURE REMOVAL

Support structures are essential components of the metal additive manufacturing process, particularly when fabricating complex geometries in laser powder bed fusion. The reason for this is that the soft bed of discrete powder particles that are allowed to freely move is unable to support the weight of the part that is being constructed.

Further details on the role of the support structures will be discussed in this section. As these support structures will end up connected to the actual product, solutions must be identified to remove them after the additive fabrication is complete. However, it is the complexity of additively manufactured components that create issues for support structure removal during post-processing (Gibson et al. 2021).

3.2.1 ROLE OF THE SUPPORT STRUCTURE

There are two purposes for support structures to aid the additive manufacturing process: (1) a support base for parts with overhanging structures to ensure fabrication accuracy, and (2) paths for heat dissipation to prevent excessive heat accumulation that affects thermal-induced residual stresses and shape distortion. In most cases, the support structure is primarily used for the first purpose, while the second is more important for metals with low thermal conductivity. A combination of these two roles has also been reported to benefit the precise fabrication of metal parts (Hintze et al. 2020). Figure 3.8 presents the importance of support structures in the manufacturing of metal parts. As an example, the inclined features that tend to overhang more (with inclination angles lesser than 30° in Figure 3.8(a)) show the deterioration of down-skin surface quality. On the other hand, the fabrication of complex geometries can be successfully achieved with the presence of support structures, as shown in Figure 3.8(b).

While the most obvious benefit of support structures is in providing a structural base for the building of parts with overhanging structures (Calignano 2014), this very support affects a multitude of other properties of the fabricated part. Microhardness, microstructure, and compressive stress distributions can be different near the part-support contact locations (Zhang, et al. 2020). These occurrences are potentially due to the complex thermal behavior of the additive manufacturing process that could result in vertical deformation of the part (Pellens et al. 2020). Contact-free heat supports can also be used to change this thermomechanical behavior of overhanging

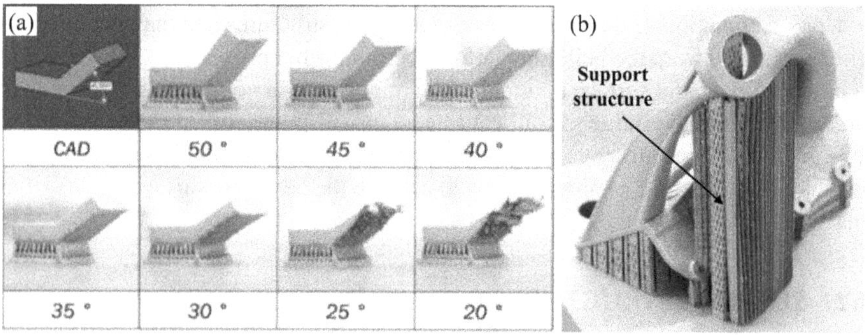

FIGURE 3.8 Additively manufactured (a) inclined features without support structures. Reprinted with permission from Springer Nature (Meng et al. 2020). (b) complex metal part with support structures. Reprinted with permission from Elsevier (Emmelmann, Herzog, and Kranz 2017).

structures to lower thermal residual stresses and distortion (Cooper et al. 2018). Enhanced thermal properties of the support structures can also be optimized through the development of algorithms that maximize heat conduction while still preserving the structural stiffness of the support (Wang et al. 2018).

3.2.2 METHODS OF REMOVING SUPPORT STRUCTURES

After fabrication is complete, the support structures are no longer needed and will be removed manually or by machining. Conventional manual removal of support structures involves operators physically removing these features with hand-held tools such as scissors, pliers, files, tweezers, and other small hand-held power tools (Bibb, Eggbeer, and Williams 2006). For example, it is common to manually remove a support structure using a pair of pliers. While the process seems simple, the quality of support structure removal is difficult to control and inconsistent as it is subject to operator workmanship. The geometry of fabricated parts remains a challenge even for operators during support structure removal. For instance, a webbed support structure is easier to remove by hand compared to tube support structures (Järvinen et al. 2014). The only benefit of this manual operation is the flexibility involved where operators can change operating configurations according to the part geometry. However, this does not outweigh the downside where low efficiency and high labor costs are involved.

As manual removal of support structures is labor intensive, it is unsuitable for integration in commercial production lines where standardization, precision, and automation are generally preferred for cost-savings and quality assurance. Hence, most support structure removal procedures turn to mechanical removal processes that can significantly improve production efficiency. At the same time, mechanical machining is widely accepted as a process to obtain high-quality and consistent surface finishing to support industrial requirements of standardizing, industrializing, and automating. Meanwhile, several downsides of mechanical machining remain prevalent in the context of AM support structure removal. Intricate and complex geometries of the fabricated part will inevitably involve machining tool paths that follow the curvatures of the part during support removal. This will require specially designed cutting tools and tool path algorithms to avoid accidental interference with the desired part.

While it may seem relatively straightforward like normal mechanical machining processes, the machining of support structures involves additional challenges. Supports are typically constructed using many individual columns to minimize material wastage and maximize time savings during the building phase. These structures are designed to maximize weight-bearing capabilities vertically. However, efficient mechanical material removal of these structures often involves cutting directions perpendicular to the support columns where the mechanical strength is theoretically weaker. Within this issue, different support geometries also show different structural integrities such as the reported cone supports that easily collapse during machining as compared to block supports, as shown in Figure 3.9. The advent of support collapse simply indicates poor machinability and is typically accompanied by higher cutting forces and severe tool wear (Cao et al. 2020; Zhang, Cao, and Lu 2022). Other support

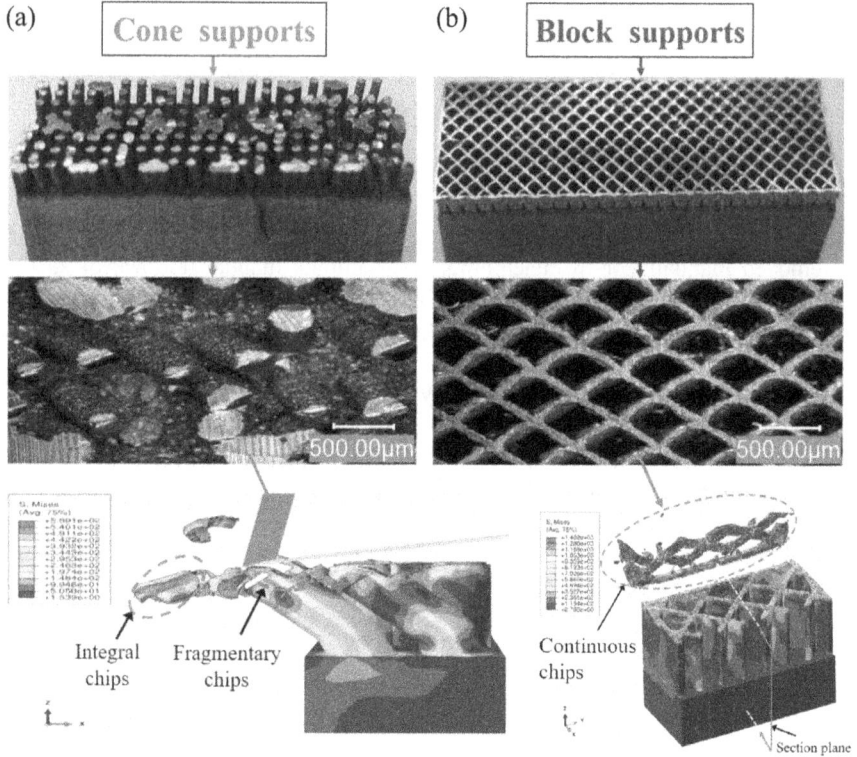

FIGURE 3.9 Machined surface morphology and corresponding simulation of cutting (a) cone support and (b) block support structures. Reprinted with permission from Elsevier (Cao et al. 2020).

structure geometries can also be employed such as hollow squarish columns and bent solid columns rather than solid structures (Hintze et al. 2020).

To mitigate the problem of support structure poor machinability, alternative processes such as wire electric discharge machining (EDM) have been tested in comparison to mechanical machining (Ameen et al. 2022). Surface finishing by mechanical machining is likely inferior due to the remnants of teeth-like features that give a roughness of 27.25 μm Ra compared to superior 4.01 μm Ra surfaces producible by wire EDM. However, heat-affected surfaces can be expected from wire EDM processes due to the high temperatures involved during the vaporization process, which changes the final processed surface properties. Moreover, material removal rates in mechanical machining far exceed that of wire EDM.

Innovative solutions have also been proposed to resolve the poor lateral mechanical integrity of support structures such as the addition of epoxy resin to fill gaps between the supports (Cao et al. 2021). This created a much stiffer composite structure that helped improve the machinability of the supports by reducing the vibrations induced during milling and, consequently, reducing cutting energy and tool wear by up to 22.6%. Effective support structure removal for thin-walled parts, without severely deforming the thin walls, was also achievable with the implementation of epoxy resin to create the rigid composite structure (Cao et al. 2022).

3.2.3 BALANCE OF SUPPORT STRUCTURES

While considering the machinability of support structures, primary considerations for support structure design currently do not include the aspect of machinability but only the optimization of part orientation to reduce material usage. These methods are used in topology optimization where minimal material usage designs can be achieved, such as the addition of a crossbeam to achieve support-free conditions (Li et al. 2017) or avoiding orientations where thin features are constructed in an overhung position (Zhang, Cheng, and Xu 2019). An example of topology optimization is given in Figure 3.10 where iterations of the overhang angle parameter were modelled to optimize the part structure that allowed larger self-support angles along the build direction, which eliminated the need for external support structures (Gaynor and Guest 2016). Another simple way of optimizing support structure design is to identify the most cost-efficient orientation of the part relative to the build direction through mathematical modeling that will employ the least volume of support structure material (Das et al. 2015).

Although support structures play an important role in additive manufacturing of complex functional metal parts, their effective removal during post-processing remains a difficult task. Therefore, when designing the support structures, it is necessary to balance AM efficacy and the support removal (cost for post-processing).

Optimizing the support structures through new designs is another effective way to ease the removal process. Vaidya and Anand (2016) set an example for support structure design while considering the support role during additive manufacturing

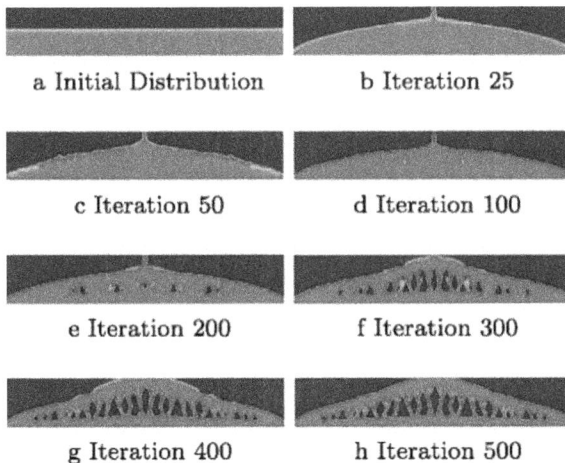

FIGURE 3.10 Design evolution of typical Messerschmidt-Bölkow-Blohm beam with 63.4-degree self-supporting angle and more aggressive algorithmic parameters. Reprinted with permission from Springer Nature (Gaynor and Guest 2016).

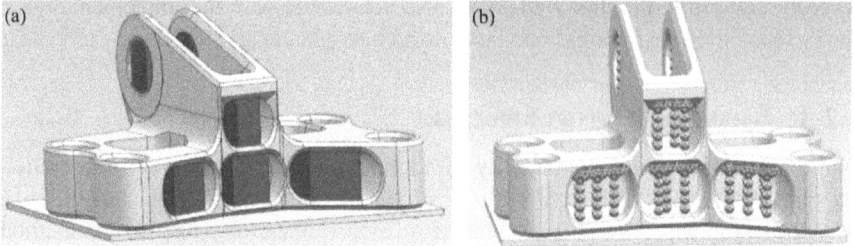

FIGURE 3.11 Support structure type: (a) fully solid support, (b) support with hollow unit cells at the interface. Reprinted with permission from Elsevier (Vaidya and Anand 2016).

and removal during post-processing. They minimized support structure usage through space-filling honeycomb structures and Dijkstra's shortest path algorithm. Meanwhile, support structure accessibility from the external surface was also made possible in the support design. Figure 3.11(a) shows the full solid support structure, which contains large volumes of materials and large contact areas between the support structures and the solid part. An improvement to the support structure design would involve unit cell voxels to build the main body of the support structure and hollow cellular unit cells at the interface between the support and the actual part, as shown in Figure 3.11(b). This configuration greatly reduced the contact area of the support structures, which would result in better surface finishing of the machined part during support structure removal. Other examples of innovative support structure designs include lattice support structures that reduce build time and support structure materials (Hussein et al. 2013) and biomimetic structures inspired by trees to reduce support structure materials by 40% and the post-processing time (Zhang et al. 2020).

3.3 MICROSTRUCTURE MODIFICATION

The process of additive manufacturing metal parts brings about unique microstructures and mechanical properties due to high cooling rates and layer-by-layer fabricating methods. It is widely known that post-processing is affected by material performance. The microstructure heterogeneity of as-built additively manufactured metal parts can influence the post-processing efficiency where surface quality and dimensional accuracy can be affected. Therefore, this section will cover one of the main processes to modulate microstructures which is the heat treatment. There are other methods to alter the microstructures such as process parameter selection during AM and surface mechanical treatment. However, changing process parameters during AM may cause other issues related to the build integrity and mechanical treatments only alter the surface properties instead of the bulk material. Hence, these alternatives will not be discussed in this chapter.

3.3.1 CONVENTIONAL HEAT TREATMENT

In metal additive manufacturing, the purpose of conventional heat treatment is to relieve residual stresses, enable microstructure homogenization, and perform aging

TABLE 3.2
Main Chemical Compositions (Mass %) of 316L Stainless Steel and Maraging Steel

Material	Fe	Ni	Cr	Co	Mo	Si	Mn	C	Ti	Al
316L	Bal.	10.00-13.00	16.50–18.50	-	2.00–2.50	<=1.00	<=2.00	<=0.03	-	-
18Ni300	Bal.	18.50	-	9.00	4.80	<=0.10	-	<=0.03	0.60	0.10

strengthening. Different types of metal alloys react differently to the heat treatment processes, which will be subsequently discussed for each type of metal (Note: for simplicity, only some representative materials are discussed in this section).

1) Iron-Based Material

The main chemical composition of 316L stainless steel and maraging steel is given in Table 3.2. 316L stainless steel is the most widely used iron-based engineering metal. Heat treatment of additively manufactured 316L stainless steels has been shown to attenuate residual stresses and microstructures at different temperatures, which thereby alter mechanical properties and corrosion resistance. Melt pool boundaries remain visible after heat treatment at 600 °C (Tascioglu, Karabulut, and Kaynak 2020) with negligible changes to the microstructure but the temperature allows for the partial release of residual stresses (Chao et al. 2021). As temperatures rise above 850 °C, polygon grains (austenite) begin to appear with clear grain boundaries as a sign of recrystallization. This variation of microstructure lowers the microhardness of the material but increases its wear resistance.

Maraging steel differs from 316L stainless steel which can be further strengthened by aging heat treatment (i.e., keep at room temperature or slightly above room temperature for a long time to improve part strength) to specifically promote precipitation. Figure 3.12(a) shows the evolution of microstructure in additively manufactured 18Ni300 maraging steel (Table 3.2) over different heat treatment procedures. Melt pool boundaries seen in as-built surfaces disappear after heat treatment at 900 °C for 1 h to be filled with fine lath martensitic grains (Bai et al. 2019; Bai et al. 2021). Aging treatment at both 440 °C and 560 °C over 6 h show negligible changes in grain morphology, but nano-scale precipitate particles (Ni_3(Ti, Al, Mo)) will form during aging treatment to cause the sharp improvement of hardness and strength (Tan et al. 2017), as shown in Figure 3.12(b). In addition, the micrometric cellular structure that is unique to additively manufactured metal alloys resulting from the thermal gradient and solidification velocity (Bai et al. 2022) can be eliminated with high-temperature heat treatment (Wang et al. 2022).

2) Titanium-Based Material

Titanium alloys have high melting points and low densities, which make them widely used in the aerospace field. Ti6Al4V alloy is the representative titanium alloy material that has been successfully produced by additive manufacturing. Ti6Al4V typically contains a dual-phase microstructure (i.e., α and β phases) when conventionally

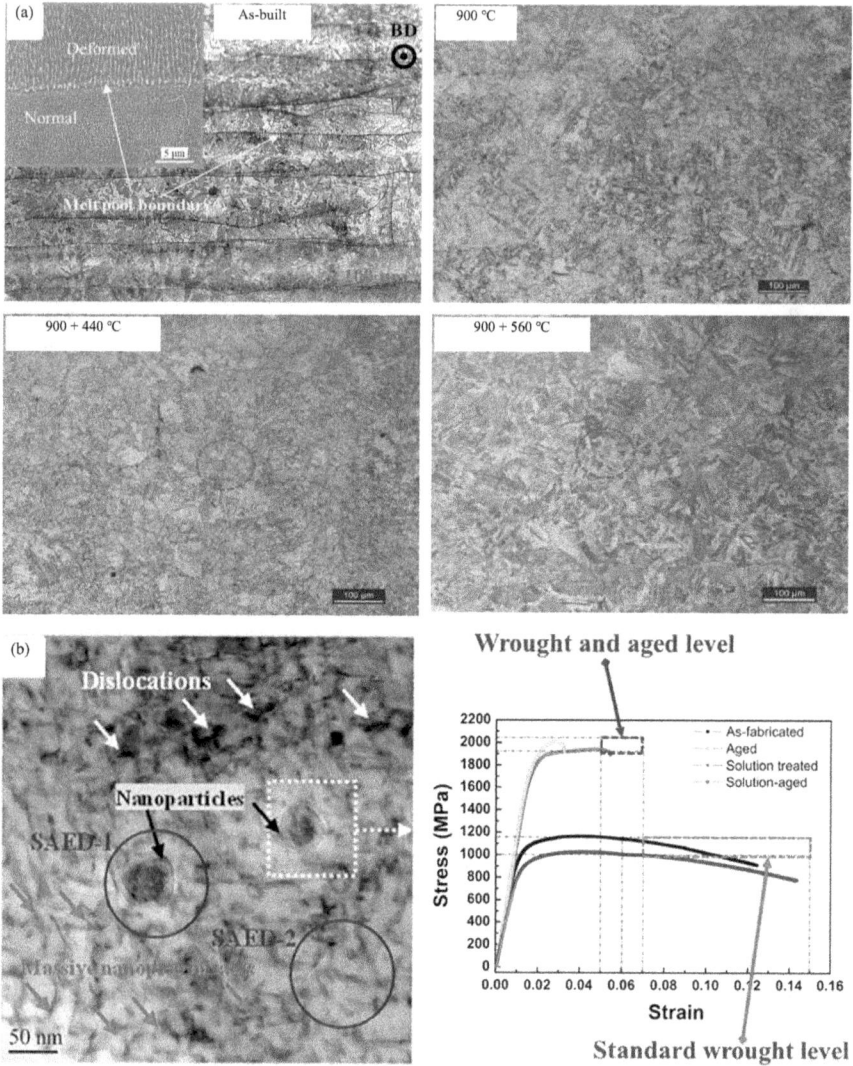

FIGURE 3.12 Maraging steel with heat treatment at different temperatures: (a) microstructure of the as-built sample, solution-treated sample at 900 °C, aging-treated sample at 440 °C after solution and aging-treated sample at 560 °C after solution. Reprinted with permission from Elsevier (Bai et al. 2019; Bai et al. 2021). (b) Dislocation and nanoparticles in the aging-treated samples, and stress–strain curves of the as-built, solution-treated, direct-aging-treated and solution+ageing-treated samples. Reprinted with permission from Elsevier (Tan et al. 2017).

produced. However, due to the high cooling rates during additive manufacturing, the as-built Ti6Al4V alloy displays α′ martensitic phase instead. Figure 3.13(a) shows the microstructure of an as-built Ti6Al4V material containing a full distribution of α′ martensitic phase characterized by the number of clearly defined α′ phase colonies (Liang et al. 2019). These boundaries are from prior β phase grains, which are large and elongated along the direction of the thermal gradient caused by the bottom-up manufacturing method. In some cases, epitaxial growth of the β phase grains can evolve into several millimeters in length (Vrancken et al. 2012).

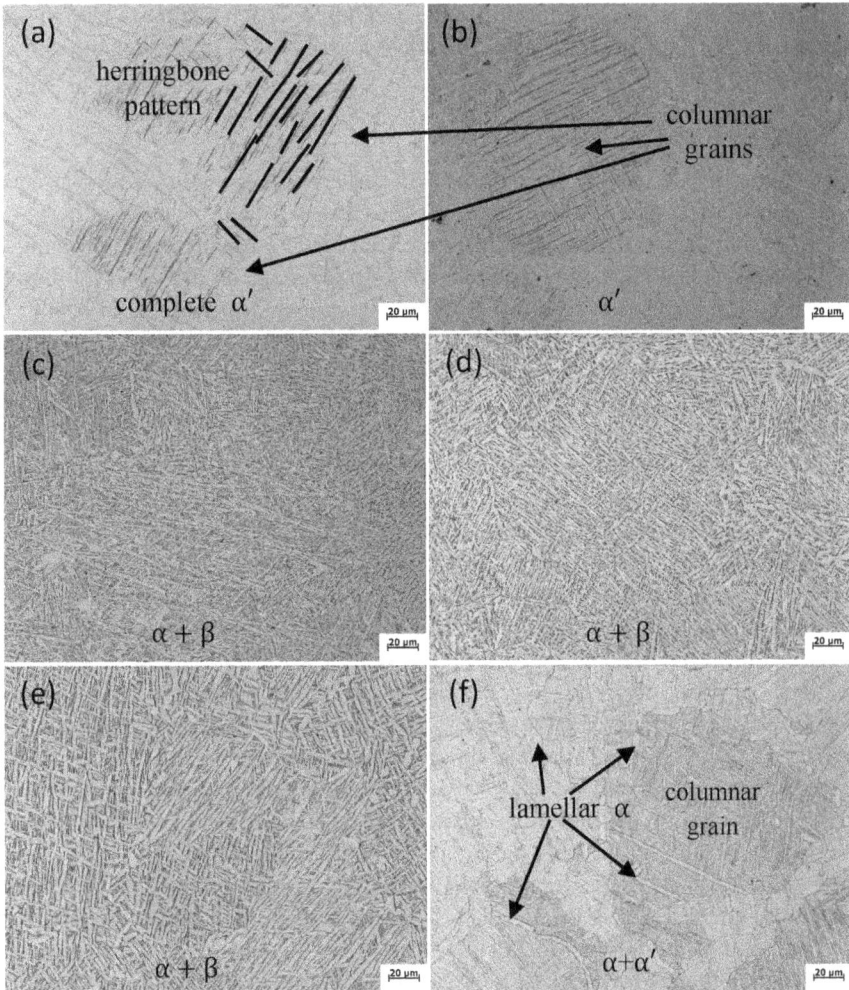

FIGURE 3.13 The microstructure on horizontal planes of as-fabricated samples and heat-treated samples. (a) as fabricated; (b) 600 °C/4 h/air cooling; (c) 800 °C/4 h/ air cooling; (d) 850 °C/4 h/ air cooling; (e) 900 °C/4 h/ air cooling; (f) 950 °C/2 h/water cooling+540 °C/4 h/air cooling. Reprinted with permission from Elsevier (Liang et al. 2019).

Annealing heat treatment to 600 °C over 4 h will show insignificant changes to the microstructure (Figure 3.13(b)) as temperatures for the formation of the α and β phases exist above 760 °C. Raising the temperature to 800 °C will cause the decomposition of the acicular α′ phase and the formation of lamellar α and irregular β phases (Figure 3.13(c)). These grains will continue to grow and form more α phase with further increase in temperature. Quenching followed by aging heat treatment of the alloy produces a mixture of lamellar α, acicular α′ and spheroidal β precipitation phases. This heat treatment is performed by holding temperatures at 950 °C for 2 h followed by water quenching, and subsequently reheating to 540 °C for 4 h and air cooled. In general, high-temperature annealing treatment will reduce the strength and enhance the ductility of the alloy. Quench-ageing treatment will increase the strength above an annealed alloy but will remain lower than the as-built counterpart.

3) Aluminum-Based Material

AlSi10Mg alloy is one of the most widely researched aluminum alloys in additive manufacturing due to its good printability and excellent mechanical performance. The alloy may be annealed, solution treated, and age treated. Annealing is performed by heating the alloy to temperatures in the range of 300–500 °C followed by furnace cooling (Butler et al. 2021). Solution treatment is performed by heating at temperatures in the range of 450–550 °C for 2 h followed by water cooling (Li et al. 2016). Artificial age treatment occurs at 180 °C for 12 h of heating followed by water cooling. The evolution of microstructures for the different heat treatments are given in Figure 3.14.

The as-built AlSi10Mg material is composed of fine cellular structures (Figure 3.14(a–c)). These structures are broken into fine Si particles and Al-Si solid solution after heat treatment at 450 °C. Further increases in temperatures result in the merging of Si particles to form larger particles. Artificial age treatment also promotes the growth of Si particles but is known to involve additional precipitation of Mg_2Si particles. Annealing also breaks the cellular structure of the Si particles and promotes grain growth. However, the temperature in annealing heat treatment is lower. Therefore, the formed Si particles are much smaller (<1 µm). Due to the much higher temperature in solution treatment, Si particles grow quickly and can be up to > 5 µm. This heat treatment results in sharp reductions of tensile strength from 434.3 MPa to 168.1 MPa while elongation increases from 5.3% to 23.7%. This sharp reduction in strength is due to the disappearance of fine grain strengthening effect caused by the breakage of fine cellular structure and growth of Si particles. The increase in elongation is due to the elimination of residual stress and the reduction of localized stress during stretching. While aging treatment promotes the formation of Mg_2Si particles, reports have shown that the presence of these precipitates has negligible impact on the strength of the AlSi10Mg alloy in comparison to the as-built state, which may be that the aging treatment not only results in the precipitation of Mg_2Si particles but also causes part of the silicon to break away from the fine cellular structure and reduce the residual stress (Park et al. 2021; Aboulkhair et al. 2016).

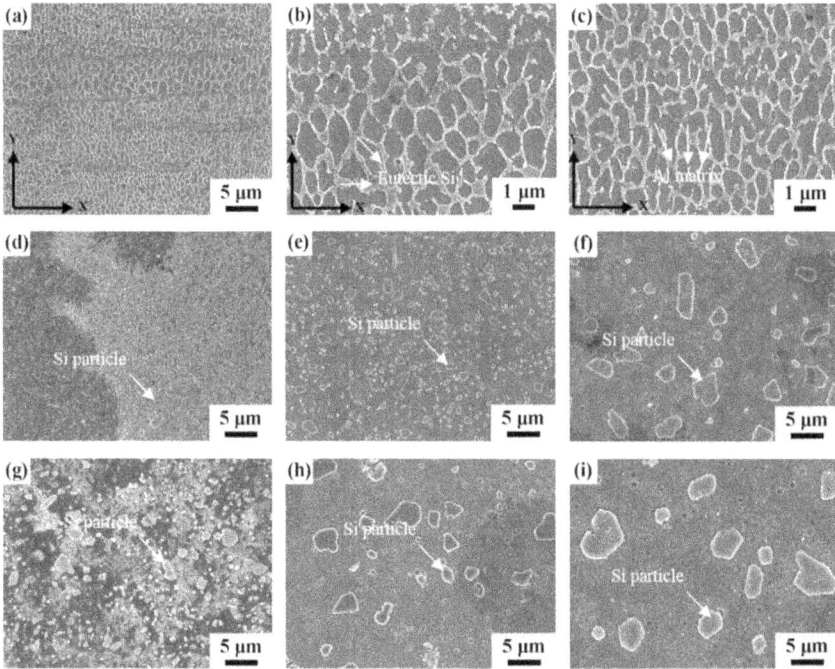

FIGURE 3.14 Microstructure of as-built and heat-treated AlSi10Mg alloy: (a) single melt track with three distinctive regions; (b) boundary of melt track with coarse cellular region and transition region; (c) interior of melt pool with fine cellular microstructure. SEM images of the eutectic microstructures of AlSi10Mg after different heat treatment conditions: (d) 450 °C for 2 h; (e) 500 °C for 2 h; (f) 550 °C for 2 h; (g) 450 °C for 2 h+180 °C for 12 h; (h) 500 °C for 2 h+180 °C for 12 h; and (i) 550 °C for 2 h+180 °C for 12 h. X-axis is the laser scanning direction and the light grey areas are Si particles. Reprinted with permission from Elsevier (Li et al. 2016).

3.3.2 HOT ISOSTATIC PRESSING

Hot isostatic pressing (HIP) is another heat treatment process employed for additively manufactured metal parts. The schematic and workflow of the process are displayed in Figure 3.15. The high temperatures and pressures involved in the process can not only modify the microstructure but also eliminate internal defects, such as micro-cracks, and pores. Therefore, it is widely applicable for additive manufacturing where these defects form easily during construction of metal parts.

There are several examples of HIP tackling the defect problem for additively manufactured metals. Figure 3.16 shows the microstructure evolution of four additively manufactured AlSi10Mg samples before and after HIP. Each sample had different relative densities (Figure 3.16(b)) corresponding to the presence of internal defects formed during building. HIP was performed with a pressure of 100 MPa and temperature of 500 °C for 2 h and demonstrated the effectiveness of the technique to

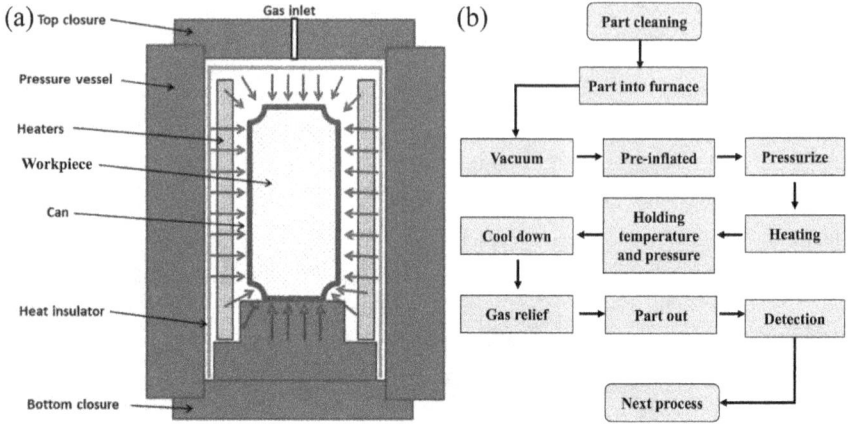

FIGURE 3.15 (a) Schematic of the hot isostatic pressuring machine and (b) the corresponding workflow. Reprinted with permission from Elsevier (Torralba 2014).

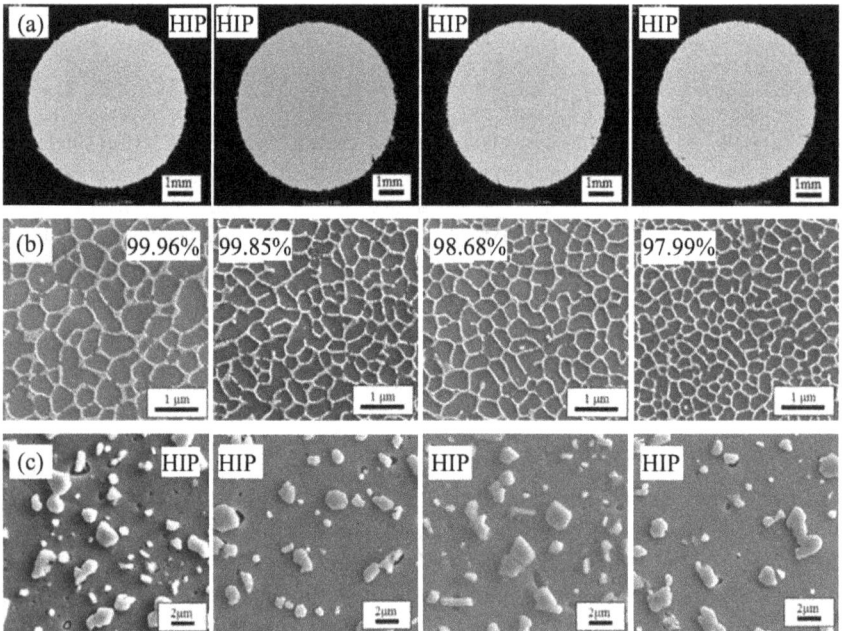

FIGURE 3.16 X-ray computed tomography scan (a) and microstructural (b, c) images of AlSi10Mg alloys: (b) before HIP and (a, c) after HIP treatment. Reprinted with permission from Elsevier (Hirata, Kimura, and Nakamoto 2020).

eliminate pores (Hirata, Kimura, and Nakamoto 2020). With the high treatment temperature, the cellular microstructure transformed into the Al matrix filled with large Si particles just as how it can be observed by normal heat treatment.

A similar increase in relative density will also be found for IN718 titanium alloy. It is also interesting that the unique microstructure composed of fine cellular structures containing Laves phase disappears after HIP and is replaced by the coarse polygonal γ phase (FCC) (Tillmann et al. 2017). HIP of Ti6Al4V with pressures of 100 MPa and 2 h of heating at 920 °C showcased a basket-like microstructure composing α and β phases, similar to normal heat treatment at 920 °C for 2 h without the applied pressure (Yu et al. 2019). While the mechanical properties between the two remain unchanged, there will be a slight increase in the width of the α phase under HIP.

HIP on 316L stainless steel requires a slightly higher temperature of 1,150 °C and pressure of 150 MPa over 3 h to effect changes to relative density and microstructure (Röttger et al. 2016). Under these conditions, grain coarsening due to recrystallization at high heating temperatures and the reduction in mechanical strength are to be expected.

HIP would produce microstructures that are like normal heat treatment conditions, but it is important to understand that the benefit of HIP is in enhancing the relative density of the fabricated parts. However, it should be mentioned that HIP has a negligible effect on densification when large internal voids are involved (Röttger et al. 2016).

3.4 CONCLUDING REMARKS

This chapter took the reader through the manufacturing process work flow that incorporates additive manufacturing technology beginning from part design considerations to the post-process step for refinement of the part properties. As the common notion for post-processes often encompass material removal processes for accurate surface and geometrical finishing, the chapter shed light on other aspects of the post-processing, which include support structure removal and microstructure modulation by heat treatment. Considerations for support structures extend beyond the function of supporting the part during additive manufacturing to the design and positioning of support structures for effective removal. Heat treatment procedures for common metals were also discussed, which would set the foundation to understand influence of microstructure on machinability in Chapter 4.

REFERENCES

Aboulkhair, Nesma T., Ian Maskery, Chris Tuck, Ian Ashcroft, and Nicola M. Everitt. 2016. "The Microstructure and Mechanical Properties of Selectively Laser Melted AlSi10Mg: The Effect of a Conventional T6-like Heat Treatment." *Materials Science and Engineering A* 667. Elsevier: 139–146. doi:10.1016/j.msea.2016.04.092.

Ali, Haider, Hassan Ghadbeigi, and Kamran Mumtaz. 2018. "Effect of Scanning Strategies on Residual Stress and Mechanical Properties of Selective Laser Melted Ti6Al4V." *Materials Science and Engineering: A* 712 (January). Elsevier: 175–187. doi:10.1016/j.msea.2017.11.103.

Ameen, Wadea, Abdulrahman Al-Ahmari, Muneer Khan Mohammed, and Osama Abdulhameed. 2022. "Evaluation of the Support Structure Removal Techniques for Additively Manufactured Ti6AL4V Parts." *Advances in Materials Science and Engineering* 2022 (Figure 2): 1–8. doi:10.1155/2022/2259974.

Attar, Hooyar, Matthias Bönisch, Mariana Calin, Lai-Chang Zhang, Sergio Scudino, and Jürgen Eckert. 2014. "Selective Laser Melting of in Situ Titanium–Titanium Boride Composites: Processing, Microstructure and Mechanical Properties." *Acta Materialia* 76 (September). Elsevier: 13–22. doi:10.1016/j.actamat.2014.05.022.

Bai, Yuchao, Cuiling Zhao, Di Wang, and Hao Wang. 2022. "Evolution Mechanism of Surface Morphology and Internal Hole Defect of 18Ni300 Maraging Steel Fabricated by Selective Laser Melting." *Journal of Materials Processing Technology* 299 (March 2021). Elsevier B.V.: 117328. doi:10.1016/j.jmatprotec.2021.117328.

Bai, Yuchao, Di Wang, Yongqiang Yang, and Hao Wang. 2019. "Effect of Heat Treatment on the Microstructure and Mechanical Properties of Maraging Steel by Selective Laser Melting." *Materials Science and Engineering A* 760 (February). Elsevier B.V.: 105–117. doi:10.1016/j.msea.2019.05.115.

Bai, Yuchao, Cuiling Zhao, Jin Yang, Ruochen Hong, Can Weng, and Hao Wang. 2021. "Microstructure and Machinability of Selective Laser Melted High-Strength Maraging Steel with Heat Treatment." *Journal of Materials Processing Technology* 288 (May). Elsevier B.V.: 116906. doi:10.1016/j.jmatprotec.2020.116906.

Bean, Glenn E, David B Witkin, Tait D McLouth, Dhruv N Patel, and Rafael J Zaldivar. 2018. "Effect of Laser Focus Shift on Surface Quality and Density of Inconel 718 Parts Produced via Selective Laser Melting." *Additive Manufacturing* 22 (August). Elsevier: 207–215. doi:10.1016/j.addma.2018.04.024.

Bibb, Richard, Dominic Eggbeer, and Robert Williams. 2006. "Rapid Manufacture of Removable Partial Denture Frameworks." *Rapid Prototyping Journal* 12 (2): 95–99. doi:10.1108/13552540610652438.

Buchbinder, D, H Schleifenbaum, S Heidrich, W Meiners, and JJPP Bültmann. 2011. "High Power Selective Laser Melting (HP SLM) of Aluminum Parts." *Physics Procedia* 12. Elsevier: 271–278. doi:10.1016/j.phpro.2011.03.035.

Butler, C., S. Babu, R. Lundy, R. O'Reilly Meehan, J. Punch, and N. Jeffers. 2021. "Effects of Processing Parameters and Heat Treatment on Thermal Conductivity of Additively Manufactured AlSi10Mg by Selective Laser Melting." *Materials Characterization* 173 (July 2020). Elsevier Inc.: 110945. doi:10.1016/j.matchar.2021.110945.

Calignano, F. 2014. "Design Optimization of Supports for Overhanging Structures in Aluminum and Titanium Alloys by Selective Laser Melting." *Materials and Design* 64. Elsevier Ltd: 203–213. doi:10.1016/j.matdes.2014.07.043.

Cao, Qiqiang, Yuchao Bai, Jiong Zhang, Zhuoqi Shi, Jerry Ying Hsi Fuh, and Hao Wang. 2020. "Removability of 316L Stainless Steel Cone and Block Support Structures Fabricated by Selective Laser Melting (SLM)." *Materials and Design* 191. The Authors: 108691. doi:10.1016/j.matdes.2020.108691.

Cao, Qiqiang, Yuchao Bai, Zhongpeng Zheng, Jiong Zhang, Jerry Ying Hsi Fuh, and Hao Wang. 2022. "Support Removal on Thin-Walled Parts Produced by Laser Powder Bed Fusion." *3D Printing and Additive Manufacturing* 00 (00): 1–14. doi:10.1089/3dp.2021.0268.

Cao, Qiqiang, Zhuoqi Shi, Yuchao Bai, Jiong Zhang, Cuiling Zhao, Jerry Ying Hsi Fuh, and Hao Wang. 2021. "A Novel Method to Improve the Removability of Cone Support Structures in Selective Laser Melting of 316L Stainless Steel." *Journal of Alloys and Compounds* 854. Elsevier B.V: 157133. doi:10.1016/j.jallcom.2020.157133.

Chao, Qi, Sebastian Thomas, Nick Birbilis, Pavel Cizek, Peter D. Hodgson, and Daniel Fabijanic. 2021. "The Effect of Post-Processing Heat Treatment on the Microstructure, Residual Stress and Mechanical Properties of Selective Laser Melted 316L Stainless Steel." *Materials Science and Engineering A* 821 (June). Elsevier B.V.: 141611. doi:10.1016/j.msea.2021.141611.

Chen, Jimin, Xubao Wang, and Tiechuan Zuo. 2003. "The Micro Fabrication Using Selective Laser Sintering Micron Metal Powder." In *Smart Sensors, Actuators, and MEMS*, edited by Jung-Chih Chiao, Vijay K. Varadan, and Carles Can%c, 5116:647. SPIE. doi:10.1117/12.501147.

Cheng, Bo, Subin Shrestha, and Kevin Chou. 2016. "Stress and Deformation Evaluations of Scanning Strategy Effect in Selective Laser Melting." *Additive Manufacturing* 12 (October). Elsevier: 240–251. doi:10.1016/j.addma.2016.05.007.

Cooper, Kenneth, Phillip Steele, Bo Cheng, and Kevin Chou. 2018. "Contact-Free Support Structures for Part Overhangs in Powder-Bed Metal Additive Manufacturing." *Inventions* 3 (1). doi:10.3390/inventions3010002.

Das, Paramita, Ramya Chandran, Rutuja Samant, and Sam Anand. 2015. "Optimum Part Build Orientation in Additive Manufacturing for Minimizing Part Errors and Support Structures." *Procedia Manufacturing* 1. The Authors: 343–354. doi:10.1016/j.promfg.2015.09.041.

Emmelmann, Claus, Dirk Herzog, and Jannis Kranz. 2017. "Design for Laser Additive Manufacturing." In *Laser Additive Manufacturing*, 259–279. Elsevier. doi:10.1016/B978-0-08-100433-3.00010-5.

Gaynor, Andrew T., and James K. Guest. 2016. "Topology Optimization Considering Overhang Constraints: Eliminating Sacrificial Support Material in Additive Manufacturing through Design." *Structural and Multidisciplinary Optimization* 54 (5). Structural and Multidisciplinary Optimization: 1157–1172. doi:10.1007/s00158-016-1551-x.

Gibson, Ian, David Rosen, Brent Stucker, and Mahyar Khorasani. 2021. *Additive Manufacturing Technologies*. Vol. 89. Cham: Springer International Publishing. doi:10.1007/978-3-030-56127-7.

Hintze, Wolfgang, Robert von Wenserski, Sebastian Junghans, and Carsten Möller. 2020. "Finish Machining of Ti6Al4V SLM Components under Consideration of Thin Walls and Support Structure Removal." *Procedia Manufacturing* 48. Elsevier B.V.: 485–491. doi:10.1016/j.promfg.2020.05.072.

Hirata, Tomotake, Takahiro Kimura, and Takayuki Nakamoto. 2020. "Effects of Hot Isostatic Pressing and Internal Porosity on the Performance of Selective Laser Melted AlSi10Mg Alloys." *Materials Science and Engineering A* 772 (June 2019). Elsevier B.V.: 138713. doi:10.1016/j.msea.2019.138713.

Hussein, Ahmed, Liang Hao, Chunze Yan, Richard Everson, and Philippe Young. 2013. "Advanced Lattice Support Structures for Metal Additive Manufacturing." *Journal of Materials Processing Technology* 213 (7). Elsevier B.V.: 1019–1026. doi:10.1016/j.jmatprotec.2013.01.020.

Järvinen, Jukka Pekka, Ville Matilainen, Xiaoyun Li, Heidi Piili, Antti Salminen, Ismo Mäkelä, and Olli Nyrhilä. 2014. "Characterization of Effect of Support Structures in Laser Additive Manufacturing of Stainless Steel." *Physics Procedia* 56 (C). Elsevier B.V.: 72–81. doi:10.1016/j.phpro.2014.08.099.

Kim, Tae Geon, and Do Sik Shim. 2021. "Effect of Laser Power and Powder Feed Rate on Interfacial Crack and Mechanical/Microstructural Characterizations in Repairing of 630 Stainless Steel Using Direct Energy Deposition." *Materials Science and Engineering: A* 828 (November). Elsevier: 142004. doi:10.1016/j.msea.2021.142004.

Lee, Eun Mi, Gwang Yong Shin, Hi Seak Yoon, and Do Sik Shim. 2017. "Study of the Effects
    of Process Parameters on Deposited Single Track of M4 Powder Based Direct Energy
    Deposition." *Journal of Mechanical Science and Technology* 31 (7). Springer: 3411–
    3418. doi:10.1007/s12206-017-0239-5.
Li, Kefeng, Xinhua Mao, Khashayar Khanlari, Kaikai Song, Qi Shi, and Xin Liu. 2020.
    "Effects of Powder Size Distribution on the Microstructural and Mechanical Properties
    of a Co–Cr–W–Si Alloy Fabricated by Selective Laser Melting." *Journal of Alloys and
    Compounds* 825 (June). Elsevier: 153973. doi:10.1016/j.jallcom.2020.153973.
Li, Wei, Shuai Li, Jie Liu, Ang Zhang, Yan Zhou, Qingsong Wei, Chunze Yan, and Yusheng
    Shi. 2016. "Effect of Heat Treatment on AlSi10Mg Alloy Fabricated by Selective Laser
    Melting: Microstructure Evolution, Mechanical Properties and Fracture Mechanism."
    Materials Science and Engineering A 663. Elsevier: 116–125. doi:10.1016/
    j.msea.2016.03.088.
Li, Zhonghua, David Zhengwen Zhang, Peng Dong, and Ibrahim Kucukkoc. 2017. "A
    Lightweight and Support-Free Design Method for Selective Laser Melting." *International
    Journal of Advanced Manufacturing Technology* 90 (9–12). The International Journal of
    Advanced Manufacturing Technology: 2943–2953. doi:10.1007/s00170-016-9509-0.
Liang, Zulei, Zhonggang Sun, Wenshu Zhang, Shikai Wu, and Hui Chang. 2019. "The Effect
    of Heat Treatment on Microstructure Evolution and Tensile Properties of Selective Laser
    Melted Ti6Al4V Alloy." *Journal of Alloys and Compounds* 782. Elsevier B.V: 1041–
    1048. doi:10.1016/j.jallcom.2018.12.051.
Liu, S.Y., H.Q. Li, C.X. Qin, R Zong, and X.Y. Fang. 2020. "The Effect of Energy Density on
    Texture and Mechanical Anisotropy in Selective Laser Melted Inconel 718." *Materials
    & Design* 191 (June). Elsevier: 108642. doi:10.1016/j.matdes.2020.108642.
Maconachie, Tobias, Martin Leary, Jianjun Zhang, Alexander Medvedev, Avik Sarker, Dong
    Ruan, Guoxing Lu, Omar Faruque, and Milan Brandt. 2020. "Effect of Build Orientation
    on the Quasi-Static and Dynamic Response of SLM AlSi10Mg." *Materials Science and
    Engineering: A* 788 (June). Elsevier: 139445. doi:10.1016/j.msea.2020.139445.
Meier, Christoph, Reimar Weissbach, Johannes Weinberg, Wolfgang A Wall, and A John Hart.
    2019. "Critical Influences of Particle Size and Adhesion on the Powder Layer Uniformity
    in Metal Additive Manufacturing." *Journal of Materials Processing Technology* 266
    (April). Elsevier: 484–501. doi:10.1016/j.jmatprotec.2018.10.037.
Meng, Liang, Weihong Zhang, Dongliang Quan, Guanghui Shi, Lei Tang, Yuliang Hou, Piotr
    Breitkopf, Jihong Zhu, and Tong Gao. 2020. "From Topology Optimization Design
    to Additive Manufacturing: Today's Success and Tomorrow's Roadmap." *Archives
    of Computational Methods in Engineering* 27 (3). Springer Netherlands: 805–830.
    doi:10.1007/s11831-019-09331-1.
Nakamoto, Takayuki, Nobuhiko Shirakawa, Yoshio Miyata, and Haruyuki Inui. 2009. "Selective
    Laser Sintering of High Carbon Steel Powders Studied as a Function of Carbon
    Content." *Journal of Materials Processing Technology* 209 (15–16). Elsevier: 5653–
    5660. doi:10.1016/j.jmatprotec.2009.05.022.
Nguyen, Quy Bau, Mui Ling Sharon Nai, Zhiguang Zhu, Chen-Nan Sun, Jun Wei, and Wei Zhou.
    2017. "Characteristics of Inconel Powders for Powder-Bed Additive Manufacturing."
    Engineering 3 (5). Elsevier: 695–700. doi:10.1016/J.ENG.2017.05.012.
Park, Tae-Hyun, Min-Seok Baek, Holden Hyer, Yongho Sohn, and Kee-Ahn Lee. 2021. "Effect
    of Direct Aging on the Microstructure and Tensile Properties of AlSi10Mg Alloy
    Manufactured by Selective Laser Melting Process." *Materials Characterization* 176
    (June). Elsevier: 111113. doi:10.1016/j.matchar.2021.111113.
Pellens, Jeroen, Geert Lombaert, Manuel Michiels, Tom Craeghs, and Mattias Schevenels.
    2020. "Topology Optimization of Support Structure Layout in Metal-Based Additive

Manufacturing Accounting for Thermal Deformations." *Structural and Multidisciplinary Optimization* 61 (6). Structural and Multidisciplinary Optimization: 2291–2303. doi:10.1007/s00158-020-02512-8.

Röttger, Arne, Karina Geenen, Matthias Windmann, Florian Binner, and Werner Theisen. 2016. "Comparison of Microstructure and Mechanical Properties of 316 L Austenitic Steel Processed by Selective Laser Melting with Hot-Isostatic Pressed and Cast Material." *Materials Science and Engineering A* 678 (April). Elsevier: 365–376. doi:10.1016/j.msea.2016.10.012.

Shi, Xuezhi, Cheng Yan, Wuwei Feng, Yulian Zhang, and Zhe Leng. 2020. "Effect of High Layer Thickness on Surface Quality and Defect Behavior of Ti-6Al-4V Fabricated by Selective Laser Melting." *Optics & Laser Technology* 132 (December). Elsevier: 106471. doi:10.1016/j.optlastec.2020.106471.

Shipley, H., D. McDonnell, M. Culleton, R. Coull, R. Lupoi, G. O'Donnell, and D. Trimble. 2018. "Optimisation of Process Parameters to Address Fundamental Challenges during Selective Laser Melting of Ti-6Al-4V: A Review." *International Journal of Machine Tools and Manufacture* 128 (January). Elsevier Ltd: 1–20. doi:10.1016/j.ijmachtools.2018.01.003.

Sun, Pei, Z Zak Fang, Yang Xia, Ying Zhang, and Chengshang Zhou. 2016. "A Novel Method for Production of Spherical Ti-6Al-4V Powder for Additive Manufacturing." *Powder Technology* 301 (November). Elsevier: 331–335. doi:10.1016/j.powtec.2016.06.022.

Tan, Chaolin, Kesong Zhou, Wenyou Ma, Panpan Zhang, Min Liu, and Tongchun Kuang. 2017. "Microstructural Evolution, Nanoprecipitation Behavior and Mechanical Properties of Selective Laser Melted High-Performance Grade 300 Maraging Steel." *Materials & Design* 134 (November). Elsevier Ltd: 23–34. doi:10.1016/j.matdes.2017.08.026.

Tascioglu, Emre, Yusuf Karabulut, and Yusuf Kaynak. 2020. "Influence of Heat Treatment Temperature on the Microstructural, Mechanical, and Wear Behavior of 316L Stainless Steel Fabricated by Laser Powder Bed Additive Manufacturing." *The International Journal of Advanced Manufacturing Technology* 107 (5–6). Springer: 1947–1956. doi:10.1007/s00170-020-04972-0.

Thijs, Lore, Frederik Verhaeghe, Tom Craeghs, Jan Van Humbeeck, and Jean Pierre Kruth. 2010. "A Study of the Microstructural Evolution during Selective Laser Melting of Ti-6Al-4V." *Acta Materialia* 58 (9). Acta Materialia Inc.: 3303–3312. doi:10.1016/j.actamat.2010.02.004.

Tillmann, W., C. Schaak, J. Nellesen, M. Schaper, M. E. Aydinöz, and K. P. Hoyer. 2017. "Hot Isostatic Pressing of IN718 Components Manufactured by Selective Laser Melting." *Additive Manufacturing* 13. Elsevier B.V.: 93–102. doi:10.1016/j.addma.2016.11.006.

Torralba, J. M. 2014. *Improvement of Mechanical and Physical Properties in Powder Metallurgy. Comprehensive Materials Processing.* Vol. 3. Elsevier. doi:10.1016/B978-0-08-096532-1.00316-2.

Vaidya, Rohan, and Sam Anand. 2016. "Optimum Support Structure Generation for Additive Manufacturing Using Unit Cell Structures and Support Removal Constraint." *Procedia Manufacturing* 5. The Author(s): 1043–1059. doi:10.1016/j.promfg.2016.08.072.

Vock, Silvia, Burghardt Klöden, Alexander Kirchner, Thomas Weißgärber, and Bernd Kieback. 2019. "Powders for Powder Bed Fusion: A Review." *Progress in Additive Manufacturing* 4 (4). Springer: 383–397. doi:10.1007/s40964-019-00078-6.

Vrancken, Bey, Lore Thijs, Jean Pierre Kruth, and Jan Van Humbeeck. 2012. "Heat Treatment of Ti6Al4V Produced by Selective Laser Melting: Microstructure and Mechanical Properties." *Journal of Alloys and Compounds* 541. Elsevier B.V.: 177–185. doi:10.1016/j.jallcom.2012.07.022.

Wagner, Marius A, Amir Hadian, Tutu Sebastian, Frank Clemens, Thomas Schweizer, Mikel Rodriguez-Arbaizar, Efrain Carreño-Morelli, and Ralph Spolenak. 2022. "Fused Filament Fabrication of Stainless Steel Structures – from Binder Development to Sintered Properties." *Additive Manufacturing* 49 (January). Elsevier: 102472. doi:10.1016/j.addma.2021.102472.

Wang, C., P. Zhu, Y. H. Lu, and T. Shoji. 2022. "Effect of Heat Treatment Temperature on Microstructure and Tensile Properties of Austenitic Stainless 316L Using Wire and Arc Additive Manufacturing." *Materials Science and Engineering A* 832 (December 2021). doi:10.1016/j.msea.2021.142446.

Wang, Di, Yongqiang Yang, Manhui Zhang, Jianbin Lu, Ruicheng Liu, and Dongming Xiao. 2013. "Study on SLM Fabrication of Precision Metal Parts with Overhanging Structures." In *2013 IEEE International Symposium on Assembly and Manufacturing (ISAM)*, 222–225. IEEE. doi:10.1109/ISAM.2013.6643532.

Wang, Junlan, Bonnie Antoun, Eric Brown, Weinong Chen, Ioannis Chasiotis, Emily Huskins-Retzlaff, Sharlotte Kramer, and Piyush R Thakre. 2018. *Mechanics of Additive and Advanced Manufacturing, Volume 9*. Edited by Junlan Wang, Bonnie Antoun, Eric Brown, Weinong Chen, Ioannis Chasiotis, Emily Huskins-Retzlaff, Sharlotte Kramer, and Piyush R. Thakre. Conference Proceedings of the Society for Experimental Mechanics Series. Cham: Springer International Publishing. doi:10.1007/978-3-319-62834-9.

Wang, L, E.L. Li, H Shen, R.P. Zou, A.B. Yu, and Z.Y. Zhou. 2020. "Adhesion Effects on Spreading of Metal Powders in Selective Laser Melting." *Powder Technology* 363 (March). Elsevier: 602–610. doi:10.1016/j.powtec.2019.12.048.

Xie, Wenqiang, Meihua Zheng, Jieqi Wang, and Xiaoyu Li. 2020. "The Effect of Build Orientation on the Microstructure and Properties of Selective Laser Melting Ti-6Al-4V for Removable Partial Denture Clasps." *The Journal of Prosthetic Dentistry* 123 (1). Elsevier: 163–172. doi:10.1016/j.prosdent.2018.12.007.

Yan, Xingchen, Shuohong Gao, Cheng Chang, Jian Huang, Khashayar Khanlari, Dongdong Dong, Wenyou Ma, Nouredine Fenineche, Hanlin Liao, and Min Liu. 2021. "Effect of Building Directions on the Surface Roughness, Microstructure, and Tribological Properties of Selective Laser Melted Inconel 625." *Journal of Materials Processing Technology* 288 (February). Elsevier: 116878. doi:10.1016/j.jmatprotec.2020.116878.

Yang, Huadong, Shiguang Li, Zhen Li, and Fengchao Ji. 2020. "Experimental and Numerical Study on the Packing Densification of Metal Powder with Gaussian Distribution." *Metals* 10 (11). MDPI: 1401. doi:10.3390/met10111401.

Yu, Hanchen, Fangzhi Li, Zemin Wang, and Xiaoyan Zeng. 2019. "Fatigue Performances of Selective Laser Melted Ti-6Al-4V Alloy: Influence of Surface Finishing, Hot Isostatic Pressing and Heat Treatments." *International Journal of Fatigue* 120 (July 2018): 175–183. doi:10.1016/j.ijfatigue.2018.11.019.

Zhang, Jiong, Qiqiang Cao, and Wen Feng Lu. 2022. "A Review on Design and Removal of Support Structures in Metal Additive Manufacturing." *Materials Today: Proceedings* 70. Elsevier: 407–411. doi:10.1016/j.matpr.2022.09.277.

Zhang, Jiong, Akshay Chaudhari, and Hao Wang. 2019. "Surface Quality and Material Removal in Magnetic Abrasive Finishing of Selective Laser Melted 316L Stainless Steel." *Journal of Manufacturing Processes* 45 (August). Elsevier: 710–719. doi:10.1016/j.jmapro.2019.07.044.

Zhang, Kaiqing, Gengdong Cheng, and Liang Xu. 2019. "Topology Optimization Considering Overhang Constraint in Additive Manufacturing." *Computers and Structures* 212. Elsevier Ltd: 86–100. doi:10.1016/j.compstruc.2018.10.011.

Zhang, Yicha, Zhiping Wang, Yancheng Zhang, Samuel Gomes, and Alain Bernard. 2020. "Bio-Inspired Generative Design for Support Structure Generation and Optimization in Additive Manufacturing (AM)." *CIRP Annals* 69 (1). Elsevier Ltd: 117–120. doi:10.1016/j.cirp.2020.04.091.

4 Machinability of AM Metals

4.1 ALUMINUM-BASED ALLOYS

Aluminum alloys commonly used in additive manufacturing include AlSi10Mg, Al6061, and AlSiMg0.75 (Vimal, Naveen Srinivas, and Rajak 2019; Bai et al. 2018; Bai et al. 2020). Most of them contain a large proportion of silicon elements as the addition of silicon can reduce the eutectic temperature, improve fluidity, and reduce the solidification range and thermal expansion coefficient, thereby reducing crack defects. Generally, due to the high conductivity, low hardness, and low strength, conventional aluminum alloys can be easy to cut with low cutting force and tool wear. Although the hardness and strength of additively manufacture aluminum alloys are higher than the conventional ones, the cutting force does not increase significantly (Zimmermann et al. 2021).

For aluminum alloys, defects, such as pores and unmelted particles caused by the high conductivity and high reflectivity of infrared lasers, are common in additively manufactured parts, and are also called laser-sintered alloys. These defects are generally not found in casted and wrought counterparts, and the difference results in burr and breach formations on the turned surface (Struzikiewicz, Zębala, and Słodki 2019). Only tool marks are observed on the turned surface of casted AlSi10Mg parts. The formation of burrs and breaches is due to the weak bonding of material in the defect regions, which tend to get pulled out from the workpiece during cutting (Aldwell et al. 2017). Even though the defects result in cutting forces comparable to the casted part and a reduction in surface roughness value can still be obtained, the existence of defects in the sample hinders the stable performance of the part. Fortunately, increasing laser energy density by reducing scanning speed can significantly reduce the pores and unmelted particles, which effectively stops the formation of burrs and breaches on the machined surface. The surface roughness of the additively manufactured metal can then be reduced by more than 20 times after milling (Matras 2019). Increasing the cutting speed and cutting tool nose radius can also have a positive impact on the machined surface roughness (i.e., a smoother surface with lower roughness). However, the increase in feed rate and cutting depths will result in poorer machined surface quality (Struzikiewicz and Sioma 2020).

DOI: 10.1201/9781003272601-4

(a)

(b)

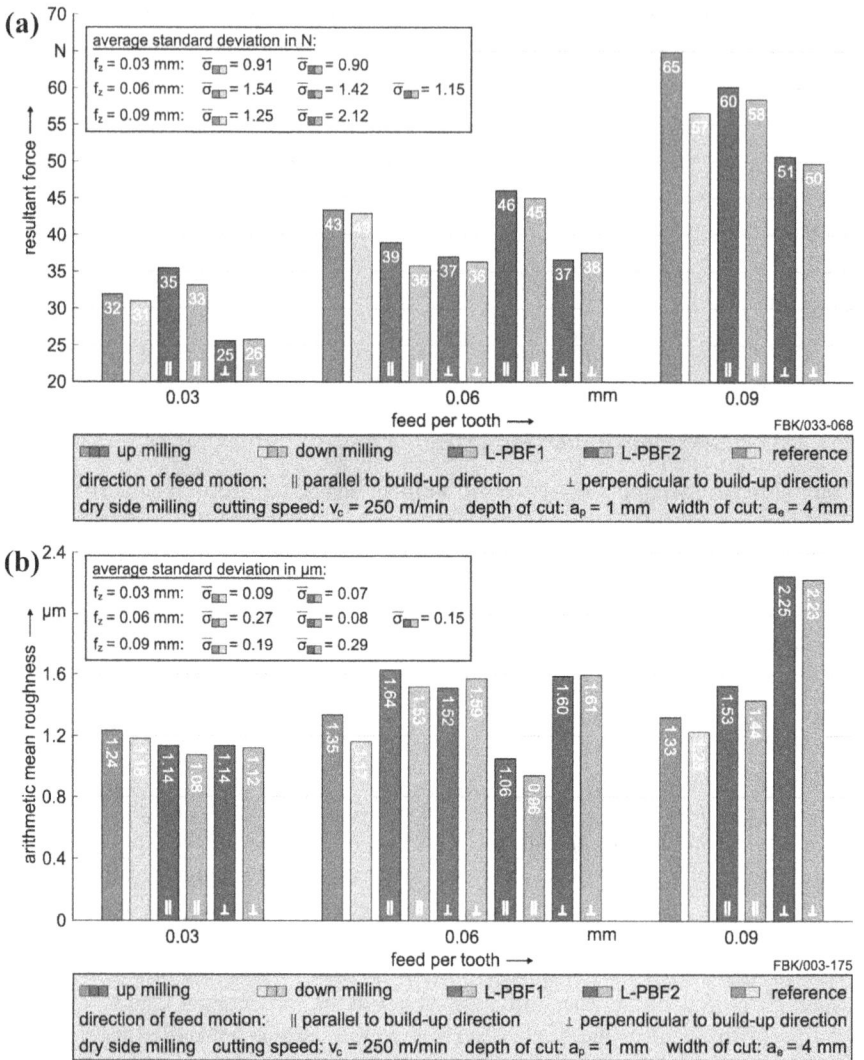

FIGURE 4.1 (a) Resultant forces for different manufacturing methods of the specimens, cutting conditions, and directions of feed motion of the tool relative to the build direction of the workpiece; (b) surface roughness according to the manufacturing method of the material and the cutting condition used during milling. Reprinted with permission from Springer Nature (Zimmermann et al. 2021).

The layer thickness and building direction in additive manufacturing can also affect the machinability of aluminum alloys as shown by the different cutting forces and machined surface roughness for AlSi10Mg in Figure 4.1 (Zimmermann et al. 2021). L-PBF1 had a layer thickness of 60 μm while L-PBF2 had a layer thickness of 30 μm. Cutting forces during milling the top surface (cutting perpendicular to the build direction) were higher than the side surface for a thinner layer thickness. However, higher layer thicknesses showed insignificant differences between the top and side surfaces, as shown by L-PBF1.

Milling forces tend to be greater along the build direction due to the larger number of melt pool boundaries. However, superior surface finishing can be obtained along these surfaces. Cutting chips tend to form more continuously in long spirals as compared to the discontinuous chips produced when milling the casted form. When compared to milling of the casted form, the machined surface quality of the additive manufactured alloy can be smoother with a roughness of approximately 1.28 μm Ra compared to the rougher surface (2.24 μm Ra) on the casted metal. The poorer machined surface quality is attributed to the formation of flaky structures with fragmented protrusions on the machined top surface. However, it is interesting to find that burrs are unlikely to form on the edges of the milled workpiece due to the fine microstructure containing Si-rich cellular structure and low ductility when compared with a casted alloy. Overall, it is understood that the machinability of a defect-free AlSi10Mg alloy is higher with low cutting force and high machined surface quality under suitable milling parameters and pre-cut surface conditions.

However, there are still challenges in machining additively manufactured aluminum alloy due to the existence of hard particles, such as oxide inclusions and silicon particles. These particles are particularly found in newly developed aluminum matrix composites fabricated by additive manufacturing. Hard particles will act as defect sources during machining that tend to get abraded through the relatively softer aluminum matrix. More details on this problem and the solutions identified to resolve them during the manufacturing of optical surfaces will be covered in Chapter 10.

4.2 IRON-BASED ALLOYS

Compared to aluminum alloys, iron-based materials such as 316L stainless steel, maraging steel, ASTM A131 steel, 17-4 stainless steel, duplex stainless steel, high-strength low-alloy steel, among others, are more difficult to cut. Additively manufactured iron-based alloys generally have higher yield strength due to the strong hindrance of dislocation movement through the fine cellular structures caused by the high cooling rate (Liu et al. 2018). These alloys can be divided into two categories that depend on the occurrence of phase changes during additive manufacturing: (1) no phase change after solidification (e.g., 316L stainless steel), and (2) phase change after solidification (e.g., maraging steel). Different material types will have different machinability when compared to the casted/wrought counterparts and considering the effect of building direction.

One of the most commonly used metals in AM is 316L stainless steel (austenite stainless steel). However, its high hardening rate, low thermal conductivity, and highly plastic properties make it a difficult-to-machine material. Although additively manufactured 316L stainless steel has a fine cellular microstructure that differs from the coarse polygonal grains in conventional counterparts, the machinability of the metal continues to remain poor as the strong epitaxial growth characteristic of the additively manufactured steel forms a strong texture in the part and results in the difference in machinability (Guo et al. 2017).

On the one hand, the surface parallel to the building direction has a uniform micro-structure, higher hardness, and higher tensile strength, which leads to higher cutting forces, more severe tool wear, and high machined-surface roughness. On the other hand, the surface perpendicular to the building direction has a number of large dendritic grains generated by epitaxial nucleation that are relatively easier to machine.

Aside from the commonly reported differences in machinability of build directions, there are also differences in the machinability of skin layers compared to bulk layers. Skin layers of 316L stainless steel tend to be easier to machine and allow longer tool life (Tapoglou and Clulow 2021). This may be due to the relatively high brittleness caused by the higher cooling rates during the building of skin layers. Down milling is also more suitable for AM 316 stainless steel compared to up milling, with results of superior surface quality and tool life extended by 330%.

Maraging steel is another commonly used material in metal additive manufacturing and undergoes a solid-state phase transition from austenite to martensite during fabrication. Moreover, the high cooling rate will lead to a fine microstructure mainly composed of the martensitic phase and high residual tensile stress, which results in differences to machinability when compared with a wrought counterpart (Du, Bai, and Zhang 2018). As-built samples have higher microhardness compared to wrought parts, which results in higher cutting forces (up to 30%). However, the machined surface roughness of the additively manufactured part is lower than the wrought part. The high hardness and relatively low plasticity reduces lateral plastic flow during milling to enable quality surface finishing.

In addition, the compressive stresses typically generated during machining would not be found on the AM metal due to the formation of residual tensile stresses near the surface during the additive manufacturing process. This, however, is only the case for low feed rates. The increase in feed rates will lead to higher magnitudes of compressive stresses according to the relationship between the increase in cutting forces with feed rate. The higher compressive stresses then lead to heavier plastic deformation in the tertiary deformation zone (i.e., on the machined surface). The analysis of residual stresses found on AM metals can also affect the machinability of thin walls due to potential part vibration and deformation during machining (Heigel et al. 2018).

Like 316L stainless steel, the surface parallel to the building direction (ABT) has poorer machinability than the surface perpendicular to the building direction (ABS) (Figure 4.2). However, the underlying mechanism is not the same as there are no large-sized epitaxially grown crystals that form in maraging steel due to the phase transition from austenite to martensite. Instead, the difference in machinability is attributed to the difference where the feed direction is more parallel to the melt tracks when milling the ABT sample as compared to milling the ABS sample (Bai, Zhao, Yang, et al. 2021). These melt tracks strengthen the material and increase the induced cutting forces and tool wear during ABT milling. The machinability of the AM maraging steel after heat treatment is illustrated in Figure 4.2. The sharp increase in strength due to the formation of nanosized particles during ageing treatment causes the tool cutting force and tool wear to sharply increase. Solution treatment (homogenization) alone will not lead to poor machinability but will minorly increase the machined surface roughness, which may be caused by the lateral flow of material

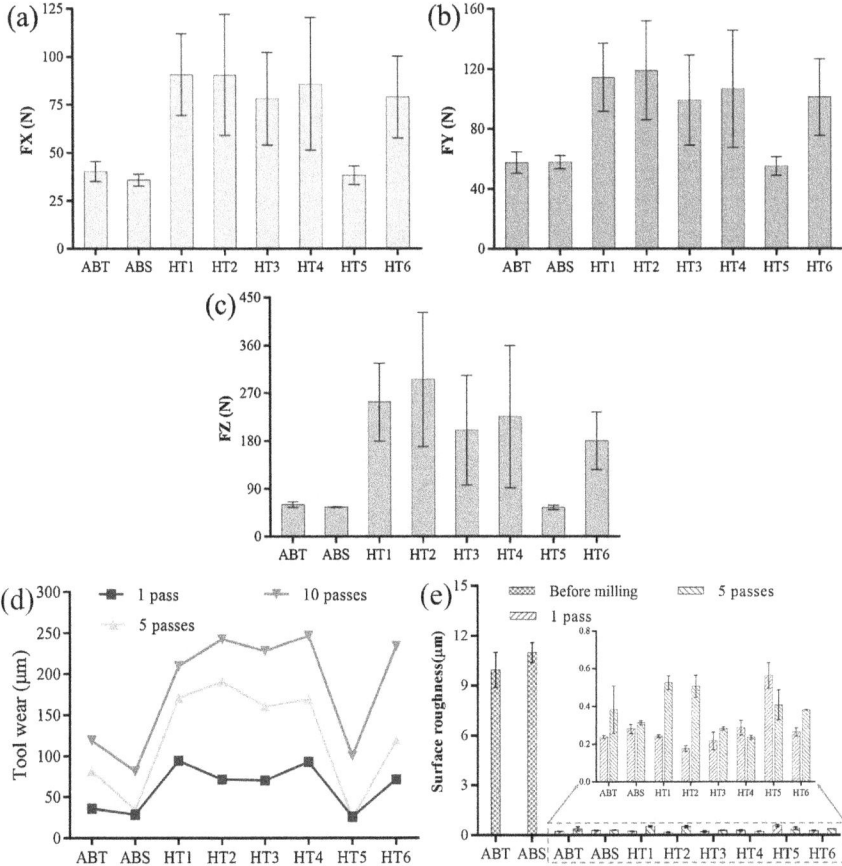

FIGURE 4.2 Machining maraging steel fabricated by SLM and under different heat treatments: (a-c) cutting force, (d) tool wear, and (e) surface roughness. Reprinted with permission from Elsevier (Bai, Zhao, Yang, et al. 2021). (ABT: surface perpendicular to the building direction, ABS: surface parallel to the building direction, 480 °C for 6 h (HT1), 520 °C for 1 h (HT2), 520 °C for 6 h (HT3), 520 °C for 12 h (HT4), 900 °C for 1 h (HT5), and 520 °C for 6 h following the treatment at 900 °C for 1 h (HT6).

during milling. In general, the surface roughness (Ra) of additively manufactured maraging steel can be reduced from ~ 10 to <0.4 μm after milling.

However, the effect of building direction on the machinability of AM metals may depend on the material types. In milling of directed energy deposited ASTM A131 steel, cutting forces on the surface parallel to the building direction is larger than that perpendicular to the building direction (Bai, Chaudhari, and Wang 2020). By comparing the forces with AM aluminum alloy in Figure 4.1, it reveals that the cutting force is much more influenced by the melt pool boundaries when the workpiece has low hardness (soft). It can be explained that small equiaxed crystals that form at the boundary of the relatively soft workpiece will contribute to the deformation resistance during machining. The fine-grain strengthening effect has a dominant influence in this scenario.

Occasionally, the solid phase transition during AM is helpful to improve machinability due to the formation of a machinable phase. An example is the machining of duplex stainless steel (SAF 2507) (Davidson, Littlefair, and Singamneni 2021), which normally consists of austenite and ferrite phases. Duplex stainless steels are considered to have poor machinability due to the high work hardening rates (mainly caused by the austenite phase), high fracture toughness, and low thermal conductivity. At high cooling rates, a predominantly ferritic microstructure forms in selective laser melted duplex stainless steel. It is known that the ferritic phase has higher machinability due to its relatively lower strength, work hardening, and tool wear characteristics as compared to duplex stainless steel. For example, when drilling wrought duplex stainless steels, work-hardening can cause periodic compression in microstructures at almost 30 μm depth from the free surface. These features would not be observed in ferritic-structured duplex stainless steels. In addition, discontinuous chips are typically obtained during the drilling of the AM duplex stainless steel due to its low ductility, which may contribute to the low cutting temperature and cutting force. The low ductility would also indicate greater tendency for built-up edge formation, resulting in increased tool damage especially when employing high cutting speeds.

4.3   TITANIUM ALLOYS

Titanium alloy is widely used in the biomedical and aerospace industries, where there are multiple parts with unique geometries. Hence, additive manufacturing is an attractive choice when it comes to fabricating titanium alloy parts (Zhang et al. 2018). However, titanium alloys are also difficult-to-machine materials due to their high thermal reactivity, low thermal conductivity, and high hardness. In most studies on the machinability of AM titanium alloy parts, Ti6Al4V has been the main material used in various metal additive manufacturing techniques. The machinability of the AM titanium alloy is quite different from wrought counterparts as solid phase transition occurs with a small difference in pre-heating temperature and cooling rates during electron beam melting (EBM), selective laser melting (SLM), and directed energy deposition (DED).

As shown in Figure 4.3, Ti6Al4V alloy parts fabricated by EBM, DMLS (also known as SLM), heat-treated, and wrought variations have different microstructures (Sartori et al. 2016). The alloy fabricated by EBM has a fine acicular microstructure composed of α phase and a fine lamella with 7% of β phase at the grain boundaries due to the high pre-heating temperature. The alloy fabricated by DMLS is characterized by a martensitic microstructure consisting of α phase and acicular α′ phase with lattice parameters similar to the HCP crystal due to the high cooling rate without high pre-heating temperature. Heat-treatment of the alloy fabricated by DMLS will form α + β lamellar structure due to the solid phase transition during heat treatment. The microstructure of the wrought sample is composed of equiaxed α grains with 8% of β grains. The alloy produced by DMLS will have the highest strength, highest hardness, and lowest thermal conductivity compared to the other three. These properties lead

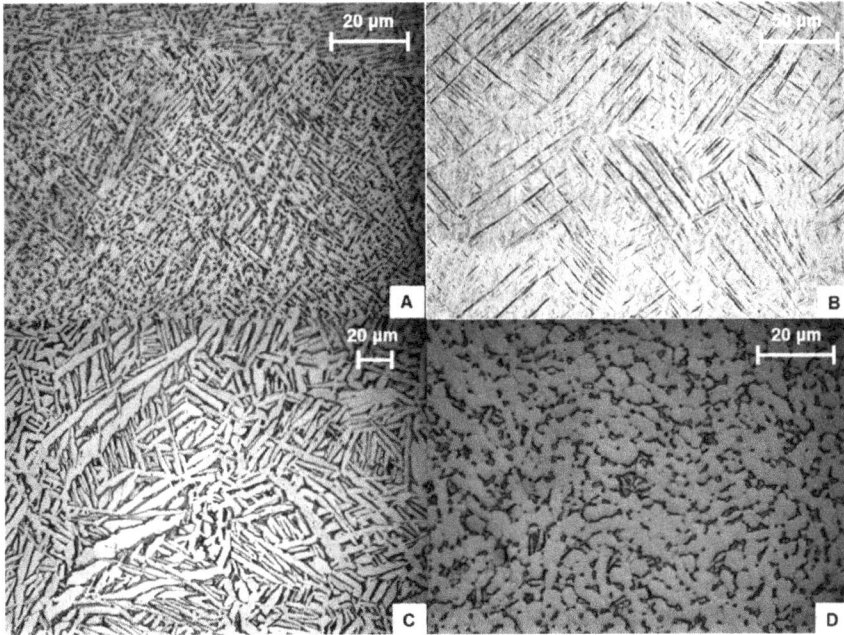

FIGURE 4.3 Microstructures of the Ti6Al4V alloys produced by: (a) EBM, (b) DMLS, (c) heat-treated and (d) wrought. Reprinted with permission from Elsevier (Sartori et al. 2016).

to the most severe tool wear during dry turning. In comparison, the alloy produced by EBM will result in lower tool wear although the wear issue cannot be entirely avoided.

Although the Ti6Al4V alloy produced by SLM has higher strength (approximately 16% higher in hardness) than a conventionally counterpart, the cutting force is 9.3% lower during micromilling (De Oliveira Campos Et Al. 2020). In addition, the formation of burr is lower along with the reduction in machined surface roughness for the SLM part. These are due to the formation of the acicular α' phase, which causes the reduction in ductility of the SLM part compared to the conventional material with equiaxed $\alpha + \beta$ grains. This prominent brittle behavior makes chip removal easier and reduces material accumulation in the cutting zone due to the heavy plastic deformation of the work material. In addition, the fine $\alpha' + \alpha$ grains have similar grain morphologies throughout the whole part compared to the $\alpha + \beta$ grains in the conventional material. Grains with similar morphology make it more conducive for dislocation movement and results in comparatively lower work hardening. Therefore, the cutting force of the alloy fabricated by SLM would be lower than that by EBM (Milton et al. 2021). It should be mentioned that the cutting force in micromilling of the Ti6Al4V alloy produced by DED is also lower than the conventional parts for the same reasons associated with the microstructure (Bonaiti et al. 2017).

The effect of building direction on machinability can also be observed in machining the additively manufactured Ti6Al4V alloy parts (Ni et al. 2020). In dry ultra-precision turning (micromachining) of the alloy produced by SLM, average cutting forces and force fluctuations on the front surfaces (i.e., surfaces parallel to the building direction) will be greater than those generated on the top surfaces (i.e., perpendicular to the building direction). The machined surface roughness (Ra) on the front surfaces will also be larger than the front surfaces by 21%–76%, depending on the scanning strategy (Ni et al. 2020). Similar results on the aniso-tropic cutting forces will also be observed in milling of the alloy fabricated by EBM (Milton et al. 2021).

Cutting force trends in machining titanium alloys differ from aluminum and iron-based alloys where the other materials are largely affected by the melt pool bound-aries. There are no observable melt pool boundaries in as-built titanium alloy parts due to extremely strong epitaxial growth of the primary β phase along the building direction and subsequent martensitic transformation. However, the reconstructed primary β phase inverse pole figure has a strong texture, which will constrain the subsequent martensitic transformation and its orientation distribution relative to the building direction. This leads to the difference in microstructure and texture between the front and top surfaces and is the most probable cause for the anisotropy in cutting force.

Although there is a difference in the machinability between the AM and conven-tional Ti6Al4V alloy parts, both types of alloys are still categorized as difficult-to-machine materials. Therefore, the methods employed to improve the machinability of conventional materials can also be applied to AM parts. For instance, cryogenic machining can be used to overcome the low thermal conductivity and improve machinability. The use of liquid nitrogen during machining of the titanium alloy fabricated by DMLS can reduce the crater wear depth in the cutting tool by 58% (Sartori et al. 2016).

Minimum quantity lubrication (MQL) has been shown to reduce the friction between the flank face and the machined surface to reduce flank wear and improve surface quality by 26.2% and 55%, respectively (Khaliq et al. 2020). Flood coolants are also helpful to improve surface quality, but the final result will still depend on the machining parameters (Bertolini et al. 2019).

4.4 SUPERALLOYS

A superalloy has the ability to operate at a high fraction of its melting point, which can be divided into three categories – iron-based, nickel-based, and cobalt-based. Among these materials, nickel-based superalloys are widely studied in additive manu-facturing, such as IN 718, IN 625, and 939 (Pratheesh Kumar et al. 2021). A super-alloy is also difficult-to-machine (more difficult than austenite stainless steel). Like most other materials, the as-built surface roughness of a superalloy is far from com-mercially acceptable conditions but a reduction by more than 90% in roughness can be achieved by post-process machining (Kaynak and Tascioglu 2020; Kaynak and Tascioglu 2018).

Generally, AM metals will have high hardness and strength derived from the high cooling rates that form fine microstructures. However, in most cases of AM superalloys, the hardness is slightly lower than the wrought counterpart (Chen et al. 2021). This is due to the low density (Moussaoui et al. 2018) and formation of Laves phase (Zhao et al. 2020), that is the absence of γ' and γ' precipitated phase particles essential for hardening. The low hardness is desirable during machining of the AM Inconel alloys due to the lower cutting forces (by ~10%), cutting temperatures (by ~6%), and cutting vibration (by ~17%) compared to the wrought counterpart. Moreover, the absence of hard particles in AM IN 718 alloy results in lower tool wear compared to the wrought one. Higher surface quality with lower surface roughness can also obtained on the machined additively manufactured part, as shown in Figure 4.4(a) (Ji et al. 2021). However, heat treatment would not further improve the machinability of the AM Inconel alloy due to the formation of precipitation particles (Careri et al. 2021). Due to the strong epitaxial growth, IN 718 alloy fabricated by SLM has large columnar dendritic grains and a strong <100> fiber texture along the building direction compared to a wrought sample with fine equiaxed crystals and weak texture (Figure 4.4(b)). In addition, the AM sample will present slight grain coarsening and texture weakening due to recrystallization caused by heat generated during cutting.

The strong texture and columnar crystals along the building direction formed in the additively manufactured parts imply anisotropy in the mechanical properties and machinability. Cutting forces on the side surface (parallel to the building direction) will be larger than that on the top surface (perpendicular to the building direction) (Pérez-Ruiz et al. 2021), which is similar to the observation in machining AM Ti6Al4V alloys. This is due to the major axis of the columnar grains being nearly parallel to the milling tool axis while milling the side surface. In addition, the grains engaged by the cutting tool during machining will be parallel to each other and will facilitate grain boundary sliding during material deformation (Pérez-Ruiz et al. 2021). The machined surface quality of the side surface will be slightly inferior compared to the top surface as observed during milling of AM IN 939 alloy (Sen et al. 2020). This can also be observed for alloys fabricated by DED, which will exhibit larger melt pools compared to SLM (Alonso et al. 2021). This would suggest that a change

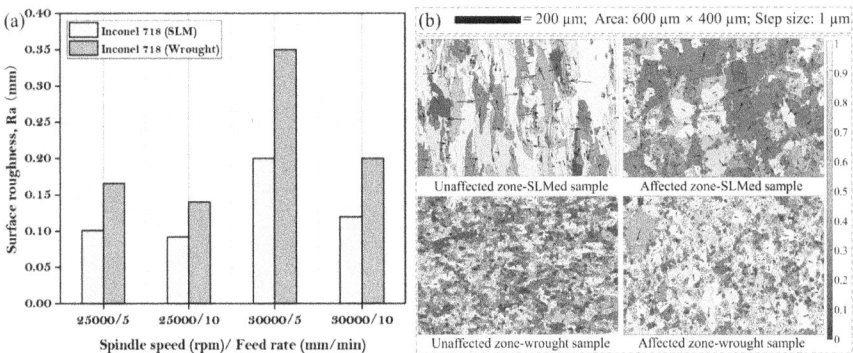

FIGURE 4.4   (a) Machined surface roughness and (b) grain scale microstructure of affected and unaffected zones in the SLM fabricated and wrought samples. Reprinted with permission from Elsevier (Ji et al. 2021). (Affected zone refers to the regions affected by the milling process).

in the size of the melt pool will have little effect on the anisotropy during machining of the Inconel alloys.

Scanning strategies will influence the anisotropy in cutting force as it will change the microstructural morphology and texture intensity of the AM workpiece (Fei et al. 2019). In general, scanning strategies that favor weakened textures and epitaxial growth can reduce the effect of building direction on machinability.

Although the machinability of as-built Inconel alloys have higher machinability than wrought counterparts, it is still a difficult-to-machine material (Fei et al. 2020). Conventional methods to improve the machinability of the material include emulsion (that is machining under a wet environment) and the use of MQL. Energy consumption during machining can be reduced by approximately 38% with emulsion (Periane et al. 2020), and superior surface quality (achievable roughness of <0.15 μm) can be obtained with MQL.

4.5   HIGH-ENTROPY ALLOYS

High-entropy alloys are formed using five or more metals in equal or approximately equal proportions. Compared to traditional alloys, these alloys have much better specific strength, fracture resistance, tensile strength, corrosion resistance, and oxidation resistance. These properties drive the interest in exploring the manufacturing of high-entropy alloys using additive manufacturing (Ostovari Moghaddam et al. 2021). However, the available understanding on the machinability of AM high-entropy alloys remains limited.

It is interesting that the superior mechanical properties of the high-entropy alloys do not affirm notions that it would be as difficult to machine as conventional alloy steels such as 316L stainless steel. On the contrary, the alloys seem to exhibit good machinability with little tool wear and achievable machined surface roughness in the nanoscale through ultra-precision machining as reported for Al80Li5Mg5Zn5Cu5 (Huang et al. 2020). The high machinability is also observed for additively manufactured high entropy alloys. The machinability of CrMnFeCoNi fabricated by SLM is better than 304L stainless steel with much lower cutting forces and tool wear, as shown in Figure 4.5 (Litwa et al. 2021). Thermal softening might be the cause for the ease in machinability of the CrMnFeCoNi alloy compared to surface hardening phenomena that occurs in machining of 304L stainless steel. In addition, the inherently lower hardness and higher ductility also contribute to the ease in machinability. However, the softness of the material may affect the machined surface quality, such as the increase in surface material plastic deformation under high cutting speed, which would increase the machined surface roughness.

4.6   OTHER METALS

Cobalt-chromium-molybdenum (CoCrMo) alloys are promising materials with excellent biocompatibility, high wear resistance, and high corrosion resistance. These properties make them applicable materials for biomedical parts such as prostheses and many other medical devices. However, the high wear resistance is a challenge for machining and often results in high tool wear. Severe tool wear can be observed when

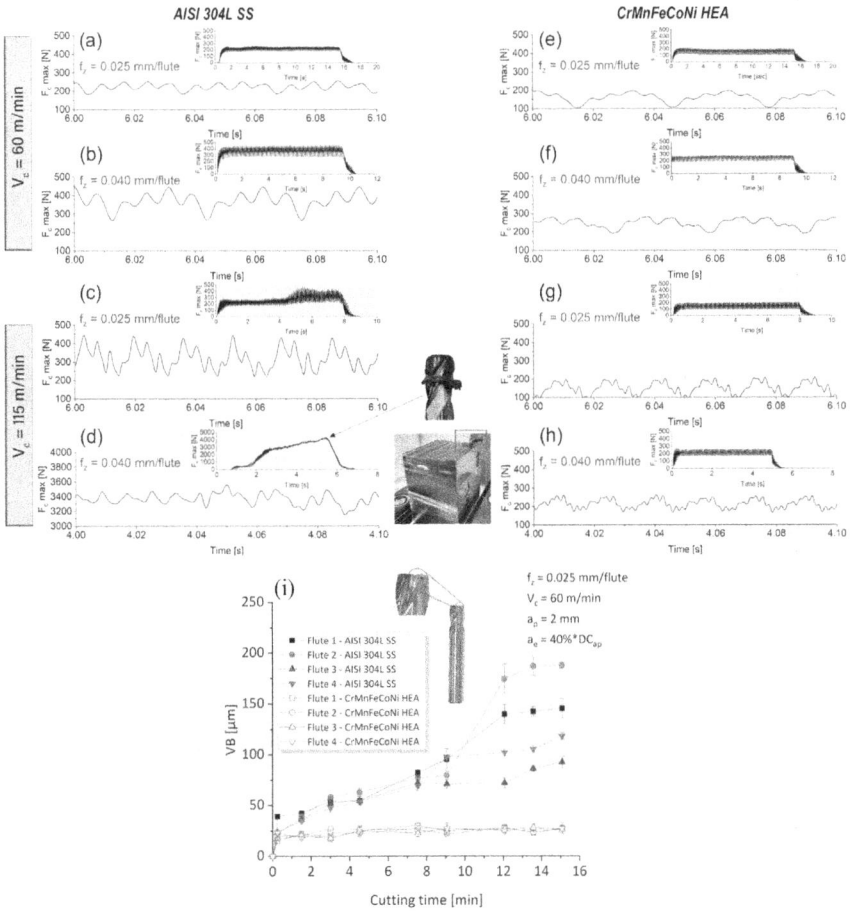

FIGURE 4.5 The resultant cutting forces as a function of cutting time in the machining of (a–d) AISI 304L stainless steel and (e–h) CrMnFeCoNi high entropy alloy; (i) Tool wear (VB) as a function of cutting time. Reprinted with permission from Elsevier (Litwa et al. 2021).

machining AM CoCrMo alloys due to the formation of equiaxed grains with massive and fine distribution of carbides (Bordin et al. 2014). This microstructure also leads to the increase in hardness and strength. As illustrated in Figure 4.6, severe tool wear is observed when turning wrought CoCrMo alloy but the wear becomes even more severe when turning the CoCrMo alloy fabricated by EBM despite the use of 5% mineral oil and water emulsion during machining. This is partly due to the harder surface of the additively manufactured alloy, which has a higher micro-hardness of > 1,000 HV compared to the wrought alloy with < 690 HV (Allegri et al. 2019). As the tools wear, the machined surface integrity of the additively manufactured alloy would be worse than the wrought material with the formation of macroscopic craters on the turned surface. These machined surface defects will increase in distribution with the increase in cutting speed and material removal rates.

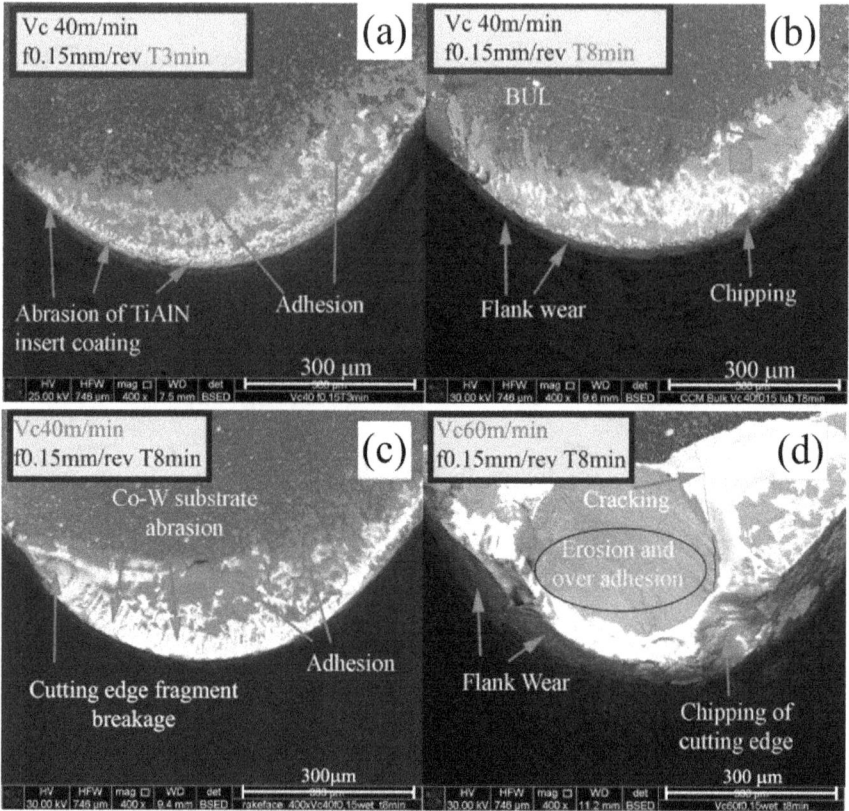

FIGURE 4.6 Tool wear during turning: (a, b) wrought alloy, and (c, d) additively manufactured CoCrMo alloys. Reprinted with permission from Elsevier (Bordin et al. 2014).

Additively manufactured copper alloys exhibit a drastic influence of the build direction on the machinability as observed for CuCrZr alloy during ultra-precision diamond turning (Bai, Zhao, Zhang, et al. 2021). The anisotropy is derived from the epitaxial grain growth during the additive manufacturing process like those observed in Inconel where the coarse columnar grains are oriented with the major axis along the building direction (Ma et al. 2020).

These grains tend to have a strong texture with the <110> direction parallel to the building direction (Wegener et al. 2021), which contribute to the anisotropy in mechanical properties. The microstructure morphology, orientation, and texture enhances the plasticity of the material along the building direction, which is also associated with the relationship between the cutting direction to facilitate grain boundary sliding during material deformation.

In this aspect, turning of the side surface (i.e., parallel to the building direction) would give a superior surface roughness compared to the top surface. The corresponding surfaces are given in Figures 4.7(a) and 4.7(b). Heat treatment of the surface would serve little impact, as shown in Figure 4.7(c) where the coarse regions

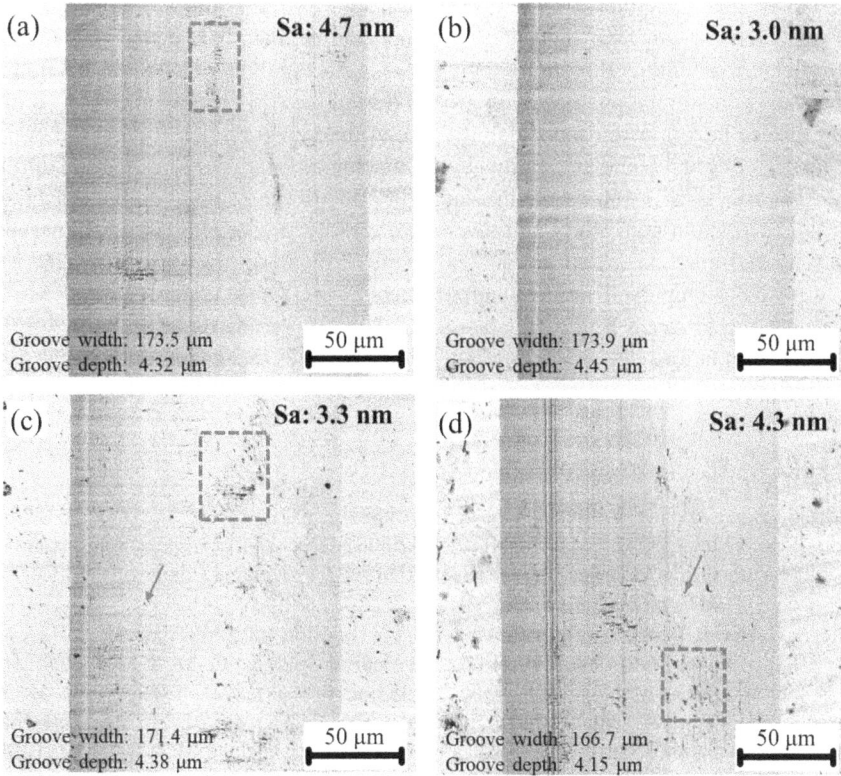

FIGURE 4.7 Groove morphology and surface roughness when cutting CuCrZr alloy manufactured by SLM on the: (a) as-built top surface, (b) as-built side surface, (c) solution-treated top surface, and (d) solution+ageing-treated top surface (cutting speed: 100 mm/min, depth of cut: 5 μm). Reprinted with permission from Elsevier (Bai, Zhao, Zhang, et al. 2021) (Red dotted box: coarse region).

on the machined surface would manifest due to the strong texture. Further heat treatment by solution and ageing treatment would result in the formation of Cr and Cu-Zr particles, which would cause the machined surface roughness to deteriorate, as shown in Figure 4.7(d). It is therefore evident that the challenge to machine AM copper alloys is due to its strong texture generated during the additive manufacturing process.

4.7 CONCLUDING REMARKS

The layer-by-layer manufacturing method and high cooling rate will induce unique microstructures and mechanical properties in additively manufactured metals, which lead to machinability that differ from conventional counterparts. Overall, AM metals have higher strength and poorer machinability compared to conventional materials except for duplex stainless steels and Inconel alloys. In addition, directionally dependent machinability is widely observed with two possible scenarios. One in softer materials with clear melt pools and boundaries that result in the side surface (parallel to the building direction) having poorer machinability, and another found in materials with strong epitaxial grain growth during the AM process that result in the side surface having better machinability due to the low grain density and the easier facilitation of grain boundary sliding during machining. Hard particles formed during the AM process will cause poor machinability due to the damage on the machined surface (e.g., aluminum alloy) and severe tool wear (e.g., CoCrMo alloy). Generally, conventional methods used to accommodate difficult-to-machine materials can be effectively employed on AM metals such as cryogenic conditions and MQL. However, the different heat histories still need to be considered when applying the different adaptations due to the different microstructures that can arise from the fabrication process.

REFERENCES

Aldwell, Barry, Elaine Kelly, Ronan Wall, Andrea Amaldi, Garret E. O'Donnell, and Rocco Lupoi. 2017. "Machinability of Al 6061 Deposited with Cold Spray Additive Manufacturing." *Journal of Thermal Spray Technology* 26 (7). Springer US: 1573–1584. doi:10.1007/s11666-017-0586-x.

Allegri, Gabriele, Alessandro Colpani, Paola Serena Ginestra, and Aldo Attanasio. 2019. "An Experimental Study on Micro-Milling of a Medical Grade Co-Cr-Mo Alloy Produced by Selective Laser Melting." *Materials* 12 (13). doi:10.3390/ma12132208.

Alonso, Unai, Fernando Veiga, Alfredo Suárez, and Alain Gil Del Val. 2021. "Characterization of Inconel 718® Superalloy Fabricated by Wire Arc Additive Manufacturing: Effect on Mechanical Properties and Machinability." *Journal of Materials Research and Technology* 14: 2665–2676. doi:10.1016/j.jmrt.2021.07.132.

Bai, Yuchao, Akshay Chaudhari, and Hao Wang. 2020. "Investigation on the Microstructure and Machinability of ASTM A131 Steel Manufactured by Directed Energy Deposition." *Journal of Materials Processing Technology* 276. Elsevier: 116410. doi:10.1016/j.jmatprotec.2019.116410.

Bai, Yuchao, Zhuoqi Shi, Yan Jin Lee, and Hao Wang. 2020. "Optical Surface Generation on Additively Manufactured AlSiMg0.75 Alloys with Ultrasonic Vibration-Assisted Machining." *Journal of Materials Processing Technology* 280. Elsevier: 116597. doi:10.1016/j.jmatprotec.2020.116597.

Bai, Yuchao, Yongqiang Yang, Zefeng Xiao, Mingkang Zhang, and Di Wang. 2018. "Process Optimization and Mechanical Property Evolution of AlSiMg0.75 by Selective Laser Melting." *Materials and Design* 140. Elsevier Ltd: 257–266. doi:10.1016/j.matdes.2017.11.045.

Bai, Yuchao, Cuiling Zhao, Jin Yang, Ruochen Hong, Can Weng, and Hao Wang. 2021. "Microstructure and Machinability of Selective Laser Melted High-Strength Maraging Steel with Heat Treatment." *Journal of Materials Processing Technology* 288 (February). Elsevier: 116906. doi:10.1016/j.jmatprotec.2020.116906.

Bai, Yuchao, Cuiling Zhao, Yu Zhang, Jie Chen, and Hao Wang. 2021. "Additively Manufactured CuCrZr Alloy: Microstructure, Mechanical Properties and Machinability." *Materials Science and Engineering: A* 819 (July). Elsevier: 141528. doi:10.1016/j.msea.2021.141528.

Bertolini, R., L. Lizzul, L. Pezzato, A. Ghiotti, and S. Bruschi. 2019. "Improving Surface Integrity and Corrosion Resistance of Additive Manufactured Ti6Al4V Alloy by

Cryogenic Machining." *International Journal of Advanced Manufacturing Technology* 104 (5–8). The International Journal of Advanced Manufacturing Technology: 2839–2850. doi:10.1007/s00170-019-04180-5.

Bonaiti, Giuseppe, Paolo Parenti, Massimiliano Annoni, and Shiv Kapoor. 2017. "Micro-Milling Machinability of DED Additive Titanium Ti-6Al-4V." *Procedia Manufacturing* 10. Elsevier B.V.: 497–509. doi:10.1016/j.promfg.2017.07.104.

Bordin, A., A. Ghiotti, S. Bruschi, L. Facchini, and F. Bucciotti. 2014. "Machinability Characteristics of Wrought and EBM CoCrMo Alloys." *Procedia CIRP* 14. Elsevier B.V.: 89–94. doi:10.1016/j.procir.2014.03.082.

Careri, Francesco, Domenico Umbrello, Khamis Essa, Moataz M. Attallah, and Stano Imbrogno. 2021. "The Effect of the Heat Treatments on the Tool Wear of Hybrid Additive Manufacturing of IN718." *Wear* 470–471 (January). Elsevier B.V.: 203617. doi:10.1016/j.wear.2021.203617.

Chen, Liyong, Qingzhong Xu, Yan Liu, Gangjun Cai, and Jichen Liu. 2021. "Machinability of the Laser Additively Manufactured Inconel 718 Superalloy in Turning." *International Journal of Advanced Manufacturing Technology* 114 (3–4). The International Journal of Advanced Manufacturing Technology: 871–882. doi:10.1007/s00170-021-06940-8.

Davidson, Karl Peter, Guy Littlefair, and Sarat Singamneni. 2021. "On the Machinability of Selective Laser Melted Duplex Stainless Steels." *Materials and Manufacturing Processes* 00 (00). Taylor & Francis: 1–17. doi:10.1080/10426914.2021.2001513.

de Oliveira Campos, Fábio, Anna Carla Araujo, André Luiz Jardini Munhoz, and Shiv Gopal Kapoor. 2020. "The Influence of Additive Manufacturing on the Micromilling Machinability of Ti6Al4V: A Comparison of SLM and Commercial Workpieces." *Journal of Manufacturing Processes* 60 (September): 299–307. doi:10.1016/j.jmapro.2020.10.006.

Du, Wei, Qian Bai, and Bi Zhang. 2018. "Machining Characteristics of 18Ni-300 Steel in Additive/Subtractive Hybrid Manufacturing." *International Journal of Advanced Manufacturing Technology* 95 (5–8). The International Journal of Advanced Manufacturing Technology: 2509–2519. doi:10.1007/s00170-017-1364-0.

Fei, Jixiong, Guoliang Liu, Kaushalendra Patel, and Tuğrul Özel. 2020. "Effects of Machining Parameters on Finishing Additively Manufactured Nickel-Based Alloy Inconel 625." *Journal of Manufacturing and Materials Processing* 4 (2). doi:10.3390/jmmp4020032.

Fei, Jixiong, Liu Guoliang, Patel Kaushalendra, and Tuğrul Özel. 2019. "Cutting Force Investigation in Face Milling of Additively Fabricated Nickel Alloy 625 via Powder Bed Fusion." *International Journal of Mechatronics and Manufacturing Systems* 12 (3/4): 196.

Guo, Peng, Bin Zou, Chuanzhen Huang, and Huabing Gao. 2017. "Study on Microstructure, Mechanical Properties and Machinability of Efficiently Additive Manufactured AISI 316L Stainless Steel by High-Power Direct Laser Deposition." *Journal of Materials Processing Technology* 240. Elsevier B.V.: 12–22. doi:10.1016/j.jmatprotec.2016.09.005.

Heigel, Jarred C., Thien Q. Phan, Jason C. Fox, and Thomas H. Gnaupel-Herold. 2018. "Experimental Investigation of Residual Stress and Its Impact on Machining in Hybrid Additive/Subtractive Manufacturing." *Procedia Manufacturing* 26. Elsevier B.V.: 929–940. doi:10.1016/j.promfg.2018.07.120.

Huang, Zhiyuan, Yuqi Dai, Zhen Li, Guoqing Zhang, Chuntao Chang, and Jiang Ma. 2020. "Investigation on Surface Morphology and Crystalline Phase Deformation of Al80Li5Mg5Zn5Cu5 High-Entropy Alloy by Ultra-Precision Cutting." *Materials and Design* 186. The Authors: 108367. doi:10.1016/j.matdes.2019.108367.

Ji, Hansong, Munish Kumar Gupta, Qinghua Song, Wentong Cai, Tao Zheng, Youle Zhao, Zhanqiang Liu, and Danil Yu Pimenov. 2021. "Microstructure and Machinability Evaluation in Micro Milling of Selective Laser Melted Inconel 718 Alloy." *Journal of Materials Research and Technology* 14. Elsevier Ltd: 348–362. doi:10.1016/j.jmrt.2021.06.081.

Kaynak, Yusuf, and Emre Tascioglu. 2018. "Finish Machining-Induced Surface Roughness, Microhardness and XRD Analysis of Selective Laser Melted Inconel 718 Alloy." *Procedia CIRP* 71. Elsevier B.V.: 500–504. doi:10.1016/j.procir.2018.05.013.

Kaynak, Yusuf, and Emre Tascioglu. 2020. "Post-Processing Effects on the Surface Characteristics of Inconel 718 Alloy Fabricated by Selective Laser Melting Additive Manufacturing." *Progress in Additive Manufacturing* 5 (2). Springer International Publishing: 221–234. doi:10.1007/s40964-019-00099-1.

Khaliq, Waqas, Chen Zhang, Muhammad Jamil, and Aqib Mashood Khan. 2020. "Tool Wear, Surface Quality, and Residual Stresses Analysis of Micro-Machined Additive Manufactured Ti–6Al–4V under Dry and MQL Conditions." *Tribology International* 151 (April). Elsevier Ltd: 106408. doi:10.1016/j.triboint.2020.106408.

Litwa, Przemyslaw, Everth Hernandez-Nava, Dikai Guan, Russell Goodall, and Krystian K. Wika. 2021. "The Additive Manufacture Processing and Machinability of CrMnFeCoNi High Entropy Alloy." *Materials and Design* 198. The Author(s): 109380. doi:10.1016/j.matdes.2020.109380.

Liu, Leifeng, Qingqing Ding, Yuan Zhong, Ji Zou, Jing Wu, Yu Lung Chiu, Jixue Li, Ze Zhang, Qian Yu, and Zhijian Shen. 2018. "Dislocation Network in Additive Manufactured Steel Breaks Strength–Ductility Trade-Off." *Materials Today* 21 (4). Elsevier Ltd: 354–361. doi:10.1016/j.mattod.2017.11.004.

Ma, Zhibo, Kaifei Zhang, Zhihao Ren, David Z. Zhang, Guibao Tao, and Haisheng Xu. 2020. "Selective Laser Melting of Cu–Cr–Zr Copper Alloy: Parameter Optimization, Microstructure and Mechanical Properties." *Journal of Alloys and Compounds* 828. Elsevier B.V: 154350. doi:10.1016/j.jallcom.2020.154350.

Matras, Andrzej. 2019. "Research and Optimization of Surface Roughness in Milling of SLM Semi-Finished Parts Manufactured by Using the Different Laser Scanning Speed." *Materials* 13 (1): 9. doi:10.3390/ma13010009.

Milton, Samuel, Olivier Rigo, Sebastien LeCorre, Antoine Morandeau, Raveendra Siriki, Philippe Bocher, and René Leroy. 2021. "Microstructure Effects on the Machinability Behaviour of Ti6Al4V Produced by Selective Laser Melting and Electron Beam Melting Process." *Materials Science and Engineering A* 823 (May). doi:10.1016/j.msea.2021.141773.

Moussaoui, K., W. Rubio, M. Mousseigne, T. Sultan, and F. Rezai. 2018. "Effects of Selective Laser Melting Additive Manufacturing Parameters of Inconel 718 on Porosity, Microstructure and Mechanical Properties." *Materials Science and Engineering A* 735 (August). Elsevier B.V.: 182–190. doi:10.1016/j.msea.2018.08.037.

Ni, Chenbing, Lida Zhu, Zhongpeng Zheng, Jiayi Zhang, Yun Yang, Jin Yang, Yuchao Bai, Can Weng, Wen Feng Lu, and Hao Wang. 2020. "Effect of Material Anisotropy on Ultra-Precision Machining of Ti-6Al-4V Alloy Fabricated by Selective Laser Melting." *Journal of Alloys and Compounds* 848 (December): 156457. doi:10.1016/j.jallcom.2020.156457.

Ostovari Moghaddam, Ahmad, Nataliya A. Shaburova, Marina N. Samodurova, Amin Abdollahzadeh, and Evgeny A. Trofimov. 2021. "Additive Manufacturing of High Entropy Alloys: A Practical Review." *Journal of Materials Science and Technology* 77. The editorial office of Journal of Materials Science & Technology: 131–162. doi:10.1016/j.jmst.2020.11.029.

Pérez-Ruiz, José David, Luis Norberto López de Lacalle, Gorka Urbikain, Octavio Pereira, Silvia Martínez, and Jorge Bris. 2021. "On the Relationship between Cutting Forces and Anisotropy Features in the Milling of LPBF Inconel 718 for near Net Shape Parts."

International Journal of Machine Tools and Manufacture 170 (February). doi:10.1016/j.ijmachtools.2021.103801.

Periane, S., A. Duchosal, S. Vaudreuil, H. Chibane, A. Morandeau, M. Anthony Xavior, and R. Leroy. 2020. "Selection of Machining Condition on Surface Integrity of Additive and Conventional Inconel 718." *Procedia CIRP* 87. Elsevier B.V.: 333–338. doi:10.1016/j.procir.2020.02.092.

Pratheesh Kumar, S., S. Elangovan, R. Mohanraj, and J. R. Ramakrishna. 2021. "A Review on Properties of Inconel 625 and Inconel 718 Fabricated Using Direct Energy Deposition." *Materials Today: Proceedings* 46. Elsevier Ltd: 7892–7906. doi:10.1016/j.matpr.2021.02.566.

Sartori, Stefano, Alberto Bordin, Lorenzo Moro, Andrea Ghiotti, and Stefania Bruschi. 2016. "The Influence of Material Properties on the Tool Crater Wear When Machining Ti6Al4 v Produced by Additive Manufacturing Technologies." *Procedia CIRP* 46: 587–590. doi:10.1016/j.procir.2016.04.032.

Sen, Ceren, Levent Subasi, Ozan Can Ozaner, and Akin Orhangul. 2020. "The Effect of Milling Parameters on Surface Properties of Additively Manufactured Inconel 939." *Procedia CIRP* 87. Elsevier B.V.: 31–34. doi:10.1016/j.procir.2020.02.072.

Struzikiewicz, G., W. Zębala, and B. Słodki. 2019. "Cutting Parameters Selection for Sintered Alloy AlSi10Mg Longitudinal Turning." *Measurement: Journal of the International Measurement Confederation* 138: 39–53. doi:10.1016/j.measurement.2019.01.082.

Struzikiewicz, Grzegorz, and Andrzej Sioma. 2020. "Evaluation of Surface Roughness and Defect Formation after the Machining of Sintered Aluminum Alloy AlSi10Mg." *Materials* 13 (7). doi:10.3390/ma13071662.

Tapoglou, Nikolaos, and Joseph Clulow. 2021. "Investigation of Hybrid Manufacturing of Stainless Steel 316L Components Using Direct Energy Deposition." *Proceedings of the Institution of Mechanical Engineers, Part B: Journal of Engineering Manufacture* 235 (10): 1633–1643. doi:10.1177/0954405420949360.

Vimal, K. E.K., M. Naveen Srinivas, and Sonu Rajak. 2019. "Wire Arc Additive Manufacturing of Aluminium Alloys: A Review." *Materials Today: Proceedings* 41. Elsevier Ltd.: 1139–1145. doi:10.1016/j.matpr.2020.09.153.

Wegener, Thomas, Julian Koopmann, Julia Richter, Philipp Krooß, and Thomas Niendorf. 2021. "CuCrZr Processed by Laser Powder Bed Fusion—Processability and Influence of Heat Treatment on Electrical Conductivity, Microstructure and Mechanical Properties." *Fatigue and Fracture of Engineering Materials and Structures* 44 (9): 2570–2590. doi:10.1111/ffe.13527.

Zhang, Lai Chang, Yujing Liu, Shujun Li, and Yulin Hao. 2018. "Additive Manufacturing of Titanium Alloys by Electron Beam Melting: A Review." *Advanced Engineering Materials* 20 (5): 1–16. doi:10.1002/adem.201700842.

Zhao, Yanan, Kai Guan, Zhenwen Yang, Zhangping Hu, Zhu Qian, Hui Wang, and Zongqing Ma. 2020. "The Effect of Subsequent Heat Treatment on the Evolution Behavior of Second Phase Particles and Mechanical Properties of the Inconel 718 Superalloy Manufactured by Selective Laser Melting." *Materials Science and Engineering A* 794 (July). Elsevier B.V.: 139931. doi:10.1016/j.msea.2020.139931.

Zimmermann, Marco, Daniel Müller, Benjamin Kirsch, Sebastian Greco, and Jan C. Aurich. 2021. "Analysis of the Machinability When Milling AlSi10Mg Additively Manufactured via Laser-Based Powder Bed Fusion." *International Journal of Advanced Manufacturing Technology* 112 (3–4). The International Journal of Advanced Manufacturing Technology: 989–1005. doi:10.1007/s00170-020-06391-7.

5 Abrasive-based Finishing Processes

5.1 DEFINITION AND CLASSIFICATION

Abrasive-based finishing processes take advantage of the high hardness of the abrasive grains to scratch or plough workpiece surfaces and smoothen out the surface to the desirable finish and dimensional accuracy. Typical processes include grinding, polishing, blasting, and mass finishing, among others. According to Hashimoto et. al (2016), abrasive-based finishing processes can be classified as bonded-abrasive finishing processes and unbonded abrasive finishing processes depending on the state of the abrasives. Common bonded-abrasive finishing processes are grinding and honing where grinding wheels or honing tools have abrasives firmly bonded to them with vitrified, resinoid, silicate, shellac, rubber, or metal. Grinding is a motion-copying finishing process where the compliance of the grinding wheel is negligible, and the material removal volume is determined by the depth of cut. The resultant geometry of the workpiece is controlled by the shape of the tool and the tool path (Malkin and Guo 2008). Due to the high efficiency and precision of the bonded abrasive finishing processes, they can control the geometrical accuracy and surface quality of parts. Honing is generally employed to finish internal cylindrical surfaces – e.g., engine cylinders, bearings, hydraulic valves (Kuchle 2009). It can also generate crosshatched patterns for lubricant storage. Unbonded abrasive finishing processes include polishing, abrasive flow machining, abrasive jet machining, and magnetic abrasive finishing. The abrasives are dispersed in a liquid (either water or oil), which deliver the abrasives toward the workpiece for material removal. These processes are also called pressure-copying finishing processes, where the normal pressure and relative velocity between the finishing tool and the workpiece determine the material removal rate. This relationship is also known as the classical Preston theory (Preston 1927) as expressed in Equation 5.1.

$$MRR = kPV \qquad (5.1)$$

where k is the Preston coefficient that corelates the properties of the polishing tool and workpiece material, and P and V are the normal contact pressure and the relative velocity between the polishing tool and workpiece, respectively.

DOI: 10.1201/9781003272601-5

Generally, unbonded abrasive finishing processes have much lower material removal rates but can generate much smoother surfaces compared to bonded abrasive finishing processes.

5.2   BONDED ABRASIVE FINISHING PROCESSES

5.2.1   GRINDING

The grinding process utilizes a grinding wheel with a high rotation speed and the hard grains fixed on the grinding wheel act as multi-edge cutting tools to remove the work material. Grinding is suitable for difficult-to-machine materials and can produce products with good geometric accuracy and smooth surface finish.

The grinding wheel is the key component in grinding. The main parameters for a grinding wheel include abrasive type (e.g., aluminum oxide, silicon carbide, cubic boron nitride, diamond), abrasive size, bonding material, hardness, and structure. The dominant factor for the achievable surface finish is the abrasive size. Smaller abrasives lead to finer surface finish while the material removal efficiency is less than that of larger abrasives. For instance, grinding of 316L samples produced by SLM using a 21.8 μm grit sandpaper can obtain a surface roughness of 0.34 μm Ra, which is a 97.7% reduction in roughness from the as-printed state of 15.03 μm Ra (Löber et al. 2013). However, the smoothening of surfaces differs between materials such as the reduction in roughness of AlSi10Mg from 7 μm Ra after SLM to 0.6 μm Ra after grinding (Teng et al. 2019). Inconel 718 produced by SLM can also be grinded to 0.8 μm Ra with a resin-bonded corundum wheel (abrasive size 250–350 μm) (Liu et al. 2019). Liu et al. compared the grindability of Inconel 718 made by selective laser melting (SLM) and casting. The layered microstructure and internal pores of the SLM Inconel 718 decrease its yield strength and tensile strength, resulting in lower power consumption and wear of the wheel during grinding. In addition, casting Inconel 718 shows better surface finish after grinding due to its homogenous microstructure.

Grinding wheels are, however, typically rigid and are usually only applicable for flat or cylindrical surface finishing. Thus, conventional grinding may not be readily applicable for complex components generated by AM. Innovative grinding techniques have to be developed. An example is the shape adaptive grinding (SAP) method using a rubber bonnet (elastic tool) covered with diamond pellets (Beaucamp et al. 2015). The rubber bonnet tool enables conformal abilities between the tool and the curved workpiece surface (either convex or concave) while the rigid pellets act as the grinding tools for material removal. These pellets are either nickel bonded (NBD) or resin bonded (RBD). The surface roughness of Ti6Al4V can be grinded down to 12.8 nm Ra with SAP from an initial roughness of 4–5 μm Ra (as-built by EBM and SLS) after sequential grinding steps with decreasing grit sizes of 40, 9, and 3 μm, respectively. The SAP is also capable of achieving shape accuracies of ±5 μm and surface finishing of 3 nm Ra on freeform artifacts fabricated by SLS on a 7-axis CNC controlled machine built by Zeeko Ltd.

5.2.2 Mass Finishing

Mass finishing refers to a variety of abrasive finishing processes in which a batch of workpieces are processed by abrasive media to improve their surface quality via deburring, descaling, surface smoothing, or edge rounding (Davidson 2001; Mediratta, Ahluwalia, and Yeo 2016). The main components in mass finishing are the container or chamber, medium with varying shapes of abrasives, workpieces, and lubricant. The key is to generate random relative motions between the abrasive medium and the workpieces that will mechanically abrade material from the workpieces over time. Due to the random motions of the mass finishing, it is applicable for parts with complex geometries. Mass finishing can be categorized into five different processes according to their vibration mode and the workpiece fixtures: (1) barrel, (2) vibratory, (3) centrifugal barrel, (4) centrifugal disk, and (5) spin/spindle finishing. Workpieces are mounted on fixtures in spin/spindle finishing while the workpieces are driven by the abrasive media where they are immersed in the other categories. A detailed comparison of these categories was summarized in a review by Davidson (2002).

Due to its effectiveness at finishing complex parts, mass finishing is commonly employed in post-processing of additively manufactured metal parts where enhanced surface quality is desired. An example is given in the processing of a Ti6Al4V impeller and an Inconel 718 manifold, as shown in Figure 5.1 (Boschetto et al. 2020). These parts possess unique surface features such as convex surfaces (A), vertical walls (B), flat top surfaces (C), internal walls (D), and inclined overhanging surfaces (E). From an initial surface roughness of 14–16 μm Ra, the mass finished surfaces reduced to < 1 μm Ra for accessible external surfaces (e.g., A) and 3 μm Ra on more challenging

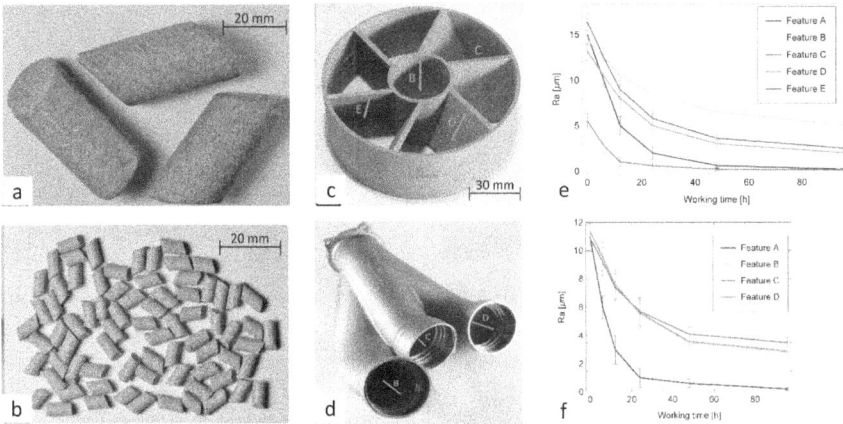

FIGURE 5.1 Barrel finishing of SLM complex parts: (a) the large media; (b) the small media; (c) the SLM Ti6Al4V impellor; (d) the Inconel 718 manifold; (e) the dependence of surface roughness Ra against the working time for the impellor finishing and (f) the dependence of surface roughness Ra against the working time for the manifold finishing. Reprinted with permission from Elsevier (Boschetto et al. 2020).

FIGURE 5.2 CDF of PBF gear structure: (a) the CDF equipment; (b) and (c) the ceramic media and its dimension; (d) and (e) the gear before and after CDF. Reprinted with permission from Authors (Fan et al. 2021).

internal and inclined surfaces (e.g., E). Despite the effectiveness to greatly improve the surface quality of the complex geometries in additively manufactured metals, a downside of conventional mass finishing is the long process running duration over days.

There are, however, operating procedure modifications and parameters that may be adjusted to overcome the disadvantage of conventional mass finishing. An example is in centrifugal disc finishing (CDF) of PBF-fabricated Ti6Al4V gears shown in Figure 5.2 (Fan et al. 2021). Rotational speed of the chamber filled with triangular ceramic abrasives (4 mm × 10 mm) was varied to find an optimal combination of 280 rpm and 30 min processing time that generates the best surface finish of 175 μinch (4.445 μm Ra) from an initial 290 μinch (7.336 μm Ra). The achievable surface quality is similar to the outcome of sandblasting.

Further reductions in surface roughness may be achieved with the appropriate selection of loading ratios, media geometry/sizes, and lubricant flow rates. However, innovative integration of vibration to the mass finishing process can further enhance surface roughness reductions.

Figure 5.3 shows the sidewall of a SLM-fabricated GRCop-84 copper waveguide post-processed by vibratory mass finishing. It is highly effective at processing external surfaces reducing the roughness values from 3.3 μm Ra to 0.45 μm Ra (Seltzman and Wukitch 2020). However, vibratory mass finishing falls short at processing internal surfaces due to the relatively smaller room for motion of the abrasive medium.

5.2.3 MAGNETICALLY DRIVEN INTERNAL FINISHING

Magnetically driven internal finishing (MDIF) is an abrasive finishing method specifically designed for internal surfaces (Zhang et al. 2018; Zhang et al. 2020; Zhang and Wang 2021). The working principle of MDIF is illustrated in Figure 5.4(a) comprising

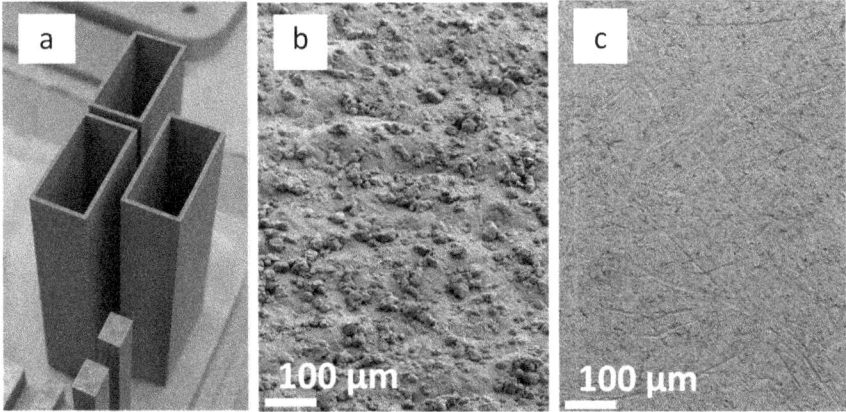

FIGURE 5.3 Vibratory mass finishing of SLM waveguides: (a) picture of the as-printed waveguides; (b) and (c) surface topography of the as-printed and processed sidewall under a microscope. Reprinted with permission from Elsevier (Seltzman and Wukitch 2020).

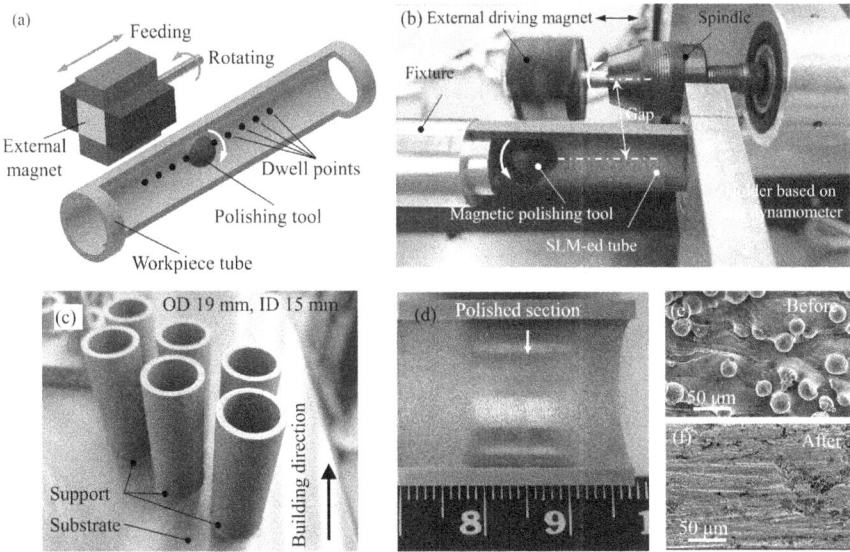

FIGURE 5.4 Post-processing of SLM 316L tubes by MDIF: (a) working principle of the MDIF; (b) experimental setup of MDIF on the SLM tubes; (c) as-printed 316L tubes; (d) the polished tube; (e) and (f) SEM images of the as-printed and polished. Reprinted with permission from Elsevier (Zhang and Wang 2022; Zhang and Wang 2021).

a stationary workpiece tube, a spherical magnet with uniformly bonded abrasives on its surface as the polishing tool, and an external magnet with four poles that rotates and drives the rotational motion of the polishing tool to remove material for each contact on the workpiece surface. The external magnet can also move along a complex

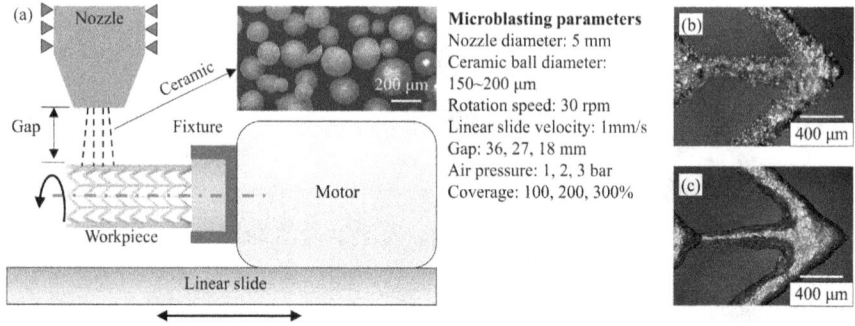

Microblasting parameters
Nozzle diameter: 5 mm
Ceramic ball diameter:
150~200 μm
Rotation speed: 30 rpm
Linear slide velocity: 1mm/s
Gap: 36, 27, 18 mm
Air pressure: 1, 2, 3 bar
Coverage: 100, 200, 300%

FIGURE 5.5 (a) Self-constructed automated micro-blasting setup and process parameters, (b) as-printed 316L stainless steel tubular lattice, (c) and the lattice after micro-blasting (Zhang 2019).

tool path to accommodate workpiece geometries and larger polishing areas. With this basic setup, MDIF can efficiently improve the surface finish of conventional tubes by more than 85% (Zhang et al. 2020; Zhang et al. 2018).

An actual MDIF setup is shown in Figure 5.4(b) where the polishing of SLM 316L tubes (Figure 5.4(c)) was being performed (Zhang and Wang 2022). Reductions in surface roughness by 92.4% can be achieved from an initial 10.629 μm Ra to 0.808 μm Ra with a shinier appearance depicted in Figure 5.4(d). Through the abrasive contact motion of the rotating polishing tool, as-printed surface features such as the unmolten particles and the staircase effect can be effectively removed, leaving behind fine scratches on the polished surface generated by the abrasives. The feasibility of MDIF for internal surfaces gives it great potential to process difficult-to-access surfaces in common machinery components such as gears or turbine blades (Krajnik et al. 2021).

5.3   UNBONDED ABRASIVE FINISHING PROCESSES

5.3.1   SANDBLASTING

Sandblasting is a mechanical process that administers a stream of abrasive material against a surface under high pressure to smooth a rough surface, roughen a smooth surface, deburr or clean a contaminated surface. The abrasive materials include metal shots, sand, glass beads, plastic beads, walnut shells, and corncobs. Generally, harder abrasives are more efficient at material removal while softer abrasives make mild modifications to the surface. The high-speed stream of the abrasives is ejected from a specially-designed nozzle via compressed air. Sandblasting can achieve a surface roughness below 5 μm Ra and is a suitable post-processing method for AM metal components before further fine finishing.

Blasting parameters include blasting pressure and nozzle-workpiece gap distance, which will affect the polishing accuracy and material removal rates. The sandblasting configuration can also vary where the nozzle travel along a stationary workpiece or the workpiece may move/rotate under a stationary nozzle as illustrated in Figure 5.5(a). Thin structures commonly found in additively manufactured metal parts such as struts in tubular lattices may be damaged by the high pressures (Zhang 2019) despite showing effective material removal and quality surface generation as seen in Figures 5.5(b) and 5.5(c). The high pressures may be more suitable for larger structures such as titanium bone scaffolds produced by SLM and can achieved a surface roughness of 0.94 µm Ra from an initial value of 3.33 µm Ra (de Wild et al. 2013). Although blasting is capable of removing partially bonded particles to enable efficient smoothening of complex AM geometries, poor selection of blasting parameters may result in abrasive embedment that can cause severe surface damage. Nonetheless, the final surface finish generated by blasting may still be inadequate for some industrial applications.

5.3.2 ABRASIVE FLOW MACHINING

Abrasive flow machining (AFM) utilizes a highly viscous abrasive-laden medium (polymeric fluid) to flow over the workpiece surface under high pressure (Uhlmann, Schmiedel, and Wendler 2015). AFM can be used for deburring, edge rounding, polishing, and recast layer removal (Bouland et al. 2019). It is suitable for finishing components with complex structures, especially internal surfaces but it can also process external surfaces with modified fixtures. An example of internal surfaces can be found in maraging steel 300 conformal cooling channels fabricated by selective laser melting where the diameter each channel measures only 3 mm shown in Figure 5.6(a–c) (Han et al. 2020). AFM can effectively reduce the surface roughness of the channel down to 1.3 µm Sa, which would enhance the cooling efficiency. AFM can also induce compressive residual stress due to negligible thermal effects, which may enhance the wear and fatigue properties (Han, Salvatore, and Rech 2019).

Media viscosity and abrasive concentration plays a critical role in AFM where high media viscosity and high concentration of abrasives are most effective on reducing surface roughness as given by AFM of SLM-fabricated maraging steel 300 (Duval-Chaneac et al. 2018). In the meantime, it is also important to control the ratio between abrasives and flowing media. Concentrations of the abrasive should go no further than 65%. A surface of approximately 2 µm Sa area roughness can be achieved using a 65% abrasive concentration media (down from an initial area roughness of 13 µm Sa) along with a significant improvement of compressive residual stress perpendicular to abrasive flow direction.

Mixture of different abrasives and sizes may also be implemented such as the abrasive mixture of #20 and #80 silicon carbide (SiC) particles employed in AFM of SLM-fabricated AlSi10Mg alloy (Peng et al. 2018). As such, larger abrasives will facilitate the removing of the partially bonded particles while the smaller abrasives contributes to generating a finer surface finish. Typical surface defects in AM (i.e., balling effects and partially bonded particles) will be effectively removed after 90

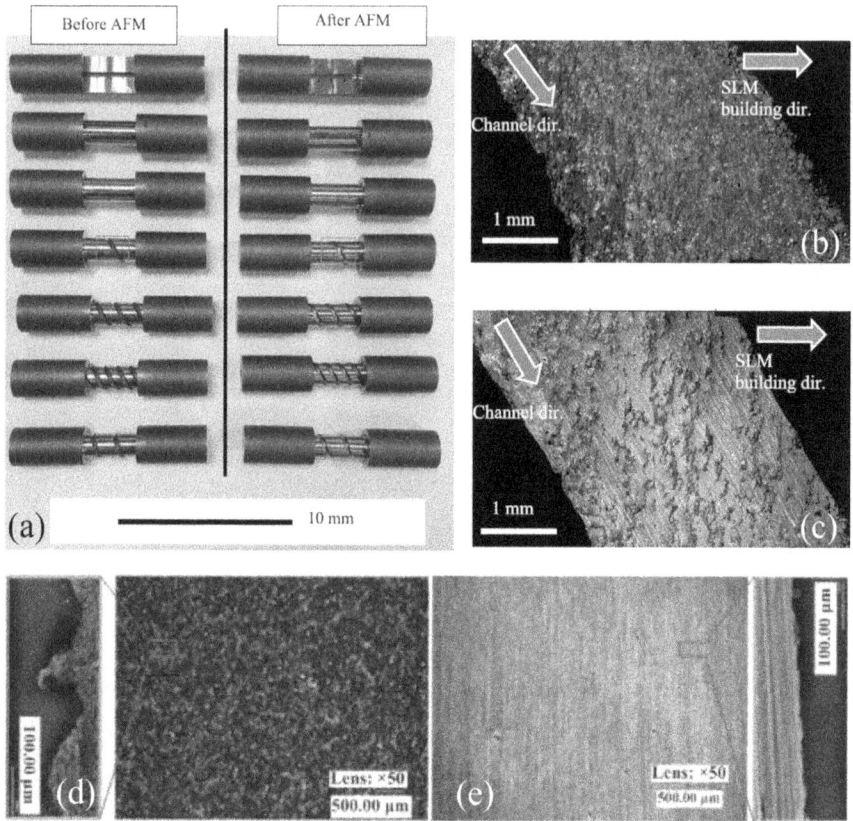

FIGURE 5.6 AM parts finished by AFM: (a) pictures of the as-printed and polished maraging steel 300 conformal cooling channels; (b) and (c) microscopic images of the as-printed and polished maraging steel 300 conformal cooling channels (Han et al. 2020); (d) and (e) microscopic images of the as-printed and the polished AlSi10Mg aluminum alloy part. Reprinted with permission from Elsevier (Peng et al. 2018).

AFM polishing cycles, as illustrated in Figures 5.6(d) and (e). However, additional cycles of up to 390 AFM cycles are necessary to reduce the surface roughness from 13 μm Sa to 1.8 μm Sa. XRD analysis also showed that compressive residual stresses were generated on the workpiece surface after AFM. The compressive stresses are beneficial to the fatigue life of the components.

Innovative integration of chemistry with AFM may also be implemented such as the combined chemical-abrasive flow polishing method (Mohammadian, Turenne, and Brailovski 2018). As an example, the roughness of the polished Inconel 625 fabricated by SLM can be reduced by 43% after an hour of polishing even on inclined surfaces. While this method is effective for difficult-to-reach areas, chemical/electrochemical-based finishing methods involve chemical hazards and are commonly identified as environmentally unfriendly. In general, normal mechanical-based AFM is sufficient as an effective post-process for additively manufactured internal surfaces.

However, AFM has certain downside that include its ineffectiveness for blind holes, low material removal rates, and non-uniform material removal characteristics.

5.3.3 Magnetic Abrasive Finishing

Magnetic abrasive finishing (MAF), also known as magnetic field-assist finishing (MFAF), is a nonconventional finishing process that employs magnetic abrasive particles (MAP) to form a flexible polishing brush on the workpiece surface (Shinmura et al. 1990). The relative motion (vibrational or rotational) between the flexible polishing brush and the workpiece enables the removal of the material via scratching or abrasion. Lubricants, such as water or barrel oil, may be applied to facilitate the generation of a smoother surface. Due to the high conformality and accessibility of the MAP, MAF has been widely applied for finishing complex components such as tubes, V-groove structures, and cutting edges (Kang, George, and Yamaguchi 2012; Wang, Wu, and Mitsuyoshi 2016; Yamaguchi et al. 2014).

MAF has also been employed to obtain nanometric surface qualities on additively manufactured metal components. With additional integration of magnetic abrasive magnetic field-assisted burnishing (MAB) with MAF, the multistage process can reduce the surface roughness of selective laser melted 316L stainless steel from 102.1 μm Rz to 0.13 μm Rz after 240 min of polishing as seen in Figures 5.7(a–c) (Yamaguchi, Fergani, and Wu 2017). Compressive residual stresses are also imparted onto the post-processed surface of the SLM components.

Meanwhile, the high conformality does not necessary indicate that the material removal is similar for different surface profiles as observed in MAF of 316L stainless steels with different slope angles fabricated by SLM (Zhang, Chaudhari, and Wang

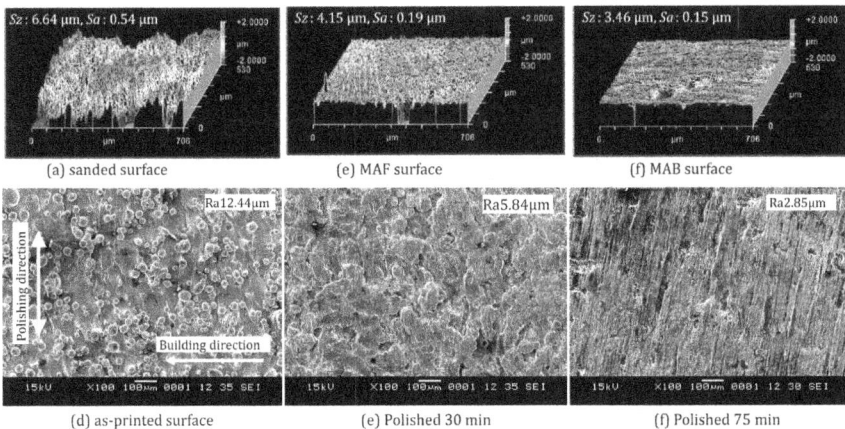

(a) sanded surface   (e) MAF surface   (f) MAB surface

(d) as-printed surface   (e) Polished 30 min   (f) Polished 75 min

FIGURE 5.7 Additively manufactured parts finished by MAF: (a–c) three-dimensional surface topography of the sanded, MAF processed and MAB processed 316L surface. Reprinted with permission from Elsevier (Yamaguchi, Fergani, and Wu 2017); (d–f) SEM images of the as-printed, 30 min MAF processed and 75 min MAF processed 316L surface. Reprinted with permission from Elsevier (Zhang, Chaudhari, and Wang 2019).

2019). The surface roughness will vary significantly between different slope angles due to the synergistic effects of the stair effect and partially bonded particles. This variation will affect the predictability of the final surface quality from MAF. Nevertheless, MAF is effective at eliminating most partially bonded particles and balling defects on as-printed parts to achieve >75% improvement in surface finish as given in Figures 5.7(d–f). MAF, however, may not be suitable for complex surfaces if working in a linear-vibration mode. Nevertheless, a rotary MAF may be employed to finish freeform surfaces with planned tool paths (Sidpara and Jain 2012).

5.3.4 CAVITATION ABRASIVE FINISHING

Cavitation abrasive finishing is an abrasive finishing technique that utilizes the synergistic effect of cavitation erosion and microparticle abrasion to remove the irregularities and defects on workpiece surface (Tan and Yeo 2017; Nagalingam, Yuvaraj, and Yeo 2020; Tan and Yeo 2020). It is specifically developed to finish additively manufactured components.

Ultrasonic cavitation abrasive finishing (UCAF) is a variation of cavitation abrasive finishing that employs an ultrasonic horn to encourage the cavitation effects at high frequencies (Tan and Yeo 2017). This was developed to process Inconel 625 flat surfaces fabricated by DMLS as given in Figures 5.8(a–c). Two material removal modes exist – (1) cavitation bubble collapse to remove the partially bonded particles, and (2) micro-abrasive impingement driven by the bubble collapse. A 0.1 mm thick hardened layer can be obtained with a surface roughness of 3.8 μm Ra can be achieved after 10 min of processing. There are several factors governing the finishing performance, which include polishing time, abrasive size, abrasive concentration, ultrasonic amplitude, and working gap. Similar to conventional polishing processes, the surface roughness decreases rapidly at the beginning of the UCAF process and will saturate after a certain polishing time. In addition, small abrasive size leads to lower surface roughness. An optimal concentration of the abrasive slurry (e.g., wt 5%) should be employed since too low concentration leads to low polishing efficiency while too high concentration inhibits bubble generation. As the ultrasonic amplitude increases from 6 μm to 120 μm, the achieved surface roughness first decreases and then increases. A high ultrasonic amplitude may cause severe erosion and excessive material removal, resulting in poor surface finish. As the working gap increases, the cavitation intensity decreases. Hence, a higher working gap leads to less material removal and poorer surface quality.

UCAF may not be suitable for complex surfaces, such as freeform surfaces and intricate internal surfaces, because cylinder ultrasonic horn cannot conform to the freeform surfaces and cannot access to the inner surfaces. Hydrodynamic cavitation abrasive finishing (HCAF) can accommodate such requirements as it involves freely suspended abrasive particles flowing through the internal surface experiencing cavitating conditions (Nagalingam and Yeo 2018). In HCAF, cavitation erosion and the microplowing of the abrasives smoothen the workpiece tube by removing the partially bonded particles and the irregularities as observed on aluminum alloy AlSi10Mg round and square tubes fabricated by DMLS (Nagalingam, Yuvaraj, and Yeo 2020) as seen in Figure 5.8(d–i). The working mechanism of HCAF equipped

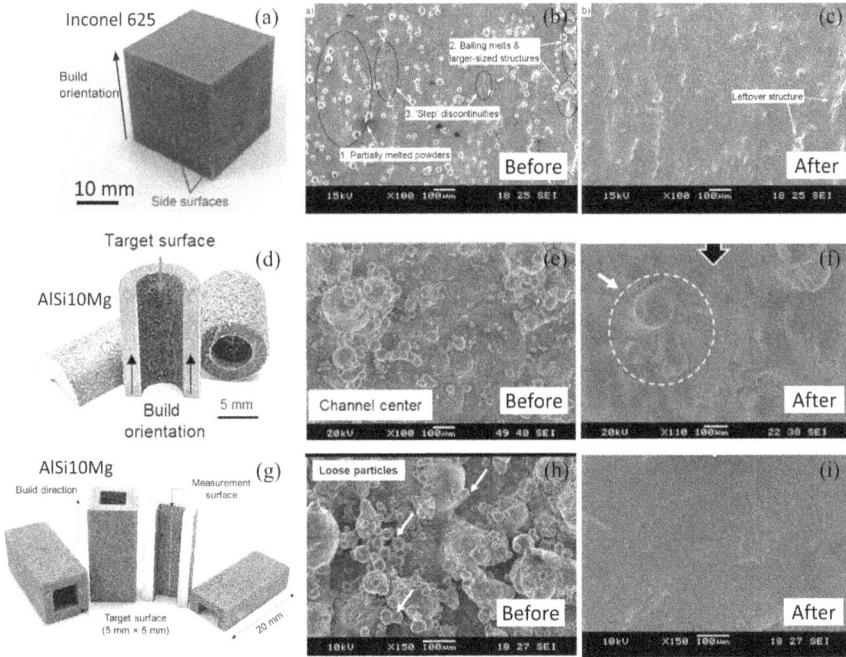

FIGURE 5.8 AM parts finished by cavitation abrasive finishing: (a–c) the as-printed Inconel 625 block and its SEM images before and after finishing. Reprinted with permission from Elsevier (Tan and Yeo 2017); (d–f) the as-printed AlSi10Mg round tubes and the SEM images before and after finishing. Reprinted with permission from Elsevier (Nagalingam and Yeo 2018); (g–i) the as-printed AlSi10Mg square tubes and the SEM images before and after finishing. Reprinted with permission from Elsevier (Nagalingam, Yuvaraj, and Yeo 2020).

with a multi-jet hydrodynamic approach can finish complex Inconel 625 internal channels with features including stepped channels and non-linear channels fabricated by LPBF (Nagalingam and Yeo 2020). Final surface roughness of the internal channels can reach < 1 μm Ra after HCAF, which corresponds to a 60%–90% reduction in surface roughness. HCAF also shows relatively good performance for internal finishing of tubes and channels with diameters < 5 mm and has been validated to finish spray nozzles with multiple branches or varying diameters (Nagalingam, Lee, and Yeo 2021). However, this method requires a complex setup, and there has not yet been any reports on finishing channels with diameters larger than 10 mm.

5.4   CONCLUDING REMARKS

Abrasive-based finishing processes are efficient and cost-effective for improving surface quality of AM parts. They can achieve nanoscale surface finish with optimal process parameters. Novel abrasive-based finishing techniques (e.g., magnetically driven internal finishing and hydrodynamic cavitation abrasive finishing) have been developed to address the challenges of the geometrical complexity of the AM parts.

These techniques can improve the surface finish of AM parts by more than 90%, showing great potential for industrial application.

REFERENCES

Beaucamp, Anthony T, Yoshiharu Namba, Phillip Charlton, Samyak Jain, and Arthur A Graziano. 2015. "Finishing of Additively Manufactured Titanium Alloy by Shape Adaptive Grinding (SAG)." *Surface Topography: Metrology and Properties* 3 (2). IOP Publishing: 024001. doi:10.1088/2051-672X/3/2/024001.

Boschetto, Alberto, Luana Bottini, Luciano Macera, and Francesco Veniali. 2020. "Post-Processing of Complex SLM Parts by Barrel Finishing." *Applied Sciences* 10 (4). Multidisciplinary Digital Publishing Institute: 1382. doi:10.3390/app10041382.

Bouland, C, V Urlea, K Beaubier, M Samoilenko, and V Brailovski. 2019. "Abrasive Flow Machining of Laser Powder Bed-Fused Parts: Numerical Modeling and Experimental Validation." *Journal of Materials Processing Technology* 273 (November). Elsevier: 116262. doi:10.1016/j.jmatprotec.2019.116262.

Davidson, David A. 2001. "Mass Finishing Processes." *Metal Finishing* 99 (January). Elsevier: 110–124. doi:10.1016/S0026-0576(01)85268-5.

de Wild, Michael, Ralf Schumacher, Kyrill Mayer, Erik Schkommodau, Daniel Thoma, Marius Bredell, Astrid Kruse Gujer, Klaus W Grätz, and Franz E Weber. 2013. "Bone Regeneration by the Osteoconductivity of Porous Titanium Implants Manufactured by Selective Laser Melting: A Histological and Micro Computed Tomography Study in the Rabbit." *Tissue Engineering Part A* 19 (23–24). Mary Ann Liebert, Inc. 140 Huguenot Street, 3rd Floor New Rochelle, NY 10801 USA: 2645–2654. doi:10.1089/ten.tea.2012.0753.

Duval-Chaneac, M.S., S Han, C Claudin, F Salvatore, J Bajolet, and J Rech. 2018. "Experimental Study on Finishing of Internal Laser Melting (SLM) Surface with Abrasive Flow Machining (AFM)." *Precision Engineering* 54 (October). Elsevier: 1–6. doi:10.1016/j.precisioneng.2018.03.006.

Fan, Foxian, Nicholas Soares, Sagar Jalui, Aaron Isaacson, Aditya Savla, Guha Manogharan, and Tim Simpson. 2021. "Effects of Centrifugal Disc Finishing for Surface Improvements in Additively Manufactured Gears." In *2021 International Solid Freeform Fabrication Symposium*. University of Texas at Austin. doi:10.26153/tsw/17678.

Han, Sangil, Ferdinando Salvatore, and Joël Rech. 2019. "Residual Stress Profiles Induced by Abrasive Flow Machining (AFM) in 15-5PH Stainless Steel Internal Channel Surfaces." *Journal of Materials Processing Technology* 267 (May). Elsevier: 348–358. doi:10.1016/j.jmatprotec.2018.12.024.

Han, Sangil, Ferdinando Salvatore, Joël Rech, and Julien Bajolet. 2020. "Abrasive Flow Machining (AFM) Finishing of Conformal Cooling Channels Created by Selective Laser Melting (SLM)." *Precision Engineering* 64 (July). Elsevier: 20–33. doi:10.1016/j.precisioneng.2020.03.006.

Hashimoto, Fukuo, Hitomi Yamaguchi, Peter Krajnik, Konrad Wegener, Rahul Chaudhari, Hans Werner Hoffmeister, and Friedrich Kuster. 2016. "Abrasive Fine-Finishing Technology." *CIRP Annals – Manufacturing Technology* 65 (2). CIRP: 597–620. doi:10.1016/j.cirp.2016.06.003.

Kang, Junmo, Andrew George, and Hitomi Yamaguchi. 2012. "High-Speed Internal Finishing of Capillary Tubes by Magnetic Abrasive Finishing." *Procedia CIRP* 1. Elsevier: 414–418. doi:10.1016/j.procir.2012.04.074.

Krajnik, Peter, Fukuo Hashimoto, Bernhard Karpuschewski, Eraldo Jannone da Silva, and Dragos Axinte. 2021. "Grinding and Fine Finishing of Future Automotive Powertrain Components." *CIRP Annals* 70 (2). Elsevier: 589–610. doi:10.1016/j.cirp.2021.05.002.

Kuchle, Aaron. 2009. *Manufacturing Processes: Grinding, Honing, Lapping*. Springer.

Liu, Zhonglei, Xuekun Li, Xiaodan Wang, Chenchen Tian, and Liping Wang. 2019. "Comparative Investigation on Grindability of Inconel 718 Made by Selective Laser Melting (SLM) and Casting." *The International Journal of Advanced Manufacturing Technology* 100 (9–12). Springer: 3155–3166. doi:10.1007/s00170-018-2850-8.

Löber, Lukas, Christoph Flache, Romy Petters, Uta Kühn, and Jürgen Eckert. 2013. "Comparison of Different Post Processing Technologies for SLM Generated 316l Steel Parts." *Rapid Prototyping Journal* 19 (3). Emerald Group Publishing Limited: 173–179. doi:10.1108/13552541311312166.

Malkin, Stephen, and Changsheng Guo. 2008. *Grinding Technology: Theory and Application of Machining with Abrasives*. Industrial Press Inc.

Mediratta, Rijul, Kunal Ahluwalia, and S H Yeo. 2016. "State-of-the-Art on Vibratory Finishing in the Aviation Industry: An Industrial and Academic Perspective." *The International Journal of Advanced Manufacturing Technology* 85 (1–4). Springer: 415–429. doi:10.1007/s00170-015-7942-0.

Mohammadian, Neda, Sylvain Turenne, and Vladimir Brailovski. 2018. "Surface Finish Control of Additively-Manufactured Inconel 625 Components Using Combined Chemical-Abrasive Flow Polishing." *Journal of Materials Processing Technology* 252 (February). Elsevier: 728–738. doi:10.1016/j.jmatprotec.2017.10.020.

Nagalingam, Arun Prasanth, Jian-Yuan Lee, and S.H. Yeo. 2021. "Multi-Jet Hydrodynamic Surface Finishing and X-Ray Computed Tomography (X-CT) Inspection of Laser Powder Bed Fused Inconel 625 Fuel Injection/Spray Nozzles." *Journal of Materials Processing Technology* 291 (May). Elsevier: 117018. doi:10.1016/j.jmatprotec.2020.117018.

Nagalingam, Arun Prasanth, and S.H. Yeo. 2018. "Controlled Hydrodynamic Cavitation Erosion with Abrasive Particles for Internal Surface Modification of Additive Manufactured Components." *Wear* 414–415 (November). Elsevier: 89–100. doi:10.1016/j.wear.2018.08.006.

Nagalingam, Arun Prasanth, and S.H. Yeo. 2020. "Surface Finishing of Additively Manufactured Inconel 625 Complex Internal Channels: A Case Study Using a Multi-Jet Hydrodynamic Approach." *Additive Manufacturing* 36 (December). Elsevier: 101428. doi:10.1016/j.addma.2020.101428.

Nagalingam, Arun Prasanth, Hemanth Kumar Yuvaraj, and S H Yeo. 2020. "Synergistic Effects in Hydrodynamic Cavitation Abrasive Finishing for Internal Surface-Finish Enhancement of Additive-Manufactured Components." *Additive Manufacturing*. Elsevier, 101110.

Peng, Can, Youzhi Fu, Haibo Wei, Shicong Li, Xuanping Wang, and Hang Gao. 2018. "Study on Improvement of Surface Roughness and Induced Residual Stress for Additively Manufactured Metal Parts by Abrasive Flow Machining." *Procedia CIRP* 71. Elsevier: 386–389. doi:10.1016/j.procir.2018.05.046.

Preston, F W. 1927. "The Theory and Design of Plate Glass Polishing Machines." *Journal of Glass Technology* 11 (44): 214–256.

Seltzman, A.H., and S.J. Wukitch. 2020. "Surface Roughness and Finishing Techniques in Selective Laser Melted GRCop-84 Copper for an Additive Manufactured Lower Hybrid Current Drive Launcher." *Fusion Engineering and Design* 160 (November). Elsevier: 111801. doi:10.1016/j.fusengdes.2020.111801.

Shinmura, T., K. Takazawa, E. Hatano, M. Matsunaga, and T. Matsuo. 1990. "Study on Magnetic Abrasive Finishing." *CIRP Annals – Manufacturing Technology* 39 (1): 325–328. doi:10.1016/S0007-8506(07)61064-6.

Sidpara, Ajay M., and V. K Jain. 2012. "Nanofinishing of Freeform Surfaces of Prosthetic Knee Joint Implant." *Proceedings of the Institution of Mechanical Engineers, Part*

B: Journal of Engineering Manufacture 226 (11). Sage Publications Sage UK: London, England: 1833–1846. doi:10.1177/0954405412460452.

Tan, K.L., and S.H. Yeo. 2017. "Surface Modification of Additive Manufactured Components by Ultrasonic Cavitation Abrasive Finishing." *Wear* 378–379 (May). Elsevier: 90–95. doi:10.1016/j.wear.2017.02.030.

Tan, K.L., and S.H. Yeo. 2020. "Surface Finishing on IN625 Additively Manufactured Surfaces by Combined Ultrasonic Cavitation and Abrasion." *Additive Manufacturing* 31 (January). Elsevier: 100938. doi:10.1016/j.addma.2019.100938.

Teng, Xiao, Guixiang Zhang, Yugang Zhao, Yuntao Cui, Linguang Li, and Linzhi Jiang. 2019. "Study on Magnetic Abrasive Finishing of AlSi10Mg Alloy Prepared by Selective Laser Melting." *The International Journal of Advanced Manufacturing Technology* 105 (5–6). Springer: 2513–2521. doi:10.1007/s00170-019-04485-5.

Uhlmann, E., C. Schmiedel, and J. Wendler. 2015. "CFD Simulation of the Abrasive Flow Machining Process." *Procedia CIRP* 31. Elsevier: 209–214. doi:10.1016/j.procir.2015.03.091.

Wang, Youliang, Yongbo Wu, and Nomura Mitsuyoshi. 2016. "A Novel Magnetic Field-Assisted Polishing Method Using Magnetic Compound Slurry and Its Performance in Mirror Surface Finishing of Miniature V-Grooves." *AIP Advances* 6 (5). AIP Publishing LLC: 56602. doi:10.1063/1.4942952.

Yamaguchi, Hitomi, Omar Fergani, and Pei Ying Wu. 2017. "Modification Using Magnetic Field-Assisted Finishing of the Surface Roughness and Residual Stress of Additively Manufactured Components." *CIRP Annals – Manufacturing Technology* 66 (1). Elsevier: 305–308. doi:10.1016/j.cirp.2017.04.084.

Yamaguchi, Hitomi, Anil K. Srivastava, Michael Tan, and Fukuo Hashimoto. 2014. "Magnetic Abrasive Finishing of Cutting Tools for High-Speed Machining of Titanium Alloys." *CIRP Journal of Manufacturing Science and Technology* 7 (4). Elsevier: 299–304. doi:10.1016/j.cirpj.2014.08.002.

Zhang, Jiong. 2019. "Micro-Blasting of 316L Tubular Lattice Manufactured by Laser Powder Bed Fusion." In *European Society for Precision Engineering and Nanotechnology, Conference Proceedings – 19th International Conference and Exhibition, EUSPEN 2019*, 254–255. Bilbao, ES. www.euspen.eu/knowledge-base/ICE19107.pdf.

Zhang, Jiong, Akshay Chaudhari, and Hao Wang. 2019. "Surface Quality and Material Removal in Magnetic Abrasive Finishing of Selective Laser Melted 316L Stainless Steel." *Journal of Manufacturing Processes* 45. Elsevier: 710–719. doi:10.1016/j.jmapro.2019.07.044.

Zhang, Jiong, Julian Hu, Hao Wang, A. Senthil Kumar, and Akshay Chaudhari. 2018. "A Novel Magnetically Driven Polishing Technique for Internal Surface Finishing." *Precision Engineering* 54. Elsevier: 222–232. doi:10.1016/j.precisioneng.2018.05.015.

Zhang, Jiong, and Hao Wang. 2021. "Generic Model of Time-Variant Tool Influence Function and Dwell-Time Algorithm for Deterministic Polishing." *International Journal of Mechanical Sciences* 211. Elsevier: 106795. doi:10.1016/j.ijmecsci.2021.106795.

Zhang, Jiong, and Hao Wang. 2022. "Magnetically Driven Internal Finishing of AISI 316L Stainless Steel Tubes Generated by Laser Powder Bed Fusion." *Journal of Manufacturing Processes* 76. Elsevier: 155–166. doi:10.1016/j.jmapro.2022.02.009.

Zhang, Jiong, Hao Wang, A. Senthil Kumar, and Mingsheng Jin. 2020. "Experimental and Theoretical Study of Internal Finishing by a Novel Magnetically Driven Polishing Tool." *International Journal of Machine Tools and Manufacture* 153. Elsevier: 103552. doi:10.1016/j.ijmachtools.2020.103552.

6 Thermal-Based Finishing Processes

6.1 DEFINITION AND WORKING PRINCIPLE

Laser polishing, also known as laser remelting, is a thermal-energy based process often employed to achieve superior surface finish by remodeling the workpiece topography through laser-material interactions as illustrated in Figure 6.1(a) (Bordatchev, Hafiz, and Tutunea-Fatan 2014). During laser remelting, the laser beam only melts a microscopic layer of material along the surface while maintaining the form accuracy of the workpiece (Lamikiz et al. 2007). Meanwhile, the irregularities and high asperities are flattened during the liquifying process. Figures 6.1(b–d) illustrate examples of metal surfaces after laser polishing. The processed surface will look shinier and more reflective under the appropriate polishing parameters. However, some irregularities and patterns will still be observed resulting from the irradiation and the mother surface. Details on the surface structures induced by laser polishing will be discussed in Section 6.1.3.

Laser polishing technology has been extensively investigated on various materials and geometries with assessments on surface finish, surface hardness, microstructure, chemical composition, and so forth. This is due to its advantageous automation capabilities using multi-axial computer numerical control (CNC) setups and excellent controllability of process parameters and "zero" tool wear (Bordatchev, Hafiz, and Tutunea-Fatan 2014). Some examples of different applications produced by laser polishing can be seen in Figure 6.2, including precision molds and biomedical implants.

6.1.1 CLASSIFICATION OF LASER POLISHING

Laser polishing has two main variants: (i) laser macro-polishing and (ii) laser micro-polishing. The former employs continuous wave (CW) laser radiation while the latter typically uses pulsed-laser radiation (Christian Nüsser et al. 2015; Bordatchev, Hafiz, and Tutunea-Fatan 2014). In laser macro-polishing, the CW laser results in a continuous melting pool where melting and re-solidification of the material occur at the same time. In laser micro-polishing, the pulsed-laser locally melts the micro-asperities. These molten regions with high surface tension transit to a state of low surface tension before re-solidification to enable the generation of a smoother surface

DOI: 10.1201/9781003272601-6

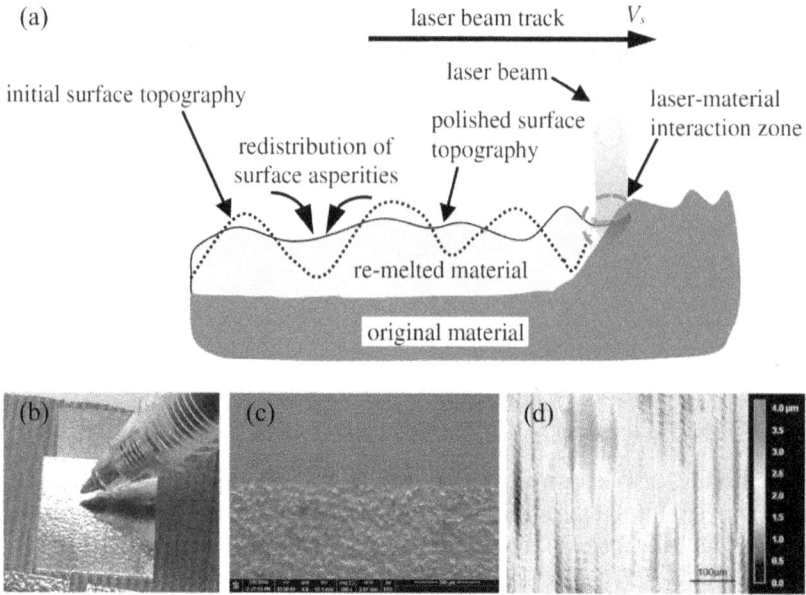

FIGURE 6.1 Laser polishing process and the resultant surfaces: (a) schematic showing the mechanism of the laser polishing process. Reprinted with permission from Elsevier (Bordatchev, Hafiz, and Tutunea-Fatan 2014); (b–d) the surface topography after laser polishing observed by a digital camera, an SEM and a confocal microscope. Reprinted with permission from Elsevier (Ma, Guan, and Zhou 2017).

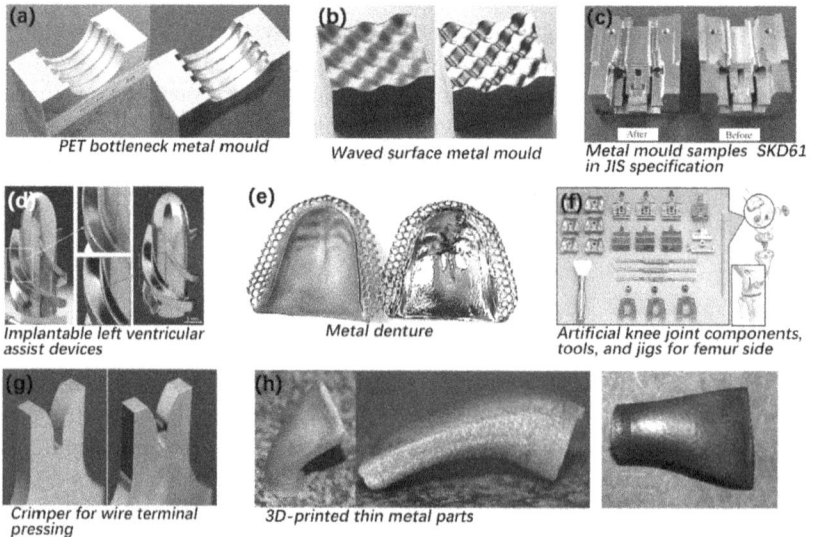

FIGURE 6.2 Applications of thermal-based finishing techniques. Reprinted with permission from Elsevier (Deng, Li, and Zheng 2020).

TABLE 6.1
Comparison of Laser Macro-Polishing and Laser Micro-Polishing for Metals and Alloys (Krishnan and Fang 2019; Pfefferkorn et al. 2013; Bordatchev, Hafiz, and Tutunea-Fatan 2014)

Factors	Laser macro polishing	Laser micro polishing
Laser type	High-power continuous-wave	Pulsed: 20–1000 ns pulse duration
Laser power (W)	30–2000	0.2–300
Process time (s/cm^2)	10–200	3
Laser spot size (μm)	40–1800	30–200
Suitable surface roughness (μm)*	1–16	0.04–3
Achievable surface roughness (μm)*	0.11–3	0.005–2
Pre-process	Not specified	Ground and milled
Re-melting depth (μm)	20–200	0.5–5
Heat affected zone (μm)	100	<10

* Evaluated by Ra or Sa

(Pfefferkorn et al. 2013). The following laser pulses will repeat the process over the full area to be polished. A comparison between parameters of laser macro-polishing and laser micro-polishing is given in Table 6.1.

The melt pool size and the remelting depths are larger under macro-polishing due to the higher laser power. This will lead to deeper heat affected zones as well. However, the deeper remelting depths benefit the processing of surfaces with high initial surface roughness such as the as-built AM metals. In contrast, laser micro-polishing is more suitable for processing surfaces with low roughness and further lower the roughness to even generate optical-grade surface finishes. In both polishing methods, the appropriate process parameters need to be selected to achieve the superior finishing. The following section will discuss the dominant factors.

6.1.2 Process Parameters in Laser Polishing

Laser polishing has several main parameters consolidated in Figure 6.3 but any factor that affects the melting and solidification processes during laser irradiation will influence the final polishing performance. The parameters comprise workpiece properties, laser subsystem parameters, and mechanical subsystem parameters (Bordatchev, Hafiz, and Tutunea-Fatan 2014).

6.1.2.1 Workpiece Properties

The workpiece properties refer to the initial surface topography (e.g., surface roughness, irregularities), optical properties (e.g., reflectivity), thermal properties (e.g., conductivity, melting point, evaporation point), mechanical properties (e.g., fracture toughness), and chemical properties (e.g., inclusions, carbon content,

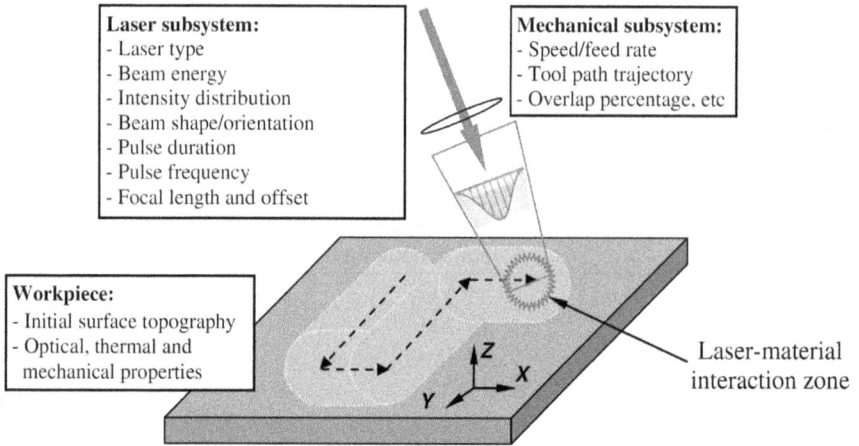

FIGURE 6.3 Dominant parameters in the laser polishing process. Reprinted with permission from Elsevier (Bordatchev, Hafiz, and Tutunea-Fatan 2014).

microstructures). They also include the properties of the material's liquid state – e.g., surface tension, viscosity, and temperature coefficient of surface tension (Mohajerani, Bordatchev, and Tutunea-Fatan 2018).

The workpiece surface is never perfectly flat. It contains low-frequency features, such as form and waviness, as well as high-frequency features such as the roughness profile and extremely short-wavelength components. The waviness and surface roughness features have a relatively similar length scale as the size of the melt pool in laser polishing. The initial surface topography of the workpiece affects the absorptivity of the laser, which influences the energy density over the laser spot (Mohajerani, Bordatchev, and Tutunea-Fatan 2018). The length scale of the long-range waviness is much larger than that of the melt pool. Hence, the waviness cannot be eliminated by laser polishing and will largely determine the final surface topography (Tian et al. 2018). A general rule is that a higher initial surface roughness will result in a higher surface roughness after laser polishing like that of the electrochemical polishing process.

6.1.2.2 Laser Subsystem Parameters

The second group of the parameters influencing the laser polishing process comes from the laser subsystem. These parameters consist of laser type, beam energy, intensity distribution, beam shape/orientation, pulse duration, pulse frequency, focal length, and offset (C. Nüsser, Sändker, and Willenborg 2013; Bordatchev, Hafiz, and Tutunea-Fatan 2014). Essential parameters, as illustrated in Figure 6.4, mainly affect the intensity and area of energy delivered to the workpiece.

Pulse duration, also known as pulse width, is the exposure time of the laser with an equivalent irradiation area of its full width half maximum (FWHM), as shown in Figure 6.4(b). The laser exposure area is the same for CW lasers with the only difference being that it provides a steady beam of irradiation. Hence, the pulse duration parameter is only applicable for laser micro-polishing where pulsed lasers are employed. The pulse duration directly affects the energy density of the laser beam. Therefore, longer pulse durations will increase the irradiation time and result in deeper melting pools. However, the pulse duration is not the only factor that affects the energy density. Generally, the laser beam shape/size, pulse duration, and the intensity distribution collectively determine the energy density (ED). This energy density is the energy radiated per surface unit and is directly dependent upon laser power, beam spot size, and the time duration the beam is applied on a surface (Chow, Bordatchev, and Knopf 2013) given by Equation (6.1).

$$ED = \frac{P \times t}{A} = \frac{4}{\pi}\left(\frac{P}{v_f \times D}\right) \tag{6.1}$$

where P is the laser power (W), t is the duration time of the laser beam, A is the area of the laser beam (μm^2), D is the laser beam spot diameter (mm), and v_f is the travel speed (mm/min).

As illustrated in Figure 6.4(b), the focal offset distance (FOD) is the distance between the laser spot size diameter at the focal point and the workpiece surface. In microscope terms, it corresponds to the focal length between the objective lens and the sample surface. FOD has great impact on the energy density. In short FOD, the energy density is high thus ablation is the main removal mechanism. In an intermediate FOD, the workpiece surface encounters remelting rather than ablation. When the FOD is large, only heat treatment effect will occur. In addition, short and long

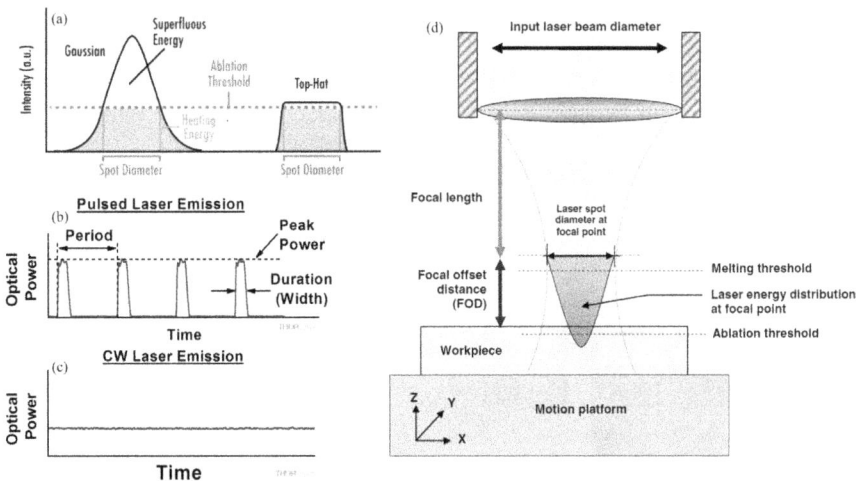

FIGURE 6.4 Schematic of the laser beam parameters. (a) Cross-section energy distribution of Gaussian and Top-Hat beams, image Courtesy of Edmund Optics, Inc. All rights reserved. (b) optical power of pulsed laser and CW laser and (c) schematic showing the focal offset distance Reprinted with permission from Springer Nature (Chow, Bordatchev, and Knopf 2013).

FODs can favorably improve the surface quality while an intermediate FOD may lead to an increase of surface roughness, when the FOD falls in the remelting stage (Chow, Bordatchev, and Knopf 2013).

6.1.2.3 Mechanical Subsystem Parameters

The third group of parameters are from the mechanical subsystem, including the laser movement speed or feed rate, tool path trajectory (Giorleo, Ceretti, and Giardini 2015), overlap percentage, and so forth. The scanning speed directly affects the energy density, as referred to the travel speed in Equation 6.1. In laser polishing, larger overlap percentage is preferred compared to conventional milling processes. Larger overlap percentages in milling essentially reduces machining productivity due to the repeated travel over "already-cut" surfaces. Overlapping in the laser path allows for reheating of the previous path. However, overlapping needs to be managed as too much overlapping may lead to overheating and surface burning (Hafiz, Bordatchev, and Tutunea-Fatan 2012).

6.1.3 SURFACE DEFECTS IN LASER POLISHING

6.1.3.1 Laser Macro-Polishing

Surface defects after laser macro-polishing can be classified into two categories as seen in Figure 6.5. The first group is represented by surface structures initiated from surface ripples and undercuts that form from the dynamics of the melt and solidification front. Ripples are usually caused by material inhomogeneity and their amplitude has a positive correlation with the scan velocity. The undercuts appear if the melting pool slides on the transitional surface between the melting and solidification fronts due to capillary forces (Nüsser et al. 2015). From the material aspect, the high heat conductivity will make the transitional surface between the melting

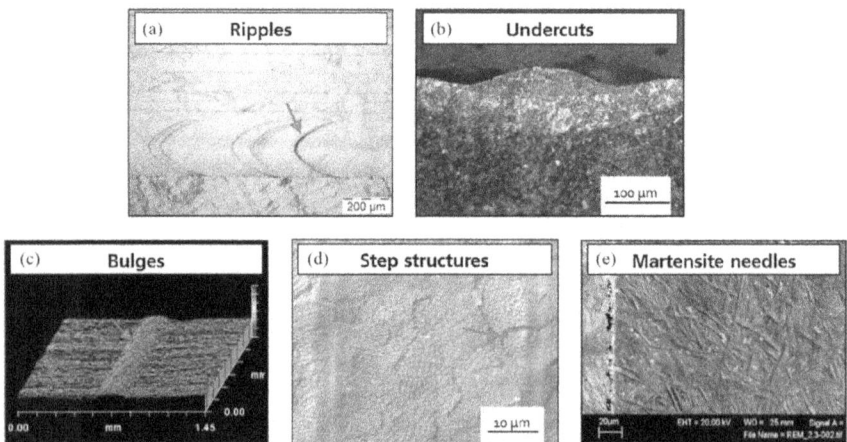

FIGURE 6.5 Typical surface defects during laser macro polishing. Reprinted with permission from Elsevier (Christian Nüsser et al. 2015).

and solidification fronts more perpendicular thus weaken the undercut. Similar to the ripples, a higher scan velocity amplifies the height of undercuts. The remelting process in laser polishing accompanies plastic deformation from the material expansion and contract due to the thermal activities.

The second group represents the plastic deformation and microstructure changes such as bulges, step structures, and martensite needles (in case of tool steels). Plastic deformation from thermal activities will lead to the formation of steps due to grain deformation (Kiedrowski 2009). These steps generally generate at the grain boundaries with heights typically < 5 µm. Higher numbers of processing passes and shorter track offsets will enlarge the height of the steps. Therefore, material properties affecting thermal responses such as the elastic modulus and coefficient of expansion will play important roles in the formation of bulges and steps. Considering the high cooling rates in the solidification process, martensitic microstructures form after laser polishing of tool steels. These formations will introduce micro-roughness to the polished surface. Decarburization will be necessary to eliminate the formation of martensitic microstructures on the surface layer.

6.1.3.2 Laser Micro-Polishing

Surface defects from laser micro-polishing consist of undercuts, micro-waviness, border-bulging, holes, step structures, and micro structures, as summarized in Figure 6.6 (Christian Nüsser et al. 2015). The first three defects are induced by the laser polishing process while the latter defects are induced by the work material properties. Undercuts in laser micro-polishing form when the melt pool is shallow due to the low laser fluence, with typical depths of 0.3–0.5 µm and widths ranging 10–30 µm.

The micro-waviness is caused by the local inhomogeneous vapor pressure above the melting pool, which results from the local errors of laser intensity distribution and uneven material properties on workpiece surface. This waviness occurs in the scale

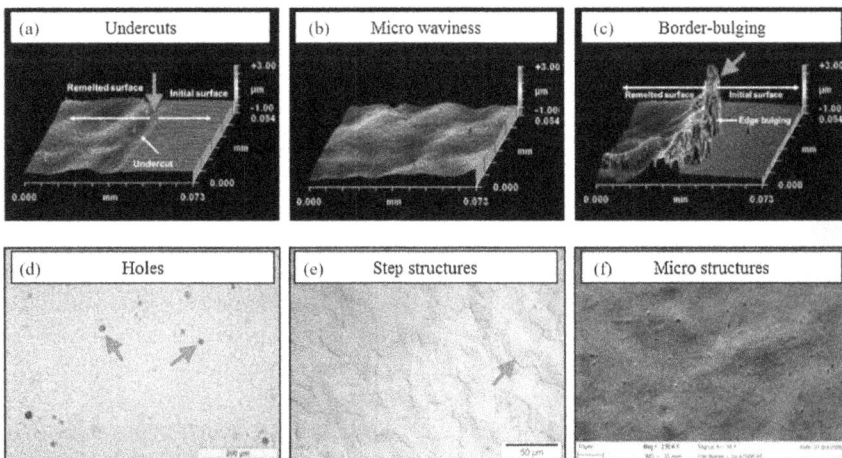

FIGURE 6.6 Typical surface defects in laser micro polishing. Reprinted with permission from Elsevier (Christian Nüsser et al. 2015).

of approximately 0.3 μm. If the laser fluence is set too high, evaporation may occur and lead to even larger micro-waviness. Additionally, evaporation may allow molten material to move from center to the edge and form border-bulges that grow up to several micrometers in height. Holes may be generated with the presence of sulfidic and oxidic inclusions on the workpiece surface as heat will accumulate in these regions and explosion-like evaporation occurs resulting in hole formation.

The holes have diameters ranging 20–100 μm and depths of 5–15 μm (Liebing 2010). Enhanced homogeneity of the work material will eliminate the occurrence of these holes. An alternative solution is to prepare the part with CW laser radiation to uniformly redistribute the iron content on the surface (Willenborg, Berichter, and Klocke 2005). The process to form step structures in laser micro-polishing follows that in laser macro-polishing.

The microstructures can also be found in laser micro-polishing of X38CrMoV5-11 when a low fluence is used. The composition of the microstructures is carbidic inclusions with melting points that are higher than the bulk material. Employing high fluence to melt these inclusions can eliminate the formation of these microstructures.

Since the workpiece properties such as melting temperature, absorption rate, heat conductivity, heat capacity, and surface tension greatly affect the performance of laser polishing, the process may also be reviewed by material categories.

6.2   LASER POLISHING OF AM METALLIC PARTS

6.2.1   CHALLENGES

Although laser polishing has shown favorable and satisfying progress on conventionally manufactured materials and components, there are several critical challenges that remain when employed for AM parts. These challenges come from the complexity of material properties, surface roughness, and complex geometries. Material properties include microstructure, porosity, mechanical, optical, thermal properties (Yung et al. 2019). High surface roughness is contributed by stair effects, rippling effects, partially bonded particles, and printing strategies (Zhang, Chaudhari, and Wang 2019). Lastly, the intricacies of part geometries include re-entrant surfaces, internal channels, internal facets (Rosa, Mognol, and Hascoët 2015; Deng, Li, and Zheng 2020). Both academic and industry partners have been pushing for innovations to apply laser polishing for AM parts over the past few decades.

The interaction between the laser and workpiece greatly depends on the thermophysical properties of the work materials, which can significantly vary (Bordatchev, Hafiz, and Tutunea-Fatan 2014). Thus, laser polishing is often described as a material-dependent finishing process. The most "laser polishable" metals are tool steel, nickel, titanium, and their alloys (e.g., Inconel 718 and Ti6Al4V). Classification of materials that can be more easily laser polished depends on the energy transfer efficiency from the laser to the specific work material without significant energy loss (Bordatchev, Hafiz, and Tutunea-Fatan 2014). The following section will discuss the effectiveness of laser polishing for different metals fabricated by AM and the progress on laser polishing of complex AM structures.

6.2.2 STEEL (STAINLESS STEEL AND TOOL STEEL)

Stainless steel and tool steel are difficult-to-machine materials in mechanical polishing processes (Ukar et al. 2010). However, they are regarded as materials that can be easily handled under laser irradiation. Tool steels are the most investigated steels in laser polishing studies due to their wide application for injection mold and die-casting components. Similarly, investigations on stainless steels and tool steels fabricated by AM have also been hotly studied under laser polishing, as summarized in Table 6.2.

As-printed AM parts have rough surfaces measuring up to tens of micrometers in surface roughness. Therefore, laser macro-polishing is usually employed as the main post-processing procedure to greatly improve the surface finish to several micrometers. Further post-processing will be required if the final surface roughness < 0.1 μm is desired. It is important to note that inappropriate selection of laser parameters can result in surfaces that are rougher than the initial as-built surfaces (Bhaduri et al. 2017). Moreover, a heat-affected zone measuring up to a hundred micrometers typically forms after laser polishing.

6.2.3 TITANIUM AND ITS ALLOYS

Titanium and its alloys are widely employed in aerospace, medical, and nuclear industries due to their high strength-weight ratio in elevated temperatures and outstanding corrosion resistance (Yang and Liu 1999). They are regarded as difficult-to-cut materials due to their low thermal conductivity and high chemical affinity with the cutting tool (Yang and Liu 1999). In this regard, the "zero tool wear" characteristic of laser polishing is a promising alternative to improve the surface finish of titanium alloys. Table 6.3 consolidates recent research on laser polishing of AM titanium alloys and classifies the material as easy to laser polish. Most literature investigated the Ti6Al4V parts. And the geometry of the parts is generally flat. Most studies show that the surface finish can be improved by more than 60% after laser polishing. An example of the most studied titanium alloy in this aspect is the reduction in surface roughness of Ti6Al4V from 6.53 μm Ra to 0.32 μm Ra (Li et al. 2019).

Surface porosity could also be significantly reduced (Liang et al. 2020). However, laser polishing of the titanium alloy may form α' martensitic phase, which increases the surface microhardness (Ma, Guan, and Zhou 2017; Li et al. 2019). Yet, there have been reports that no α' martensitic phase can be found after laser polishing (Marimuthu et al. 2015). High energy density can lead to undesirable surface oxidation and carbonization (Marimuthu et al. 2015), but the high energy density is also preferable for better surface finish (Zhang et al. 2020). Thus, the final application of the part will determine the desirable process parameters.

6.2.4 OTHER ALLOYS (INCONEL, COCR, AL)

Cobalt Chromium (CoCr), Inconel, and Al alloys have important applications in the medical and aerospace industries. Similar to the other metals discussed, the as-printed materials show high surface roughness and require post-finishing. Laser polishing of these AM alloys is summarized in Table 6.4. In all cases, laser polishing can

TABLE 6.2
Summary of Laser Polishing on Stainless Steel and Tool Steel Fabricated by AM

Ref.	Laser System	Workpiece Material	Range of Process Parameters Tested	Best Surface Quality Results	Prerequisites for Best Results	Remarks
(Poprawe and Schulz 2003)	Q-switched mode of slab laser (max. 700 W)	SLM steel (stainless steel) 316 L or tool steel H11)	• Pulse duration: 15 ns*	• Ra initial = 2.7 µm • Ra final = 0.2 µm • 92.7% decrease		* Not explicitly stated.
(Marinescu, Ghiculescu, and Anger 2008)	CO_2 laser (max. 2500 W)	LaserForm ST-100 steel (stainless steel)	• Power: 1.2 kW • Feed rate: 1 – 1.4 m/min • Standoff distance: 22 mm	• $R_{a\,initial}$ = 7.51 µm • $R_{a\,final}$ = 0.574 µm • 92.4% decrease	• Standoff distance: 1.2 m/min	Higher energy density due to lower absorption of stainless steel component.
(Lamikiz et al. 2007)	CO_2 slab laser (max. 2500 W, wavelength: 1060 nm)	LaserForm ST-100 / AISI 420 (bronze infiltrated stainless steel)	Line polishing • Power: 550 – 1200 W • Feed rate: 800 – 2000 mm/min • Focal offset: 20 – 40 mm • Spot diameter: 0.54 – 1.3 mm • Energy density: 1935 – 16666 J/cm²	Line polishing • $R_{a\,initial}$ = 7.5 µm • $R_{a\,final}$ ≈ 0.77 µm* • 89.7% decrease Planar area polishing • $R_{a\,initial}$ = N/A • $R_{a\,final}$ < 2.5 µm** • 68.2% decrease 3D area polishing • $R_{a\,initial}$ = N/A • $R_{a\,final}$ = N/A*** • 74.2% decrease	Line polishing • Power: 900 W • Feed rate: 1000 mm/min • Focal offset: 35 mm • Spot diameter: 1.2 mm • Energy density: 4500 J/cm²	* not explicitly stated; **"below 1 µm most areas"; *** between 1.25 and 2.5 µm.

Reference	Laser system	Material	Parameters	Results	Polishing settings	Remarks
(Yasa, Deckers, and Kruth 2011)	Nd:YAG CW laser (max. 100 W wavelength: 1064 nm)	AISI 316L (stainless steel)	Planar area polishing • Spot diameter: 0.8 – 1.2 mm • Stepover: 0.7 – 1.5 mm • Overlapping index: 15-30% • Power: 1100 W • Feed rate: 1100 mm/min 3D area polishing • Slope angles: 15°, 30°, 45° • Spot diameter: 53 – 133 µm • Feed rate: 50 – 800 mm/s • Scan spacing factor: 0.1 • Power: 60 – 100 W	• $R_{a\,\text{initial}}$ = 15 µm • $R_{a\,\text{final}}$ = 1.5 µm • 90% decrease	Planar area polishing • Overlapping index: 25% 3D area polishing • Same settings • Feed rate: 200 mm/s • Power: 98 W	Melting of materials with dissimilar melting points requires increased energy densities® reduced LP performance Best performance for medium Laser power and low feed rate
(Ramos et al. 2001)	CO_2 laser	LaserForm ST-100 / AISI 420 (bronze infiltrated stainless steel)	• Power: 220 – 420 W • Spot size: 350 ± 50 µm • Feed rate: 1.8 – 4.5 mm/s	• $R_{a\,\text{initial}}$ = 2.38 µm • $R_{a\,\text{final}}$ = 0.82 µm • 65.5% decrease	• Power: 420 W • Feed rate: 4.5 mm/s	An optimal feed rate is a prerequisite for polishing

(continued)

TABLE 6.2 (Continued)
Summary of Laser Polishing on Stainless Steel and Tool Steel Fabricated by AM

Ref.	Laser System	Workpiece Material	Range of Process Parameters Tested	Best Surface Quality Results	Prerequisites for Best Results	Remarks
	Nd:YAG laser	LaserForm ST-100 / AISI 420 (bronze infiltrated stainless steel)	• Power: 220 W • Spot size: 250 ± 50 μm • Feed rate: 1.46 – 1.76 mm/s	• $R_{a\,\text{initial}}$ = 9.0 μm • $R_{a\,\text{final}}$ = 3.0 μm • 66.7% decrease	• Feed rate: 1.46 mm/s	Slightly better results obtained at 1.7 mm/s ($R_{a\,\text{final}}$ = 2.4 μm)
(Ramos, Bourell, and Beaman 2003)	CO_2/Nd:YAG laser	LaserForm ST-100 / AISI 420 (bronze infiltrated stainless steel)	• Power: 320 W • Feed rate: 101.3 – 692.7 mm/s	• $R_{a\,\text{initial}}$ = 2.98 μm • $R_{a\,\text{final}}$ = 2.05 μm* • 31.2% decrease	• Feed rate: 177 mm/s	$R_{a\,\text{initial}}$ and energy density are important * Not explicitly stated
(Bhaduri et al. 2017)	Yb-doped fiber nanosecond (ns) laser (Max. 50 W, wavelength: 1064 nm)	Digital Metal® SS316L Size: 10 × 10 × 10 mm artifact	• Power: 37.2 W • Pulse duration: 220 ns • Fluence: 5 – 9 J/cm² • Overlap: 82 – 95% • Pulse frequency: 30 000 – 50 000 Hz • Spot size: 59 μm • Hatching pitch: 3 – 11 μm • Scanning speed: 900 – 5500 mm/s	• $S_{a\,\text{initial}}$ = 3.8 μm • $S_{a\,\text{final}}$ = 0.23 μm 94% decrease	• Fluence: 9 J/cm² • Overlap: 88 – 91%	Heat-affect zone around 100 μm Ra will increase if using 20 J/cm² in air

| (Yung et al. 2019) | CW laser (a low-power SPI fiber laser system with max. 100 W and a high-power IPG fiber laser system with max. 1000 W) | Concept Laser M2 maraging tool steel (Flat and free form) | Laser macro polishing (cw) • Power: 100 – 300 W • Scanning speed: 254 – 1207 mm/s • Hatching pitch: 64 – 508 μm • Spot size: 3000 ± 50 μm | • $R_{a\,initial}$ = 12 μm • $R_{a\,final}$ = 0.74 μm • 94% decrease | • Coarse (200 W, 507 mm/s, 127 μm hatching) and delicate (100 W, 1016 mm/s, 508 μm hatching) polishing are combined • Energy density are 3.1 and 0.2 J/cm², respectively | For freeform surface: • $R_{a\,initial}$ = 5.01 μm • $R_{a\,final}$ = 0.44 μm • 91% decrease |
| | | | Laser micro polishing (pulse) • Power: 15 – 100 W • Scanning speed: 1000 – 3000 mm/s • Hatching pitch: 2 – 16 μm • Spot size: 2500 ± 50 μm | • $R_{a\,initial}$ = 12 μm • $R_{a\,final}$ = 4.67 μm • 61% decrease | • Coarse (40 W, 1500 mm/s, 8 μm hatching) and delicate (25 W, 2000 mm/s, 50 μm hatching) polishing are combined • Energy density are 3.3 and 0.3 J/cm², respectively | Surface hardness increased by 14.6% |

(continued)

TABLE 6.2 (Continued)
Summary of Laser Polishing on Stainless Steel and Tool Steel Fabricated by AM

Ref.	Laser System	Workpiece Material	Range of Process Parameters Tested	Best Surface Quality Results	Prerequisites for Best Results	Remarks
(Muhannad A. Obeidi et al. 2019)	CO_2 laser (max. 1500 W)	SLM 316L SS	• Power: 90 – 130 W • Scanning speed*: 628 – 2512 mm/min • Overlap: -20 – 20% • Focus offset: -0.5 – 0.5 mm • Hatching pitch: 2 – 16 μm Spot size: 2500 ± 50 μm	• $R_{a\,initial}$ = 10.4 μm • $R_{a\,final}$ = 2.7 μm • 74% decrease	• Scanning speed*: 628 mm/min • Overlap: 20% • Focus offset: 0 mm	* Converted from workpiece rotation speed
(Braun et al. 2019)	Solid State Laser (max. 4000 W)	SLM 316L SS Size: 60 ×40 × 30 mm artifact	• Power: 150 – 300 W • Spot size: 870 μm • Hatching pitch: 50 – 200 μm	• $R_{a\,initial}$ = 6.9 – 11.8 μm • $R_{a\,final}$ = 0.54 μm • 92% decrease	• Power: 350 W • Hatching pitch: 50 μm • Feed rate: 200 mm/s • Fluence: 35 J/mm² • Area rate: 6 cm²/min	A robot was used to provide the motion
(Rosa, Mognol, and Hascoët 2015)	CW fiber laser (max. 800 W and wavelength: 1070 nm)	DED 316L SS flat surface	• Power: 210 W • Scanning speed: 3000 mm/min • Overlap: 60% • Focus offset: 0 mm • Spot size: 800 μm	• R_a = 21 μm • $R_{a\,final}$ = 0.79 μm • 96% decrease	• 5 polishing passes	Use the same laser as the printing process

Reference	Laser source	Sample	Parameters	Results	Optimal/Notes	Remarks
		DED 316L SS thin-wall part	• Power: 100 W • Scanning speed: 3000 mm/min • Overlap: 60% • Focus offset: 0 – 7mm • Spot size: 800 μm	• $R_{a\ initial}$ = 14 μm • $R_{a\ final}$ = 5.39 μm • 61% decrease	• 5 polishing passes • Focus offset varies according to the workpiece profile	Main goal to eliminate geometrical deviation but not roughness
(Alrbaey et al. 2014)	Fiber laser (max. 400 W)	SLM 316L SS Size: 30 ×10 × 4 mm flat samples	• Power: 160, 180, 200 W • Scanning speed: 400, 500, 600 mm/min • Hatching pitch: 400, 500, 600 μm • Standoff distance: 128 mm • Spot size: ~1 mm	• $R_{a\ initial}$ = 10 μm • $R_{a\ final}$ = 1.3 μm • 87% decrease	• Power: 178 W • Scanning speed: 400 mm/ min • Hatching pitch: 400 μm	NA
		SLM 316L SS Size: 30 ×10 × 4 mm with inclined angles	• Power: 180 W • Scanning speed: 500 mm/min • Hatching pitch: 400 μm • Standoff distance: 128 mm • Spot size: ~1 mm	• $R_{a\ initial}$ = 8 – 16 μm • $R_{a\ final}$ = 1.5 μm • 80–90% decrease	• Same as before	Best Ra obtained at 0° sample ($R_{a\ final}$ = 1.43 μm) and worst at 60° sample ($R_{a\ final}$ = 1.55 μm)
(D. Zhang et al. 2020)	CW fiber laser (max. 400 W, wavelength: 1060 nm)	SLM 316L SS Size: 40 × 40 × 5 mm flat	• Laser power: 100 – 400 W • Feed rate: 500– 1000 mm/s • Hatching distance: 80 μm • Spot size: 100 μm	• Sa initial = 14.59 μm • Sa final = 3.91 μm • 73% decrease	• Laser power: 400 W • Feed rate: 500 mm/s	

TABLE 6.3
Summary of Laser Polishing on AM Ti Alloy

Ref.	Laser System	Workpiece Material	Range of Process Parameters Tested	Best Surface Quality Results	Prerequisites for Best Results	Remarks
(Yasa, Deckers, and Kruth 2011)	Yb:YAG CW fiber laser (max. 300 W, wavelength: 1085 nm)	SLM Ti6Al4V	• Spot diameter: 50 μm • Feed rate: 20 – 1280 mm/s • Power: 16 – 256 W • Power density: $1.3{\times}10^5$–$2.72{\times}10^7$ w/cm^{2}*	• $R_{a\ initial} = 15$ μm • $R_{a\ final} = 2.9$ μm • 80.7% decrease	• Feed rate: 640 mm/s • Power: 256 w	Best performance for high power and medium feed rate * not explicitly stated
(Marimuthu et al. 2015)	CW fiber laser (max. 200 W, wavelength: 1070 – 1090 nm)	SLM Ti6Al4V Size: 40 × 10 × 5 mm	Standoff distance: 5 mm Spot size: 500 μm Line polishing: • Power: 50 – 200 W • Feed rate: 200 – 2400 mm/min Area polishing: • Power: 150 – 180 W • Feed rate: 500 – 1000 mm/min • Hatching distance: 0.3 – 0.55 mm	• Ra initial = 10.2 μm • Ra final = 2.4 μm • 76% decrease	• Feed rate: 750 mm/s • Power: 160 w • Hatching distance: 0.35 mm	No alpha case formation No thermal cracking Higher energy density results in surface oxidation and carbonization

Reference	Laser	Material	Parameters	Surface roughness	Optimized parameters	Notes
(Ma, Guan, and Zhou 2017)	Ns pulsed fiber laser (wavelength: 1060 nm)	AM Ti6Al4V (TC4) flat	• Pulse duration: 220 ns • Repetition rate: 500 khz • Spot size: 44 μm • Scanning speed: 200 mm/s • Energy density: 1.2×10^7 w/cm^2 • Overlap: 50%	• Ra initial = 5.23 μm • Ra final = 0.38 μm* • 92% decrease	• Previous parameters already optimized	* Wire-EDMed before laser polishing A' martensitic formed leading to 32% increase in surface micro hardness Remelting depth around 170 μm
		AM Ti-6.5Al-3.5Mo-1.5Zr-0.3Si (TC11) flat		• Ra initial = 7.21 μm • Ra final = 0.73 μm* • 89% decrease		* Wire-EDMed before laser polishing α' martensitic formed leading to 42% increase in surface micro hardness remelting depth around 90 μm
(Li et al. 2019)	CW fiber laser (wavelength: 1060 nm)	SLM Ti6Al4V	• Laser power: 70 – 100 W • Feed rate: 50 – 250 mm/s • Overlap: 10%	• Ra initial = 6.53 μm • Ra final = 0.32 μm • 95% decrease	• Feed rate: 150 mm/s • Power: 90 W	Surface microhardness increased 25% after laser polishing New α' martensitic phase formed HAZ up to 160 μm

(continued)

TABLE 6.3 (Continued)
Summary of Laser Polishing on AM Ti Alloy

Ref.	Laser System	Workpiece Material	Range of Process Parameters Tested	Best Surface Quality Results	Prerequisites for Best Results	Remarks
(Liang et al. 2020)	Ns pulsed fiber laser (wavelength: 1064 nm)	SLM Ti6Al4V	• Spot size: 150 μm • Pulse duration: 270 nm • Repetition rate: 50 kHz • overlap: 50% • Feed rate: 3750 mm/s • Energy density: 0.853×10^5 W/cm^2 – 2.503×10^5 W/cm^2	• Ra initial = 10.2 μm • Ra final = 2.1 μm • 79% decrease	• Energy density: 1.703×10^5 W/cm^2	Porosity decreased from 0.68% to 0.02% after laser polishing HAZ around 120 μm
(D. Zhang et al. 2020)	CW fiber laser (max. 400 W, wavelength: 1060 nm)	SLM Ti6Al4V Size: 40 × 40 × 5 mm flat	• Laser power: 100 – 400 W • Feed rate: 500–1000 mm/s • Hatching distance: 120 μm • Spot size: 100 μm	• Sa initial = 8.8 μm • Sa final = 3.32 μm • 62% decrease	• Laser power: 400 W • Feed rate: 500 mm/s	Under the selected parameters, the higher the energy density, the larger the improvement of the surface finish

Reference	Laser	Material/Sample	Parameters	Results		Remarks
(Tian et al. 2018)	Fiber laser (wavelength: 1070 nm)	EBM Ti6Al4V Size: 30 × 70 × 50 mm flat	• Laser power: 100 W • Spot size: 400 μm • Feed rate: 300 mm/min • Hatching distance: 30 μm	• Sa initial = 21.46 μm • Sa final = 5.5 μm • 75% decrease	• Previous parameters already optimized	If measured by inferometer within 0.65 mm× 0.86 mm, the surface roughness Sa is 1.65 μm
(Muhannad Ahmed Obeidi et al. 2022)	CO_2 laser (max. 1.5 kW)	SLM Ti6Al4V Size: 50 × 10 × 2 mm flat	• Laser power: 45 – 107 W • Scanning speed: 4500 mm/s • Spot size: 200 μm • Overlap: 25 – 75%	• Sa initial = 17.15 μm • Sa final = 3.31 μm • 81% decrease	• Laser power: 70 W • Scanning speed: 4500 mm/s	Surface hardness increased by 7% in a depth of 100 to 150 μm from the sample surface

TABLE 6.4
Summary of Laser Polishing on Other AM Alloys (Inconel, CoCr, Al)

Ref.	Laser System	Workpiece Material	Range of Process Parameters Tested	Best Surface Quality Results	Prerequisites for Best Results	Remarks
(Dadbakhsh, Hao, and Kong 2010)	CO_2 laser (max. 1.8 kW HQ CO_2)	LMD Inconel 718 (Ni-Cr alloy)	• Power: 268–532 W • Feed rate: 737–1263 mm/min • Spot size: 0.5–1.1 mm	• $R_{a\,initial} \approx$ 9.85 μm* • $R_{a\,final} \approx$ 2 μm • 80% decrease	• Power: 500 W • Feed: 800 – 850 mm/min • Spot size: 0.7 mm • Power density: $1.3 \cdot 10^5$ W/cm²**	* Not explicitly stated Energy density and power are critical for R_a ** Not explicitly stated
(Witkin et al. 2016)	CW laser (max. 200 W wavelength: 1064 nm)	SLM Inconel 625	• Power: 5–10 W • Scanning speed: 240 mm/min • Hatching space: 0.01 mm • Spot size: 0.03 mm	As received: • $R_{a\,initial} \approx$ 10.04 μm • $R_{a\,final} \approx$ 6.51 μm • 35% decrease Milled: • $R_{a\,initial} \approx$ 1.47 μm • $R_{a\,final} \approx$ 0.79 μm • 46% decrease	• Power: 5–10 W • Scanning speed: 240 mm/min • Hatching space: 0.01 mm • Spot size: 0.03 mm	Microstructure remains unchanged after laser polishing
(Yung et al. 2018)	Pulsed fiber laser (max. 70 W wavelength: 1064 nm)	SLM CoCr alloy Slant, convex and concave surfaces	• Power: 30–70 W • Scanning speed: 15–300 mm/min • Hatching space: 0.02–0.05 mm	• $R_{a\,initial} \approx$ 4.51 μm • $R_{a\,final} \approx$ 0.28 μm • 93% decrease	• Power: 70 W • Scanning speed: 300 mm/min • Hatching space: 0.03 mm	The surface hardness of the laser polished samples was enhanced by 8%

Reference	Laser	Material	Parameters	Results	Optimal parameters		Remarks
(Richter et al. 2019)	CW fiber laser (max. 200 W wavelength: 1070 nm)	SLM Co-Cr-Mo alloy Size: 20.5 × 9.4 × 3.4 mm	• Power: 53–200 W • Scanning speed: 200–300 mm/min • Hatching space: 0 • Spot size: 0.1–0.15 mm	• $R_{a\,initial} \approx$ 20.19 μm • $R_{a\,final} \approx$ 6.51 μm • 68% decrease	• Power: 200 W • Scanning speed: 200 mm/min • Spot size: 0.15 mm	• Spot size: 0.05 mm • Defocusing distance: 6 mm	Single-track polishing and no hatching
(Schanz et al. 2016)	Solid state disk laser (max. 4 kW)	SLM AlSi10Mg Size: 100 × 30 × 2 mm	• Power: 1700 W • Scanning speed: 40 mm/min • Overlap: 93% • Spot size: 1 mm	• $R_{a\,initial} \approx$ 8.7 μm • $R_{a\,final} \approx$ 0.66 μm • 92% decrease	• Power: 1700 W • Scanning speed: 40 mm/min • Overlap: 93% • Spot size: 1 mm		NA

efficiently reduce the surface roughness of these alloys for both flat surfaces and complex surfaces. For example, the surface roughness of CoCr fabricated by SLM can be reduced to 0.28 μm Ra and surface hardness is slightly increased (Yung et al. 2018). Surface roughness of 0.79 μm Ra can be obtained on Inconel 625 fabricated by SLM through milling followed by laser polishing (Witkin et al. 2016). However, the pre-milling process will increase extra cost and labor.

6.3   CONCLUDING REMARKS

This chapter discussed the most employed thermal-based finishing method – laser polishing. Dominant process parameters and common surface defects of laser polishing were covered along with the consolidation of literature on laser polishing of various AM metals These include steels, titanium, Inconel, CoCr, Al, and their alloys.

Through these studies, laser macro-polishing is deemed suitable for as-printed surfaces and can achieve final surface measuring sub-micrometers. On the other hand, laser micro-polishing is more suitable as a secondary or fine finishing of AM metals where nanoscale surface finishing may be obtained with the optimal parameters.

With the integration of multi-axial CNC capabilities, laser polishing would be able to polish the complex geometries of AM components such as inclined (Lamikiz et al. 2007; Yung et al. 2018; Alrbaey et al. 2014), cylinder (Muhannad A. Obeidi et al. 2019), concave (Yung et al. 2018), convex (Yung et al. 2018), freeform surfaces (Yung et al. 2019; Braun et al. 2019), and thin wall parts (Rosa, Mognol, and Hascoët 2015). However, common internal features in AM – such as internal channels, lattice structures, and internal facets – pose significant challenges for laser polishing. Additionally, thin-walled parts that have wide application in the aerospace industry may be affected by laser polishing where the applied heat to the workpiece may lead to thermal distortion of the thin-walled parts (Rosa, Mognol, and Hascoët 2015). Nonetheless, laser polishing is an effect post-process for additive manufacturing to greatly reduce surface roughness without incurring any tool wear.

REFERENCES

Alrbaey, K., D. Wimpenny, R. Tosi, W. Manning, and A. Moroz. 2014. "On Optimization of Surface Roughness of Selective Laser Melted Stainless Steel Parts: A Statistical Study." *Journal of Materials Engineering and Performance* 23 (6). Springer: 2139–2148. doi:10.1007/s11665-014-0993-9.

Bhaduri, Debajyoti, Pavel Penchev, Afif Batal, Stefan Dimov, Sein Leung Soo, Stella Sten, Urban Harrysson, Zhenxue Zhang, and Hanshan Dong. 2017. "Laser Polishing of 3D Printed Mesoscale Components." *Applied Surface Science* 405. Elsevier: 29–46. doi:10.1016/j.apsusc.2017.01.211.

Bordatchev, Evgueni V., Abdullah M K Hafiz, and O. Remus Tutunea-Fatan. 2014. "Performance of Laser Polishing in Finishing of Metallic Surfaces." *International Journal of Advanced Manufacturing Technology* 73 (1–4): 35–52. doi:10.1007/s00170-014-5761-3.

Braun, Markus, Markus Hofele, Jochen Schanz, Simon Ruck, Mario Pohl, Rainer Börret, and Harald Riegel. 2019. "Time-Based Offline Track Planning of Robot Guided Laser

Applications for Complex 3D Freeform Surfaces." In *Laser 3D Manufacturing VI*, 10909:30. International Society for Optics and Photonics. doi:10.1117/12.2508340.

Chow, Michael T.C., Evgueni V. Bordatchev, and George K. Knopf. 2013. "Experimental Study on the Effect of Varying Focal Offset Distance on Laser Micropolished Surfaces." *International Journal of Advanced Manufacturing Technology* 67 (9–12). Springer: 2607–2617. doi:10.1007/s00170-012-4677-z.

Dadbakhsh, Sasan, Liang Hao, and Choon Yen Kong. 2010. "Surface Finish Improvement of LMD Samples Using Laser Polishing." *Virtual and Physical Prototyping* 5 (4). Taylor & Francis: 215–221. doi:10.1080/17452759.2010.528180.

Deng, Tiantian, Jianjun Li, and Zhizhen Zheng. 2020. "Fundamental Aspects and Recent Developments in Metal Surface Polishing with Energy Beam Irradiation." *International Journal of Machine Tools and Manufacture* 148. Elsevier: 103472. doi:10.1016/j.ijmachtools.2019.103472.

Giorleo, L., E. Ceretti, and C. Giardini. 2015. "Ti Surface Laser Polishing: Effect of Laser Path and Assist Gas." *Procedia CIRP* 33. Elsevier: 446–451. doi:10.1016/j.procir.2015.06.102.

Hafiz, Abdullah M Khalid, Evgueni V Bordatchev, and Remus O Tutunea-Fatan. 2012. "Influence of Overlap between the Laser Beam Tracks on Surface Quality in Laser Polishing of AISI H13 Tool Steel." *Journal of Manufacturing Processes* 14 (4). Elsevier: 425–434. doi:10.1016/j.jmapro.2012.09.004.

Kiedrowski, Thomas. 2009. „ *Oberflächenstrukturbildung Beim Laserstrahlpolieren von Stahlwerkstoffen* ". Shaker.

Krishnan, Arun, and Fengzhou Fang. 2019. "Review on Mechanism and Process of Surface Polishing Using Lasers." *Frontiers of Mechanical Engineering* 14 (3). Springer: 299–319. doi:10.1007/s11465-019-0535-0.

Lamikiz, A., J. A. Sánchez, L. N. López de Lacalle, and J. L. Arana. 2007. "Laser Polishing of Parts Built up by Selective Laser Sintering." *International Journal of Machine Tools and Manufacture* 47 (12–13). Elsevier: 2040–2050. doi:10.1016/j.ijmachtools.2007.01.013.

Li, Yu Hang, Bing Wang, Cheng Peng Ma, Zhi Hao Fang, Long Fei Chen, Ying Chun Guan, and Shou Feng Yang. 2019. "Material Characterization, Thermal Analysis, and Mechanical Performance of a Laser-Polished Ti Alloy Prepared by Selective Laser Melting." *Metals* 9 (2). Multidisciplinary Digital Publishing Institute: 112. doi:10.3390/met9020112.

Liang, Chunyong, Yazhou Hu, Ning Liu, Xianrui Zou, Hongshui Wang, Xinping Zhang, Yulan Fu, and Jingyun Hu. 2020. "Laser Polishing of Ti6Al4V Fabricated by Selective Laser Melting." *Metals* 10 (2). MDPI: 191. doi:10.3390/met10020191.

Liebing, Christiane. 2010. *Beeinflussung Funktioneller Oberflächeneigenschaften von Stahlwerkstoffen Durch Laserpolieren*. na.

Ma, C. P., Y. C. Guan, and W. Zhou. 2017. "Laser Polishing of Additive Manufactured Ti Alloys." *Optics and Lasers in Engineering* 93. Elsevier: 171–177. doi:10.1016/j.optlaseng.2017.02.005.

Marimuthu, S., A. Triantaphyllou, M. Antar, D. Wimpenny, H. Morton, and M. Beard. 2015. "Laser Polishing of Selective Laser Melted Components." *International Journal of Machine Tools and Manufacture* 95. Elsevier: 97–104. doi:10.1016/j.ijmachtools.2015.05.002.

Marinescu, N I, D Ghiculescu, and C Anger. 2008. "Correlation between Laser Finishing Technological Parameters and Materials Absorption Capacity." *International Journal of Material Forming* 1 (S1). Springer: 1359–1362. doi:10.1007/s12289-008-0116-y.

Mohajerani, Shirzad, Evgueni V Bordatchev, and O Remus Tutunea-Fatan. 2018. "Recent Developments in Modeling of Laser Polishing of Metallic Materials." *Lasers in Manufacturing and Materials Processing* 5 (4). Springer: 395–429. doi:10.1007/s40516-018-0071-5.

Nüsser, C., H. Sändker, and E. Willenborg. 2013. "Pulsed Laser Micro Polishing of Metals Using Dual-Beam Technology." *Physics Procedia* 41. Elsevier: 346–355. doi:10.1016/j.phpro.2013.03.087.

Nüsser, Christian, Judith Kumstel, Thomas Kiedrowski, Andrei Diatlov, and Edgar Willenborg. 2015. "Process- and Material-Induced Surface Structures during Laser Polishing." *Advanced Engineering Materials* 17 (3). Wiley Online Library: 268–277. doi:10.1002/adem.201400426.

Obeidi, Muhannad A., Eanna McCarthy, Barry O'Connell, Inam Ul Ahad, and Dermot Brabazon. 2019. "Laser Polishing of Additive Manufactured 316L Stainless Steel Synthesized by Selective Laser Melting." *Materials* 12 (6). Multidisciplinary Digital Publishing Institute: 991. doi:10.3390/ma12060991.

Obeidi, Muhannad Ahmed, Andre Mussatto, Merve Nur Dogu, Sithara P. Sreenilayam, Eanna McCarthy, Inam Ul Ahad, Shane Keaveney, and Dermot Brabazon. 2022. "Laser Surface Polishing of Ti-6Al-4V Parts Manufactured by Laser Powder Bed Fusion." *Surface and Coatings Technology* 434. Elsevier: 128179. doi:10.1016/j.surfcoat.2022.128179.

Pfefferkorn, Frank E., Neil A. Duffie, Xiaochun Li, Madhu Vadali, and Chao Ma. 2013. "Improving Surface Finish in Pulsed Laser Micro Polishing Using Thermocapillary Flow." *CIRP Annals – Manufacturing Technology* 62 (1). Elsevier: 203–206. doi:10.1016/j.cirp.2013.03.112.

Poprawe, Reinhart, and Wolfgang Schulz. 2003. "Development and Application of New High-Power Laser Beam Sources." *RIKEN Review: Focused on Laser Precision Microfabrication (LPM2002)* 50 (50). Citeseer: 3–10.

Ramos, J. A., D. L. Bourell, and J. J. Beaman. 2003. "Surface Over-Melt during Laser Polishing of Indirect-SLS Metal Parts." *Materials Research Society Symposium – Proceedings* 758. Cambridge University Press: 53–61. doi:10.1557/proc-758-ll1.9.

Ramos, J A, J Murphy, K Wood, D L Bourell, and J J Beaman. 2001. "Surface Roughness Enhancement of Indirect-SLS Metal Parts by Laser Surface Polishing 28." In *2001 International Solid Freeform Fabrication Symposium*. doi:10.26153/tsw/3233.

Richter, B., N. Blanke, C. Werner, F. Vollertsen, and F. E. Pfefferkorn. 2019. "Effect of Initial Surface Features on Laser Polishing of Co-Cr-Mo Alloy Made by Powder-Bed Fusion." *Jom* 71 (3). Springer: 912–919. doi:10.1007/s11837-018-3216-2.

Rosa, Benoit, Pascal Mognol, and Jean-yves Hascoët. 2015. "Laser Polishing of Additive Laser Manufacturing Surfaces." *Journal of Laser Applications* 27 (S2): S29102. doi:10.2351/1.4906385.

Schanz, Jochen, Markus Hofele, Leonhard Hitzler, Markus Merkel, and Harald Riegel. 2016. "Laser Polishing of Additive Manufactured Alsi10mg Parts with an Oscillating Laser Beam." In *Advanced Structured Materials*, 61:159–169. Springer. doi:10.1007/978-981-10-1082-8_16.

Tian, Yingtao, Wojciech S. Gora, Aldara Pan Cabo, Lakshmi L. Parimi, Duncan P. Hand, Samuel Tammas-Williams, and Philip B. Prangnell. 2018. "Material Interactions in Laser Polishing Powder Bed Additive Manufactured Ti6Al4V Components." *Additive Manufacturing* 20. Elsevier: 11–22. doi:10.1016/j.addma.2017.12.010.

Ukar, E., A. Lamikiz, L. N. López de Lacalle, D. del Pozo, and J. L. Arana. 2010. "Laser Polishing of Tool Steel with CO2 Laser and High-Power Diode Laser." *International Journal of Machine Tools and Manufacture* 50 (1). Elsevier: 115–125. doi:10.1016/j.ijmachtools.2009.09.003.

Willenborg, Edgar, Emsland Berichter, and Fritz Klocke. 2005. *Polieren von Werkzeugstählen Mit Laserstrahlung*. Shaker Aachen.

Witkin, David, Henry Helvajian, Lee Steffeney, and William Hansen. 2016. "Laser Post-Processing of Inconel 625 Made by Selective Laser Melting." In *Laser 3D Manufacturing III*, 9738:97380W. International Society for Optics and Photonics. doi:10.1117/12.2213745.

Yang, Xiaoping, and C. Richard Liu. 1999. "Machining Titanium and Its Alloys." *Machining Science and Technology* 3 (1). Taylor & Francis: 107–139. doi:10.1080/10940349908945686.

Yasa, Evren, Jan Deckers, and Jean Pierre Kruth. 2011. "The Investigation of the Influence of Laser Re-Melting on Density, Surface Quality and Microstructure of Selective Laser Melting Parts." *Rapid Prototyping Journal* 17 (5). Emerald Group Publishing Limited: 312–327. doi:10.1108/13552541111156450.

Yung, K. C., T. Y. Xiao, H. S. Choy, W. J. Wang, and Z. X. Cai. 2018. "Laser Polishing of Additive Manufactured CoCr Alloy Components with Complex Surface Geometry." *Journal of Materials Processing Technology* 262. Elsevier: 53–64. doi:10.1016/j.jmatprotec.2018.06.019.

Yung, K. C., S. S. Zhang, L. Duan, H. S. Choy, and Z. X. Cai. 2019. "Laser Polishing of Additive Manufactured Tool Steel Components Using Pulsed or Continuous-Wave Lasers." *International Journal of Advanced Manufacturing Technology* 105 (1–4). Springer: 425–440. doi:10.1007/s00170-019-04205-z.

Zhang, Dongqi, Jie Yu, Hui Li, Xin Zhou, Changhui Song, Chen Zhang, Shengnan Shen, Linqing Liu, and Chengyuan Dai. 2020. "Investigation of Laser Polishing of Four Selective Laser Melting Alloy Samples." *Applied Sciences (Switzerland)* 10 (3). MDPI: 760. doi:10.3390/app10030760.

Zhang, Jiong, Akshay Chaudhari, and Hao Wang. 2019. "Surface Quality and Material Removal in Magnetic Abrasive Finishing of Selective Laser Melted 316L Stainless Steel." *Journal of Manufacturing Processes* 45. Elsevier: 710–719. doi:10.1016/j.jmapro.2019.07.044.

7 Chemical-based Finishing Processes

7.1 DEFINITION AND CLASSIFICATION

7.1.1 CHEMICAL ETCHING (CE)

Chemical etching is a cost-effective finishing method that has been widely utilized in many industries (Han and Fang 1998). Also known as chemical polishing, it is a surface treatment aimed at improving the surface finish of a part. Generally, the workpiece will be submerged in a specially designed chemical solution over time. The material removal mechanism is illustrated in Figure 7.1 (Basha et al. 2022; Ishizawa et al. 2008). Material on the surface will dissolve into metal ions that diffuse into the solution. Metal ions diffuse faster at convex regions such as peaks of asperities due to the high density of cations (H^+). Hence, material dissolves faster from asperities. On the other hand, dissolved metal ions accumulate in concave regions and the density of the cations (H^+) is relatively lower, which leads to slower dissolution of the metals (i.e., lower material removal rates) in the concave regions.

7.1.2 ELECTROCHEMICAL POLISHING (ECP)

Electrochemical polishing (ECP), or electrolytic polishing (EP), is a chemical finishing process that gradually removes material by anodic dissolution as presented in Figure 7.2(a). The workpiece will assume the role of an anode immersed in an electrolyte and connected to a power supply, while the tool as the cathode will be positioned at a controlled distance from the anode otherwise known as the inter-electrode gap. A sequence of chemical reactions will occur as the voltage increases, beginning with etching followed by passivating, polishing, and, finally, gas evolution – as illustrated in Figure 7.2(b). Like CE, ions from high asperities of the anode will dissolve quicker than the valleys during ECP, which enables smoothening of the workpiece surface.

The material removal mechanism of ECP involves the viscous layer theory (Landolt 1987), passive film theory (Hoar and Mowat 1950), mass transport theory (West et al. 1992; Matlosz, Magaino, and Landolt 1994), and enhanced oxidation-dissolution

DOI: 10.1201/9781003272601-7

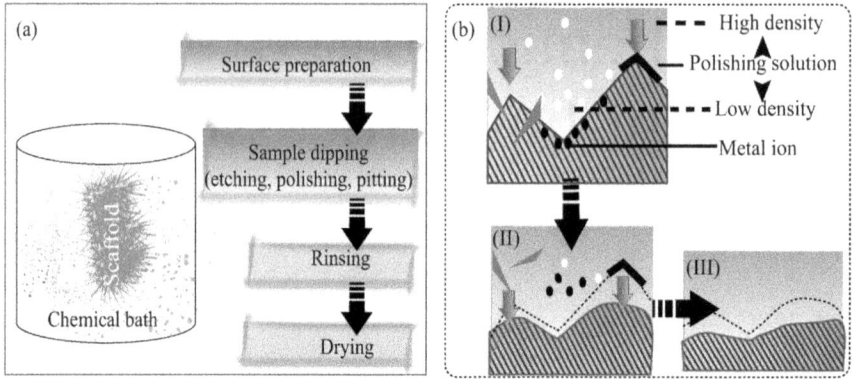

FIGURE 7.1   (a) Workflow of chemical polishing and (b) material removal mechanism. Reprinted with permission from Elsevier (Basha et al. 2022; Ishizawa et al. 2008).

theory (Rokicki and Hryniewicz 2012). However, the intrinsic material removal mechanism of ECP remains unclear (Han and Fang 2019; Yi et al. 2020).

ECP is suitable for finishing parts with complex geometries such as sharp corners that can be accessed by the liquid electrolyte. The main process parameters include electrolyte temperature, electrolyte composition, polishing time, initial surface roughness, electrode rotation speed, inter-electrode gap, microstructure of workpiece, and current density. A detailed description of these process parameters can be found in Han and Fang 2019.

7.1.3   HYBRID ELECTROCHEMICAL POLISHING

Hybrid electrochemical polishing refers to the simultaneous action of electrochemical material removal and another removal mechanism. They include chemical mechanical polishing, dry electrolytic polishing, and electrochemical mechanical polishing. The goals are to increase the polishing efficiency and achieve better surface finish. Chemical mechanical polishing, also known as chemical mechanical planarization (CMP), employs a rotational polishing pad to improve the surface finish and flatness using a mixture of fine abrasive particles and corrosive chemical slurry. It is widely applied in wafer production and metallographic sample preparation. Dry electrolytic polishing refers to the electrochemical polishing process without liquid electrolyte (GPAINNOVA 2020). In this polishing process, porous particles holding chemicals are employed as the polishing media. The workpiece moves inside the polishing media to ensure its entire surface is polished while providing adequate abrasion for mechanical removal. Electrochemical mechanical polishing also integrates the mechanical removal mechanism and it can be applied for both flat surfaces and curved surfaces. Both free abrasives and bonded abrasives can be used. Detailed examples of the hybrid electrochemical polishing processes are discussed in Section 7.4.

(a)

(b)

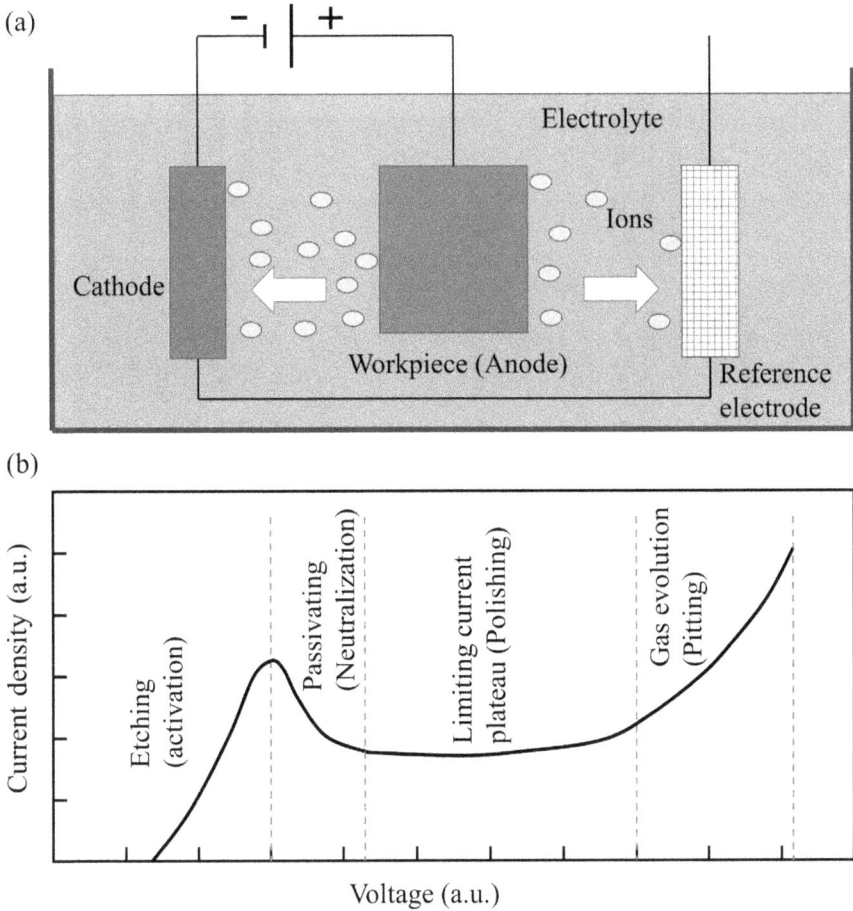

FIGURE 7.2 Schematics of electrochemical polishing and the voltage-current curve: (a) schematic setup showing the electrolyte, cathode, workpiece (anode), and ions; (b) voltage-current curve showing various stages in electrochemical polishing, i.e., etching, passivating, limiting current plateau and gas evolution.

7.2 CHEMICAL ETCHING OF AM PARTS

Studies on chemical polishing of AM metals (mostly Ti6Al4V by SLM) are summarized in Table 7.1. HF and HNO_3 are typical electrolytes used for CE. The concentration of the electrolyte also plays a critical role in the chemical reactions. Low concentration of HF will have insignificant impact on the surface finish (Pyka et al. 2012) regardless of the heat applied to the electrolyte. High HF concentration may lead to the formation of oxide layers after polishing. Recommended electrolyte concentrations in a mix of the two acids are approximately 10 vol.% HNO_3 and 10 vol.% HF (Balyakin, Shvetcov, and Zhuchenko 2018). However, a solution comprising

TABLE 7.1

Summary of Chemical Polishing on AM Materials

Ref.	Workpiece Material	Electrolyte solution	Range of Process Parameters Tested	Best Surface Quality Results	Prerequisites for Best Results	Remarks
(Lyczkowska et al. 2014)	SLM Ti6Al7Nb Size: cylinder lattice with a 6.2-mm diameter and a 6.0 mm height. Strut thickness 0.25 mm	Bath No. 1:80 vol.% H_2O, 6 vol.% HF and 14 vol.% HNO_3 Bath No. 2:99 vol.% H_2O and 1 vol.% HF	• Temperature: not specified • Polishing time: $60-900$ s • Stirring speed: not specified	• $R_{a\,initial}$ = not specified • $R_{a\,final}$ = not specified	• Polishing time: 600 s • Bath No. 2	A magnetic stirrer is better to control the solution flow than the ultrasonic stirrer
(Pyka et al. 2012)	SLM Ti6Al4V Size: cylinder lattice with a 6-mm diameter and a 12-mm height. Strut thickness 0.1 mm	HF	• Temperature: not specified • Polishing time: 10 min • Stirring speed: not specified • HF (v%): 1%	• $R_{a\,initial}$ = 7 μm • $R_{a\,final}$ = 6 μm • 14% decrease	• 10 min	NA
(Balyakin, Shvetcov, and Zhuchenko 2018)	SLM Ti6Al4V Size: 35 mm × 10 mm × 2 mm flat samples	HF + HNO_3	• Temperature: 25 °C • Polishing time: $5-15$ min • Stirring speed: not specified HNO_3 (v%): 3% – 20% • HF (v%): 3% – 10%	• R_a $_{initial}$ = 3.99 μm • $R_{a\,final}$ = 1.69 μm • 58% decrease	• HNO_3 (v%): 10% • HF (v%): 10% • Polishing time: 5 min	HNO_3 accelerates etching, eliminates hydrogen gas production, and produces a smoother surface. Higher HF concentrations result in the generation of titanium oxide on the surface;

Reference	Material & Size	Chemical solution	Process parameters	Surface roughness	Optimized parameters	Remarks
(Bezuidenhout et al. 2020)	SLM Ti6Al4V Size: dog bone test specimen with 7.03-mm diameter and 92-mm length.	HF + HNO$_3$	• Temperature: 25 °C • Polishing time: 60 min • Stirring speed: 400 rpm • HNO$_3$: 3.17 M • HF: 1–4 M	• $R_{a\,initial}$ = 10.93 μm • $R_{a\,final}$ = 0.90 μm • 92% decrease	• HF: 4 M	Specimens polished in 2 M solutions exhibited better fatigue performance
(Soro et al. 2020)	SLM Ti6Al4V Size: TPMS cylinder with a 10-mm diameter and 10-mm length	HF + HNO$_3$	• Temperature: 25 °C • Polishing time: 60 min • Stirring speed: 120 rpm • HNO$_3$: 3.3% • HF: 11.1%	• $R_{q\,initial}$ = 72 μm • $R_{q\,final}$ = 47 μm • 35% decrease	• Same as stated	NA
(Tyagi et al. 2019)	SLM 316L SS Size: complex artifacts with both internal and external features	DS-9-314 solution (10–30% H$_3$PO$_4$, 1–10% HCl, 1–10% HNO$_3$, and 1–10% surfactants)	• Temperature: 75 ± 2 °C • Polishing time: 3 0–90 min • Stirring speed: not specified	• $S_{a\,initial}$ = 13.88 μm • $S_{a\,final}$ = 5.22 μm • 62% decrease	• Temperature: not specified • Polishing time: 45 min • Stirring speed: not specified	The as-printed surface was sandblasted before CP Both external and internal surfaces were efficiently polished

(continued)

TABLE 7.1 (Continued)
Summary of Chemical Polishing on AM Materials

Ref.	Workpiece Material	Electrolyte solution	Range of Process Parameters Tested	Best Surface Quality Results	Prerequisites for Best Results	Remarks
(Hooreweder et al. 2017)	SLM CoCr Size: cylinder lattice with a 6-mm diameter and an 8-mm length	$HCl + H_2O_2$	• Temperature: not specified • Polishing time: 4 min • Stirring speed: not specified • HCl (v%): 20% – 30% • H_2O_2 (v%): 0% – 10%	• $R_{q\,initial} = 8\ \mu m$ • $R_{q\,final} = 5.5\ \mu m$ • 31% decrease	• HCl (v%): 27% • H_2O_2 (v%): 8%	The quasi-static and fatigue performance retained after etching
(Scherillo 2019)	SLM AlSi10Mg Size: 20 mm × 20 mm square	Step 1: $HNO_3 + HF$ Step 2: $H_2O + H_3PO_4 + H_2SO_4 + HNO_3 + HF + CuSO_4$	• Temperature: 85 °C • Polishing time: 75 min • Stirring speed: not specified • HNO_3 (v%): not specified • HF (v%): not specified	• $S_{q\,initial} = 24\ \mu m$ • $S_{q\,final} = 10\ \mu m$ • 58% decrease	• Same as stated	Two steps, i.e., chemical machining and chemical brightening

three acids is also effective such as H_3PO_4, HCl, and HNO_3 for 316L stainless steel (SS) as-printed by SLM. Under optimal conditions, the surface roughness can be reduced to as low as 0.9 μm Ra (Bezuidenhout et al. 2020). In general, the surface roughness of chemically etched metal surfaces reaches approximately 5 μm Sa from as-printed surfaces with roughness of approximately 13 μm Sa (Tyagi et al. 2019). Chemical polishing can also be used to polish internal surfaces of the AM part. Fatigue life of the polished parts was also reported since AM components usually have a low fatigue limit due to various surface defects. The chemically etched parts generally show better fatigue life because there are fewer surface crack initiations.

A mixture of HCl and H_2O_2 is useful to polish cylindrical lattice structures CoCr fabricated by SLM and can obtain a final surface roughness of 5.5 μm Ra after 4 min of polishing (Van Hooreweder et al. 2017). The authors also noted that the fatigue life of the etched lattice structure was retained though not improved. This abnormal result may be due to the low notch sensitivity of the CoCr parent material. Another reason may be that the internal porosity of the SLM CoCr dominates the fatigue performance rather than the surface defects. AlSi10Mg could be processed by chemical polishing in a mix of HNO_3 and HF followed by chemical brightening in a solution of water, H_3PO_4, H_2SO_4, and $CuSO_4$ (Scherillo 2019). Chemical polishing removes the large irregularities, and the following chemical brightening process smoothens the surface to obtain a final surface with 6.8 μm Ra roughness.

It is, therefore, evident that chemical polishing is suitable for processing a variety of materials and AM components (that arrive with both external and internal surfaces), but the achievable surface finish is relatively poorer in comparison to machining, abrasive-based methods, and laser micro-polishing.

7.3 ELECTROCHEMICAL POLISHING OF AM PARTS

Chemical polishing has its advantages in processing internal surfaces but is still unsuitable for producing uniformly polished surfaces. ECP is a good alternative to meet this demand and has been widely investigated across different AM metals such as Ti6Al4V, Inconel, stainless steel, NiTi, CoCr, and AlSi10Mg. A summary of ECP on AM components is listed in Table 7.2, which shows the superior surface finishing achievable of ECP compared to CE. Moreover, less hazardous solutions are used in ECP compared to high concentrations of acids in CE. Instead, gentler solutions such as the mixture of ethanol-isopropyl alcohol, $ZnCl_2$, and $AlCl_3$ can be employed (Yang, Lassell, and Paiva 2020; García-Blanco et al. 2015). ECP can also generate more uniform surfaces on simple plates, internal surfaces, and complex freeform surfaces by employing specially designed cathode geometries.

Like laser polishing, ECP can reduce the surface roughness of the investigated AM materials from an initial surface roughness that exceeds 20 μm to sub-micrometric finishing. Additional processing of as-printed components prior to ECP is also known to augment the finishing process and achieve superior surface quality. These processes include sandblasting, machining, or laser polishing.

TABLE 7.2
Summary of Electrochemical Polishing on AM Materials

Ref.	Workpiece Material	Electrolyte solution	Range of Process Parameters Tested	Best Surface Quality Results	Prerequisites for Best Results	Remarks
(Urlea and Brailovski 2017)	SLM Ti6Al4V Size: cylinder, V-shape and stave-shape with different dimensions	60% $HClO_4$ and CH_3COOH mixed in a 1:9 vol ratio	• Cathode: SLM Ti6Al4V • Temperature: 23 °C • Stirring speed: 100 rpm • Polishing time: 0 – 27 min • Gap: 5 mm • Voltage: 20 – 25 V • Current density: 160 – 320 mA/cm²	• $R_{a\ initial} =$ 11.25 µm • $R_{a\ final} = 0.87$ µm • 92% decrease	• Polishing time: 27 min • Current density: 240 mA/cm²	Various cathodes were designed and printed With substantial details of experiments
(Chang et al. 2019)	SLM 316L SS Size: various shapes including plate, lattice structure and curved pipes	60 vol.% H_3PO_4, 30 vol.% H_2SO_4 sulfuric acid, 0.3 vol.% $C_3H_8O_3$ and 9.7 vol.% DI water	• Cathode: SLM 316L • Temperature: 50 – 60 °C • Polishing time: 0 – 60 min • Stirring speed: 100 rpm • Voltage: 1 – 2.4 V • Current density: 0 – 300 mA/cm²	• $R_{a\ initial} = 8$ µm • $R_{a\ final} =$ 0.18 µm • 98% decrease	• 20-min overpotential ECP + 20-min ECP • Voltage: 1.5 – 1.9 V	Various cathodes were designed and printed The ECP lattice showed enhanced energy absorption ability
(Dong, Marleau-Finley, and Zhao 2019)	DMLS Ti6Al4V Size: lattice structure	995 ml ethanol, 100 ml n-butyl alcohol, 109 g of $(Al(H_2O)_6Cl_3)$, 250 g $(ZnCl_2)$	• Cathode: stainless steel • Temperature: 50 – 60 °C • Polishing time: 10 – 15 min • Stirring speed: not specified	• $R_{a\ initial} =$ 10.314 µm • $R_{a\ final} =$ 3.309 µm • 68% decrease	• Polishing time: 15 min • Voltage: 50 V	ECP is more suitable for external surfaces

Reference	Material / size	Electrolyte	Process parameters	Results	Additional conditions	Remarks
(Y. Zhang, Li, and Che 2018)	SLM Ti6Al4V Size: plate with a dimension of 20 mm × 20mm × 2 mm	$HClO_4$, CH_3COOH and distilled water with a volume ratio of 1:10:1.2	• Voltage: 40 – 50 V • Gap: 5 mm • Cathode: graphite • Temperature: 30 °C • Polishing time: 5 – 20 min • Stirring speed: not specified • Gap: 50 mm • Current density: 0.3 A/cm²		• Polishing time: 15 min	After the electropolishing process, the corrosion resistance of the Ti-6Al-4V alloy enhanced
(Rotty et al. 2016)	SLM 316L SS Size: not specified	45% H_3PO_4, 35% H_2SO_4 and 20% H_2O	• Cathode: a platinum-coated titanium mesh • Temperature: 35 – 70 °C • Polishing time: 10 min • Stirring speed: 500 rpm • Current density: 120 – 360 mA/cm²	• $R_{a\,initial}$ = 6.33 μm • $R_{a\,final}$ = 1.132 μm • 82% decrease	• Temperature: 70 °C • Voltage: 2 V	*sandpaper ground The best electropolishing potential is preferred at the end of this plateau
(Liu et al. 2022)	SLM AlSi10Mg Size: 25 mm × 30 mm × 3 mm	aqueous sodium hydroxide (NaOH) with pH of 13	• Cathode: not specified • Temperature: 70 °C • Polishing time: 1000 s • Stirring speed: not specified • Current density: 120 – 360 mA/cm² • Voltage: 5 V	• $R_{a\,initial}$ = 1.54 μm* • $R_{a\,final}$ = 0.103 μm • 93% decrease	• Intermittent electrochemical polishing (IECP) with 50-s interval and a total polishing time of 1000 s	ECP of AlSi10Mg was dominated by the dissolution and diffusion of the aluminum element

• $S_{a\,initial}$ = 14.9 μm
• $S_{a\,final}$ = 1.83 μm
• 88% decrease

7.3.1 EXTERNAL SURFACES

It is expected that ECP can process external surfaces, but it may also be employed to remove support structures. An example is the processing of SLM 316L stents using a solution mixture of H_3PO_4 and H_2SO_4 as the electrolyte and a titanium spiral cathode (Grad et al. 2021). The example shown in Figure 7.3 requires solution temperatures to be kept constant at 65 °C and an applied voltage of 6 V to maintain the trans-passive potential region. These conditions facilitate the removal of support structure and the generation of shinier surfaces with a roughness of 0.5 μm Sa. ECP conditions will vary between workpiece materials. Surface preparation prior to ECP can also help the process such as laser polishing of 316L stainless steel parts fabricated by SLM, as shown in Figure 7.4(a). The stainless steel will require choline chloride (ChCl) and ethylene glycol in a 1:2 weight ratio in oxalic acid as the electrolyte (Alrbaey et al. 2016). A surface with 0.34 μm Ra roughness can be achieved with a potential of 4–5.5 V, polishing time of 60 min, and electrolyte temperature of 40 °C, as shown in Figure 7.4(b).

ECP of a few AM metals may be easier such as Inconel 718, as seen in Figure 7.5. With 20 vol.% H_2SO_4 in absolute methanol as the electrolyte and a stainless steel cathode, partially bonded particles can be fully removed after only 1 min of polishing and a smooth surface with surface roughness of 3.66 μm Ra can be achieved after 5 min of polishing (Zhang et al. 2017).

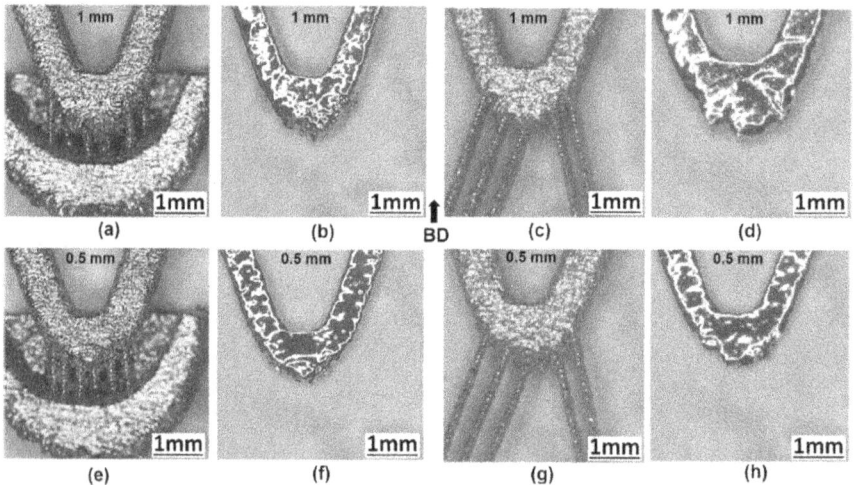

FIGURE 7.3 Electrochemical polishing of stent structure fabricated by SLM: (a) as-printed stent with 1-mm cross support structure; (b) electrochemical polished stent; (c) as-printed stent with 1-mm line support structure; (d) electrochemical polished stent; (e) as-printed stent with 0.5-mm cross support structure; (f) electrochemical polished stent; (g) as-printed stent with 0.5-mm line support structure; (h) electrochemical polished stent. Reprinted with permission from MDPI (Grad et al. 2021).

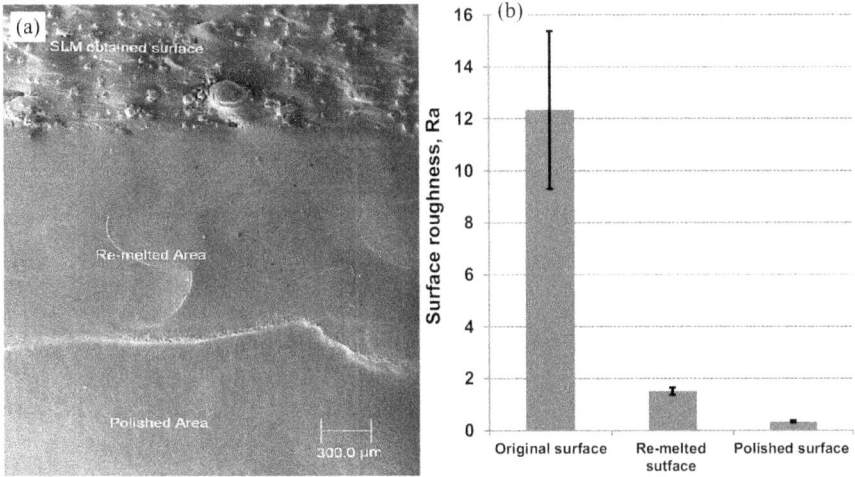

FIGURE 7.4 Electrochemical polishing of SLM 316L stainless steel: (a) SEM image showing the as-printed, re-melted and polished surfaces; (b) surface roughness of the original, re-melted and polished surfaces. Reprinted with permission from Springer Nature (Alrbaey et al. 2016).

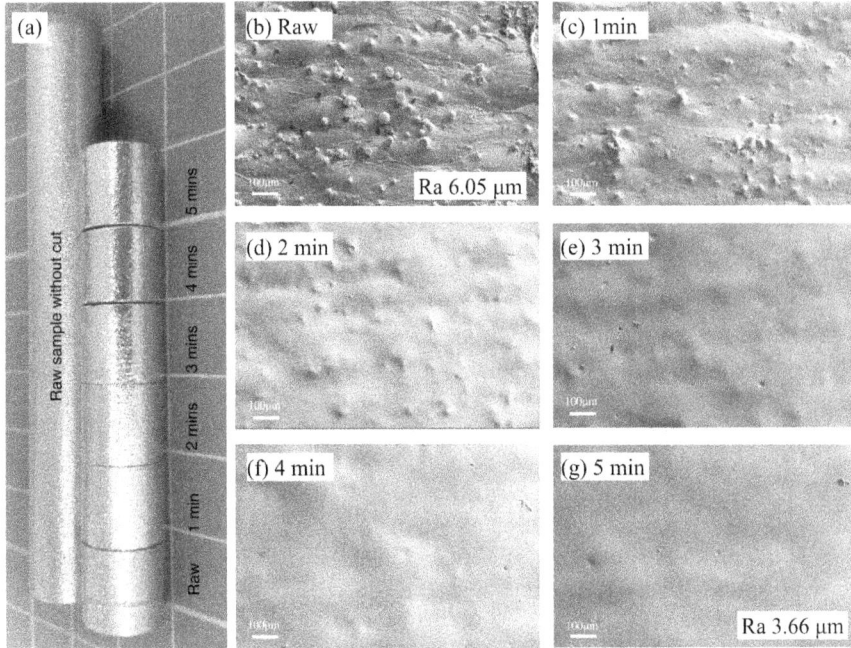

FIGURE 7.5 Electrochemical polishing of cylinder surfaces fabricated by SLM: (a) as-printed sample and electrochemically polished sample; (b) SEM image of the as-printed sample; (c–g) SEM images of the surfaces after 1 min, 2 min, 3 min, 4 min and 5 min polishing, respectively. Reprinted with permission from Elsevier (Zhang et al. 2017).

FIGURE 7.6 Electrochemical polishing of SLM AlSi10Mg: (a) unpolished surface (UP); (b) surface processed by continuous electrochemical polishing (CECP); (c) surface processed by intermittent electrochemical polishing (IECP); (d–f) cross-section images of the UP, CECP and IECP surfaces, respectively. Reprinted with permission from Elsevier (Liu et al. 2022).

Customized cathode geometries which can conform the shape of the targeted surfaces may also be employed to polish external surfaces with irregular geometries (Urlea and Brailovski 2017; Chang et al. 2019). The material removal principle remains the same while the inter-electrode gap controls the chemical reactivity for the work material removal to conform to the cathode shape.

As ECP is driven by the applied voltage, material removal may be controlled to be continuous or intermittent, otherwise coined as continuous electrochemical polishing (CECP) and intermittent electrochemical polishing (IECP) (Liu et al. 2022). The frequency of changes in the application of a voltage is relatively low at intervals in tens of seconds up to 100 s. Interestingly, IECP of AM AlSi10Mg plates using a less corrosive electrolyte, is capable of obtain superior surface finishing compared to CECP, as shown in Figure 7.6. This is because IECP can maintain a small thickness of the polishing product, which keeps a high electrochemical current and also facilitates the dispersion of the dissolved ions. CECP can eliminate most partially bonded particles on the AM surface but will leave behind raised fold-like features. On the other hand, IECP will remove these features to produce a surface with minor waviness. This superior surface eliminates the electrochemical reaction sites thus the parts after IECP have higher anti-corrosion performance than that of the CECP parts.

7.3.2 INTERNAL SURFACES

In general, ECP is capable of processing internal surfaces such as channels and lattice structures to a certain extent. For example, CoCr stents fabricated by SLM can be processed by ECP as presented in Figure 7.7. The final surface roughness can reach 1.45 μm along internal struts using 45 vol.% sulfuric acid (H_2SO_4), 50 vol.% phosphoric acid (H_3PO_4), and 5 vol.% water as the electrolyte solution (Demir and Previtali 2017). Internal channels additively manufactured by nickel-titanium shape memory alloy (Figure 7.8) can be processed by ECP with a mixture of acetic (CH_3COOH) and perchlorate Acid ($HClO_4$) as the electrolyte and nickel foil as the cathode (Mingear et al. 2019). The internal walls of the channels were smoothened out with a roughness of 25 μm Sa after ECP.

Innovative techniques can be introduced to ECP to further enable uniform material removal on internal surfaces. One solution is overpotential electrochemical polishing (OECP) to process lattice structures fabricated by SLM, as shown in Figure 7.9 (Chang et al. 2019). Generally, ECP is conducted in the current plateau region (refer to Figure 7.2(b)). A benchmarking experiment has shown that an as-printed 316L SS flat sample demonstrates higher current density than a mechanically polished 316L SS flat sample. The maximum difference of the current density occurs near the over potential region that is slightly above the current plateau region in the voltage-current curve. This may be caused by the partially bonded particles which provide more sites for electron/ion migration. Hence, using an overpotential may fasten the dissolving

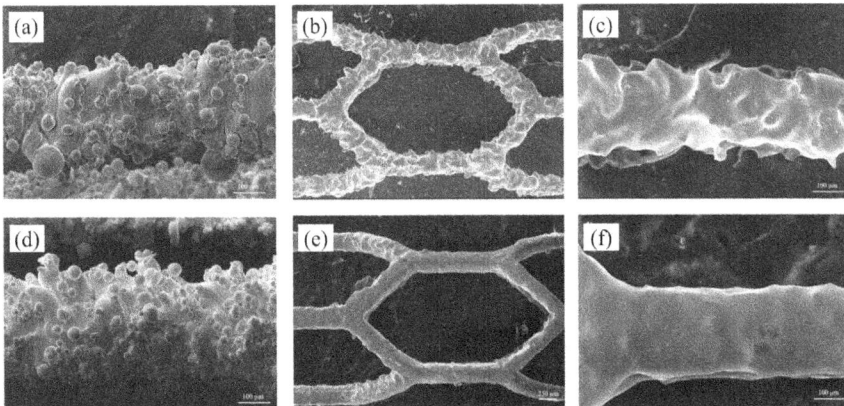

FIGURE 7.7 Electrochemical polishing of CoCr stents produced by SLM: (a) the stent printed with hatching scanning method; (b) the polished stent; (c) zoom-in image of the polished strut; (d) the stent printed with concentric scanning method; (e) the polished stent; (f) zoom-in image of the polished strut. Reprinted with permission from Elsevier (Demir and Previtali 2017).

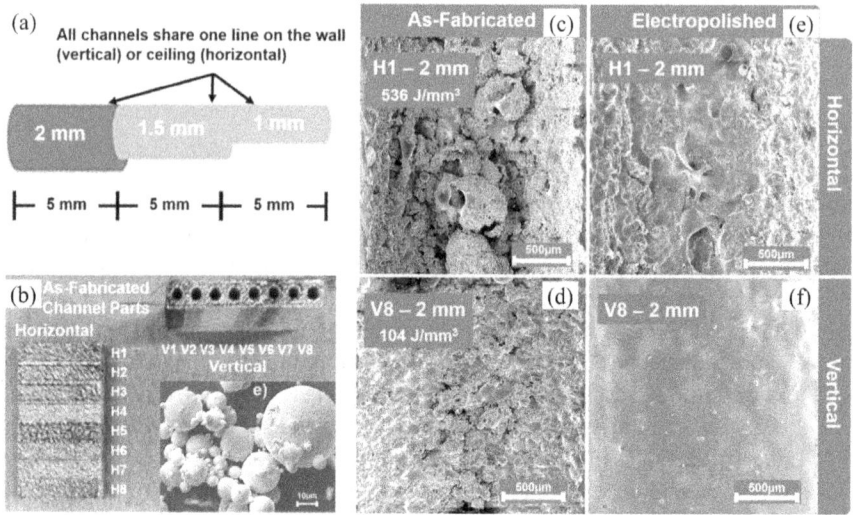

FIGURE 7.8 Electrochemical polishing of NiTi grooves fabricated by SLM: (a) schematic showing the channel dimensions; (b) the as-fabricated channel parts and the metal powders used for printing; (c) SEM image of the channel printed horizontally; (d) SEM image of the channel printed vertically; (e) SEM image of the polished channel which is printed horizontally and (f) SEM image of the polished channel which is printed vertically. Reprinted with permission from Elsevier (Mingear et al. 2019).

process, which serves the foundation of OECP process. OECP can be employed as the roughing process to smoothen out the surface. Then conversational ECP is used as finishing process to produce surfaces with 0.18 μm Ra roughness on 316L SS. For OECP, the preferable electrolyte for 316L SS comprises a mixture of 60 vol.% phosphoric acid, 30 vol.% sulfuric acid, 0.3 vol.% glycerol, and 9.7 vol.% deionized water. The optimal temperature range is within 50–60 °C.

When comparing ECP and OECP, Figure 7.10 illustrates the superiority of OECP at handling as-built surface topographies. These polished surfaces double the energy absorption capacity to further enhance the surface quality. Counter electrodes with different geometries can be designed and fabricated by AM to be inserted into channels and served as the cathodes to process internal surfaces.

7.4   HYBRID POLISHING

Figure 7.11 shows the schematic of a hybrid electrochemical-mechanical process (PEMEC). This is the combination of mass finishing with electrochemical polishing (Rech et al. 2022). It has the advantage of efficiently processing freeform surfaces. Figure 7.12 illustrates the polishing setup to process a freeform surface produced by SLM using 2 mm Al_2O_3 pyramidal-shaped grains as the abrasive media and a mixture of phosphoric acid H_3PO_4 (85 wt.%) and deionized water as the electrolyte. With the

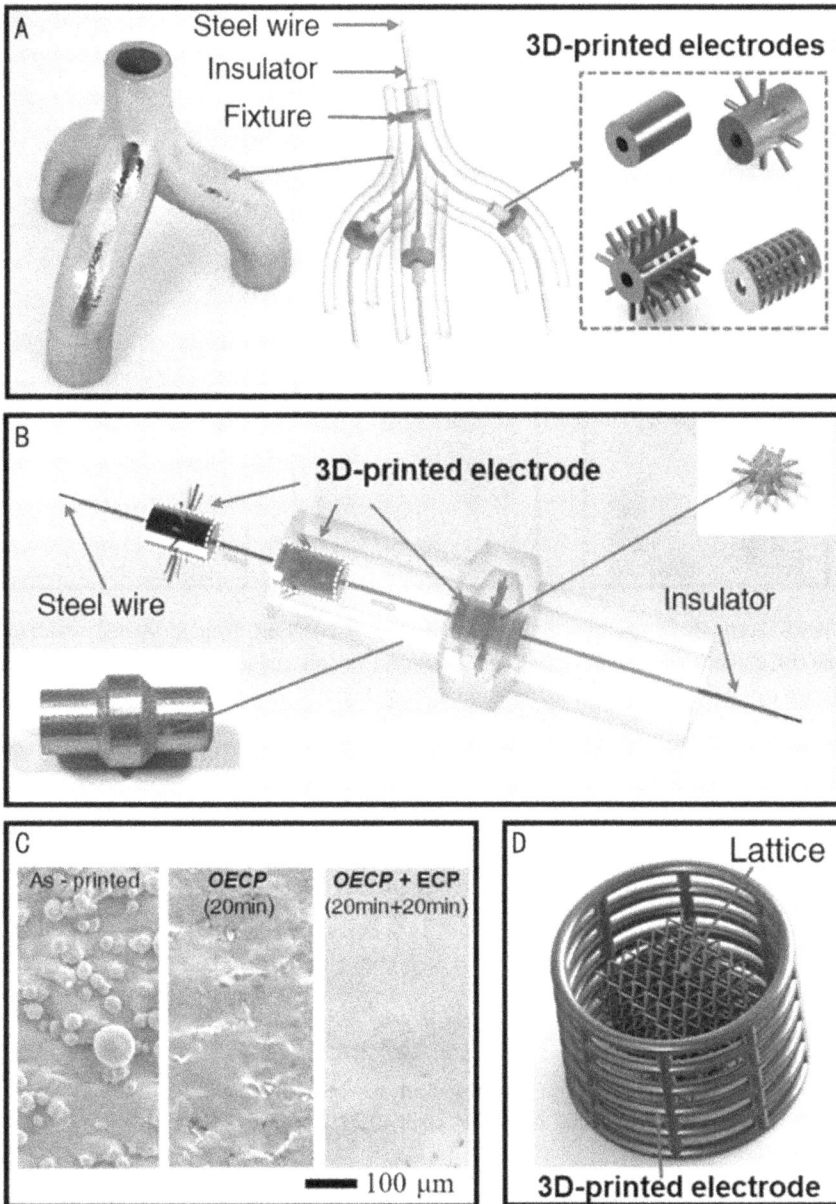

FIGURE 7.9 Electrochemical polishing of complex internal structures by SLM: (a) polished branch tube and the 3d-printed electrodes; (b) polished step tube and the 3D-printed electrode; (c) SEM images of the as-printed, OECP and OECP+ECP surfaces; (d) setup showing the polishing of lattice structure with a 3D-printed electrode. Reprinted with permission from Taylor & Francis (Chang et al. 2019).

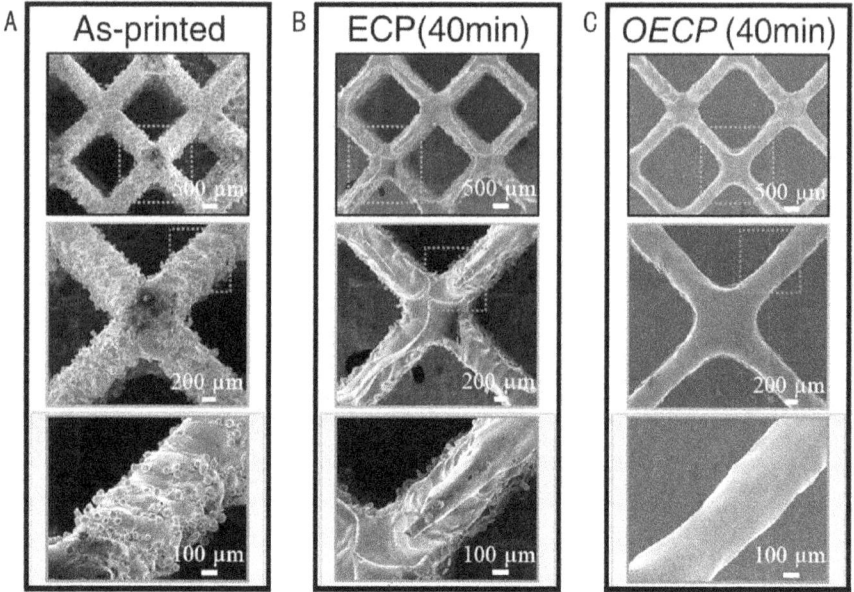

FIGURE 7.10 Benchmarking of conventional ECP and OECP on SLM body-centered-cubic (BCC) lattice structure: (a) as-printed lattice; (b) lattice polished by ECP and (c) lattice polished by OECP. Reprinted with permission from Taylor & Francis (Chang et al. 2019).

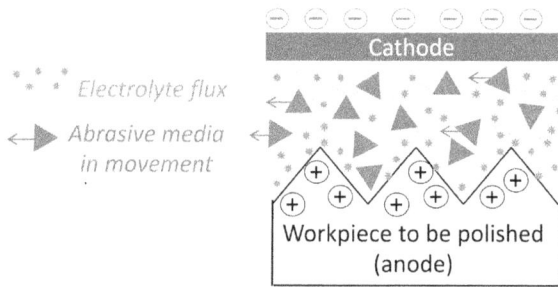

FIGURE 7.11 Schematic of the hybrid electrochemical-mechanical process (PEMEC). Reprinted with permission from Elsevier (Rech et al. 2022).

voltage setting at 12 V corresponding to a current density of 20 A/dm^2 and the work-piece being dragged through the abrasives with a velocity of 0.5 m/s, the processed workpiece gave a reflective appearance and an achievable surface roughness of 1.2 μm Sa after 60 mins of processing.

Considering that conventional ECP involves hazardous electrolytes such as HF, H_2SO_4, HNO_3, and HCl, dry mechanical-electrochemical polishing (DMECP) can be employed to replace the liquid electrolyte with porous solid particles containing the electrolyte below the saturation level (MILLET 2018; MILLET et al. 2019).

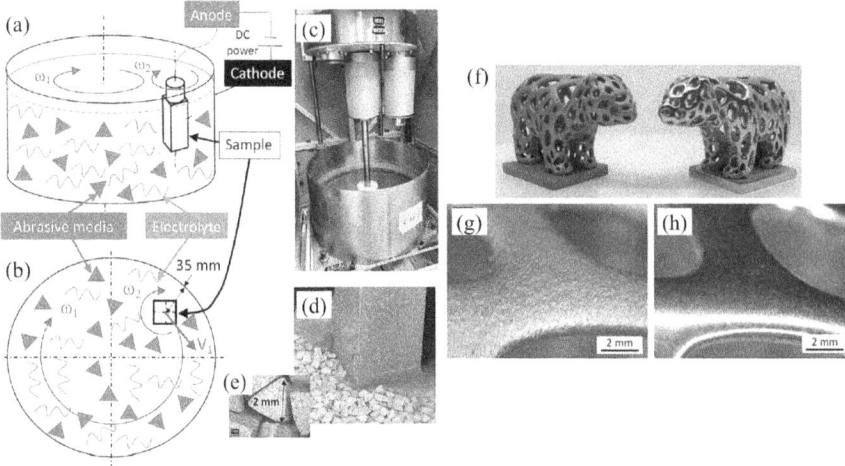

FIGURE 7.12 Setup of PEMEC and its performance on SLM 316L SS freeform surface: (a) oblique view of the schematic working principle; (b) top view of the schematic working principle; (c) image of the built setup; (d) the mixture of abrasive media and the electrolyte; (e) the pyramidal alumina abrasive; (f) the freeform surface before and after PEMEC; (g) the zoom-in image of the initial surface and (h) the zoom-in image of the polished surface. Reprinted with permission from Elsevier (Rech et al. 2022).

The workpiece is dragged inside a barrel of these porous particles. The solid particles also serve as abrasives to facilitate mechanical material removal. DMECP has shown great potential in processing complex metal parts produced by additive manufacturing – e.g., dental implants, medical device, and industrial components (GPAINNOVA 2020). An example of DMECP on AM 316L SS is given in Figure 7.13. The DMECP, however, performs more than just material removal and includes oxide growth, oxide crack, oxide fragmentation, oxide shedding, and new surface formations (Bai et al. 2020). Samples polished by DMECP will have surface roughness below 1 μm Ra. However, challenges arise for this method in processing internal surfaces where the porous particles are unable to freely access. Particles may also get stuck in small corners of the workpiece, which will eventually deteriorate the geometric accuracy of the parts.

Another technique to incorporate both mechanical and chemical aspects of polishing is in electrochemical mechanical combined polishing (EMCP), shown in Figure 7.14 (An, Wang, and Zhu 2022). The polishing tool is modified where a cylinder cathode is sandwiched between two grinding wheels. One wheel is for coarse grinding while the other is for fine grinding. The major material removal mechanism is primarily from electrochemical dissolution while the coarse grinding wheel removes large irregularities on the as-printed parts, and the fine grinding wheel eliminates layers of passivation films. With the setup configuration in Figure 7.14, EMCP would be able to uniformly polish curved 316L SS tubes with a maximum internal diameter of 10 mm.

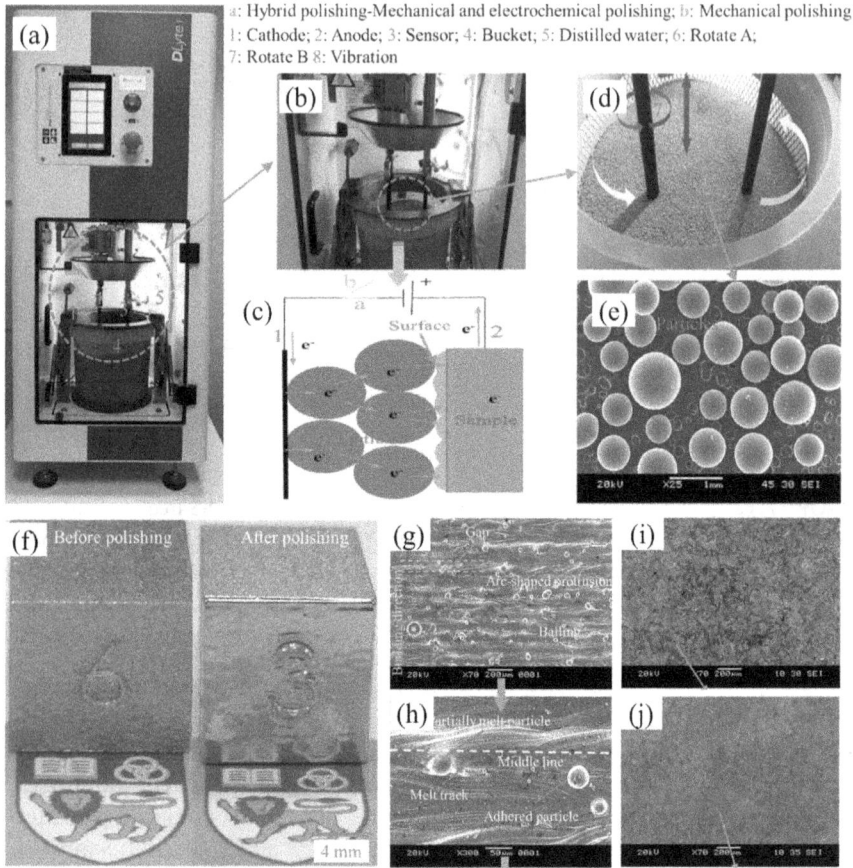

FIGURE 7.13 Setup of DMECP and its performance on SLM 316L SS workpiece: (a) overall image of the polishing machine; (b) functional region of the polishing machine; (c) schematic of the DMECP process; (d) motion of the workpiece in the media; (e) SEM image of the polishing media; (f) image of the workpiece before and after DMECP; (g) SEM image of the as-printed surface at ×70 magnification; (h) SEM image of the as-printed surface at ×300 magnification; (i) SEM image of the DMECP surface and (h) SEM image of the DMECP surface followed by a pure mechanical polishing process. Reprinted with permission from Elsevier (Bai et al. 2020).

7.5 CONCLUDING REMARKS

In this chapter, three chemical-based finishing methods for AM metal parts were introduced – chemical etching, electrochemical polishing, and hybrid electrochemical polishing. The key process parameters and their influence on the final surface finish are discussed. Chemical etching has high accessibility to complex surfaces, but has inadequate impact on the surface finishing quality. Electrochemical polishing has higher material removal rates, superior surface finish, and a more uniform surface processing. It can also be applied for external and internal surfaces. Hybrid methods

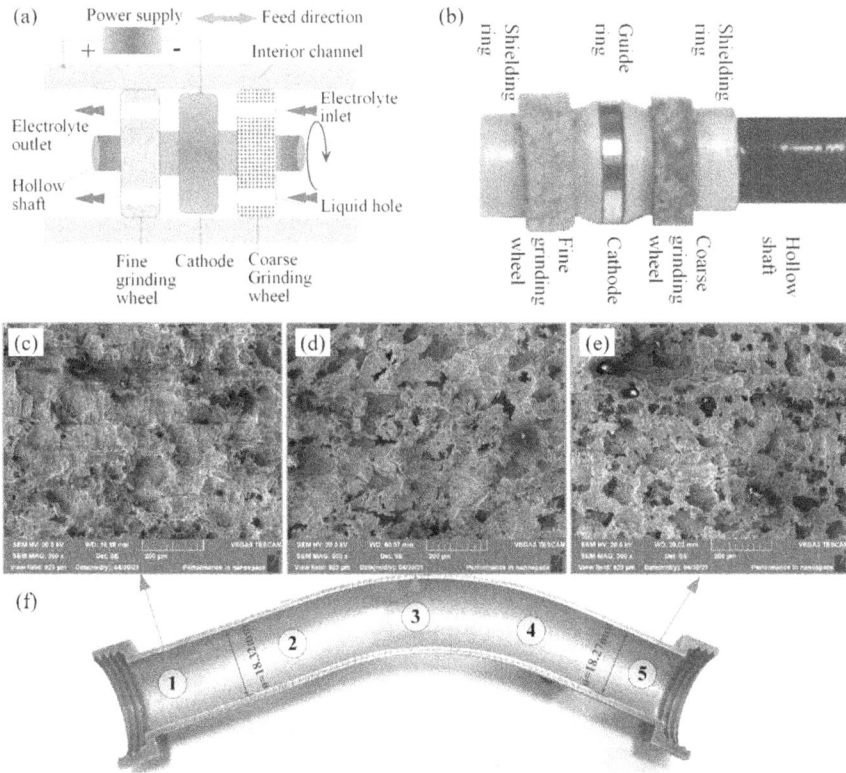

FIGURE 7.14 Working principle of electrochemical mechanical combined polishing (EMCP) and its performance on an SLM 316L SS curve tube: (a) schematic of EMCP; (b) image of the polishing tool; (c–e) SEM images of the polished surfaces in the inlet, middle and the outlet (f) image of the polished tube. Reprinted with permission from Elsevier (An, Wang, and Zhu 2022).

match together the advantages of mechanical polishing and electrochemical polishing to maximize material removal rates. Nevertheless, more research is needed on freeform surface finishing and the inner surfaces. And more chemical-based finishing techniques need be developed to conform to the rapid development of AM technology and its industrial applications.

REFERENCES

Alrbaey, K., D. I. Wimpenny, A. A. Al-Barzinjy, and A. Moroz. 2016. "Electropolishing of Re-Melted SLM Stainless Steel 316L Parts Using Deep Eutectic Solvents: 3 × 3 Full Factorial Design." *Journal of Materials Engineering and Performance* 25 (7). Springer: 2836–2846. doi:10.1007/s11665-016-2140-2.

An, Linchao, Dengyong Wang, and Di Zhu. 2022. "Combined Electrochemical and Mechanical Polishing of Interior Channels in Parts Made by Additive Manufacturing." *Additive Manufacturing* 51. Elsevier: 102638. doi:10.1016/j.addma.2022.102638.

Bai, Yuchao, Cuiling Zhao, Jin Yang, Jerry Ying Hsi Fuh, Wen Feng Lu, Can Weng, and Hao Wang. 2020. "Dry Mechanical-Electrochemical Polishing of Selective Laser Melted 316L Stainless Steel." *Materials and Design* 193. Elsevier: 108840. doi:10.1016/j.matdes.2020.108840.

Balyakin, Andrey V., Aleksei N. Shvetcov, and Evgeny I. Zhuchenko. 2018. "Chemical Polishing of Samples Obtained by Selective Laser Melting from Titanium Alloy Ti6Al4V." In *MATEC Web of Conferences*, 224:1031. EDP Sciences. doi:10.1051/matecconf/201822401031.

Basha, M. M., S. M. Basha, V. K. Jain, and M. R. Sankar. 2022. "State of the Art on Chemical and Electrochemical Based Finishing Processes for Additive Manufactured Features." *Additive Manufacturing* 58. Elsevier: 103028. doi:10.1016/j.addma.2022.103028.

Bezuidenhout, Martin, Gerrit Ter Haar, Thorsten Becker, Sabrina Rudolph, Oliver Damm, and Natasha Sacks. 2020. "The Effect of HF-HNO3 Chemical Polishing on the Surface Roughness and Fatigue Life of Laser Powder Bed Fusion Produced Ti6Al4V." *Materials Today Communications* 25. Elsevier: 101396. doi:10.1016/j.mtcomm.2020.101396.

Chang, Shuai, Aihong Liu, Chun Yee Aaron Ong, Lei Zhang, Xiaolei Huang, Yong Hao Tan, Liping Zhao, Liqun Li, and Jun Ding. 2019. "Highly Effective Smoothening of 3D-Printed Metal Structures via Overpotential Electrochemical Polishing." *Materials Research Letters* 7 (7). Taylor & Francis: 282–289. doi:10.1080/21663831.2019.1601645.

Demir, Ali Gökhan, and Barbara Previtali. 2017. "Additive Manufacturing of Cardiovascular CoCr Stents by Selective Laser Melting." *Materials and Design* 119. Elsevier: 338–350. doi:10.1016/j.matdes.2017.01.091.

Dong, Guoying, Julien Marleau-Finley, and Yaoyao Fiona Zhao. 2019. "Investigation of Electrochemical Post-Processing Procedure for Ti-6Al-4V Lattice Structure Manufactured by Direct Metal Laser Sintering (DMLS)." *International Journal of Advanced Manufacturing Technology* 104 (9–12). Springer: 3401–3417. doi:10.1007/s00170-019-03996-5.

García-Blanco, M. B., M. Díaz-Fuentes, O. Garrido, G. Vara, and J. A. Díez. 2015. "Electropolishing Process to Reduce Surface Roughness of a Ti Alloy Fabricated by SLM." In *Proceedings Euro PM 2015: International Power Metallurgy Congress and Exhibition*. Reims, France: The European Powder Metallurgy Association.

GPAINNOVA. 2020. "The New Concept of Polishing." Accessed February 25. www.dlyte.es/.

Grad, Marius, Naresh Nadammal, Ulrich Schultheiss, Philipp Lulla, and Ulf Noster. 2021. "An Integrative Experimental Approach to Design Optimization and Removal Strategies of Supporting Structures Used during L-PBF of SS316L Aortic Stents." *Applied Sciences (Switzerland)* 11 (19). MDPI: 9176. doi:10.3390/app11199176.

Han, Keping, and Jingli Fang. 1998. "Study on Chemical Polishing for Stainless Steel." *Transactions of the Institute of Metal Finishing* 76 (1). Taylor & Francis: 24–25. doi:10.1080/00202967.1998.11871186.

Han, Wei, and Fengzhou Fang. 2019. "Fundamental Aspects and Recent Developments in Electropolishing." *International Journal of Machine Tools and Manufacture* 139. Elsevier: 1–23. doi:10.1016/j.ijmachtools.2019.01.001.

Hoar, T. P., and J. A. S. Mowat. 1950. "The Electropolishing of Nickel in Urea-Ammonium Chloride Melts." *Transactions of the IMF* 26 (1). Taylor & Francis: 7–25. doi:10.1080/00202967.1950.11869532.

Hooreweder, Brecht Van, Karel Lietaert, Bram Neirinck, Nicholas Lippiatt, and Martine Wevers. 2017. "CoCr F75 Scaffolds Produced by Additive Manufacturing: Influence of Chemical Etching on Powder Removal and Mechanical Performance." *Journal of the Mechanical Behavior of Biomedical Materials* 70. Elsevier: 60–67. doi:10.1016/j.jmbbm.2017.03.017.

Ishizawa, K., H. Kurisu, S. Yamamoto, T. Nomura, and N. Murashige. 2008. "Effect of Chemical Polishing in Titanium Materials for Low Outgassing." In *Journal of Physics: Conference Series*, 100:92023. IOP Publishing. doi:10.1088/1742-6596/100/9/092023.

Landolt, D. 1987. "Fundamental Aspects of Electropolishing." *Electrochimica Acta* 32 (1). Elsevier: 1–11. doi:10.1016/0013-4686(87)87001-9.

Liu, Han, Minheng Ye, Zuoyan Ye, Lili Wang, Guowei Wang, Xianfeng Shen, Ping Xu, and Chao Wang. 2022. "High-Quality Surface Smoothening of Laser Powder Bed Fusion Additive Manufacturing AlSi10Mg via Intermittent Electrochemical Polishing." *Surface and Coatings Technology* 443. Elsevier: 128608. doi:10.1016/j.surfcoat.2022.128608.

Lyczkowska, E., P. Szymczyk, B. Dybała, and E. Chlebus. 2014. "Chemical Polishing of Scaffolds Made of Ti-6Al-7Nb Alloy by Additive Manufacturing." *Archives of Civil and Mechanical Engineering* 14 (4). Elsevier: 586–594. doi:10.1016/j.acme.2014.03.001.

Matlosz, M., S. Magaino, and D. Landolt. 1994. "Impedance Analysis of a Model Mechanism for Acceptor-Limited Electropolishing." *Journal of The Electrochemical Society* 141 (2). IOP Publishing: 410–418. doi:10.1149/1.2054741.

MILLET, Pau SARSANEDAS. 2018. "Method for Smoothing and Polishing Metals Via Ion Transport Free Solid Bodies and Solid Bodies for Performing the Method." Google Patents.

MILLET, Pau SARSANEDAS, Pau Manuel NARCIS GUASCH PIRIZ, Arnau GARRELL BUNUEL, and Gerard TORDERA XANDRI. 2019. "Device for Burnishing and Smoothing Metal Parts." Google Patents.

Mingear, Jacob, Bing Zhang, Darren Hartl, and Alaa Elwany. 2019. "Effect of Process Parameters and Electropolishing on the Surface Roughness of Interior Channels in Additively Manufactured Nickel-Titanium Shape Memory Alloy Actuators." *Additive Manufacturing* 27. Elsevier: 565–575. doi:10.1016/j.addma.2019.03.027.

Pyka, Grzegorz, Andrzej Burakowski, Greet Kerckhofs, Maarten Moesen, Simon Van Bael, Jan Schrooten, and Martine Wevers. 2012. "Surface Modification of Ti6Al4V Open Porous Structures Produced by Additive Manufacturing." *Advanced Engineering Materials* 14 (6). Wiley Online Library: 363–370. doi:10.1002/adem.201100344.

Rech, J., D. Krzak, F. Roy, F. Salvatore, A. Gidon, and S. Guérin. 2022. "A New Hybrid Electrochemical-Mechanical Process (PEMEC) for Polishing Complex and Rough Parts." *CIRP Annals* 71 (1). Elsevier: 173–176. doi:10.1016/j.cirp.2022.03.011.

Rokicki, R., and T. Hryniewicz. 2012. "Enhanced Oxidation-Dissolution Theory of Electropolishing." *Transactions of the Institute of Metal Finishing* 90 (4). Taylor & Francis: 188–196. doi:10.1179/0020296712Z.00000000031.

Rotty, C., A. Mandroyan, M. L. Doche, and J. Y. Hihn. 2016. "Electropolishing of CuZn Brasses and 316L Stainless Steels: Influence of Alloy Composition or Preparation Process (ALM vs. Standard Method)." *Surface and Coatings Technology* 307. Elsevier: 125–135. doi:10.1016/j.surfcoat.2016.08.076.

Scherillo, Fabio. 2019. "Chemical Surface Finishing of AlSi10Mg Components Made by Additive Manufacturing." *Manufacturing Letters* 19. Elsevier: 5–9. doi:10.1016/j.mfglet.2018.12.002.

Soro, Nicolas, Nicolas Saintier, Hooyar Attar, and Matthew S. Dargusch. 2020. "Surface and Morphological Modification of Selectively Laser Melted Titanium Lattices Using a Chemical Post Treatment." *Surface and Coatings Technology* 393. Elsevier: 125794. doi:10.1016/j.surfcoat.2020.125794.

Tyagi, Pawan, Tobias Goulet, Christopher Riso, Robert Stephenson, Nitt Chuenprateep, Justin Schlitzer, Cordell Benton, and Francisco Garcia-Moreno. 2019. "Reducing the Roughness of Internal Surface of an Additive Manufacturing Produced 316 Steel

Component by Chempolishing and Electropolishing." *Additive Manufacturing* 25. Elsevier: 32–38. doi:10.1016/j.addma.2018.11.001.

Urlea, V., and V. Brailovski. 2017. "Electropolishing and Electropolishing-Related Allowances for IN625 Alloy Components Fabricated by Laser Powder-Bed Fusion." *International Journal of Advanced Manufacturing Technology* 92 (9–12). Elsevier: 4487–4499. doi:10.1007/s00170-017-0546-0.

West, A. C., R. D. Grimm, D. Landolt, C. Deslouis, and B. Tribollet. 1992. "Electrohydrodynamic Impedance Study of Anodically Formed Salt Films on Iron in Chloride Solutions." *Journal of Electroanalytical Chemistry* 330 (1–2). IOP Publishing: 693–706. doi:10.1016/0022-0728(92)80337-4.

Yang, Li, Austin Lassell, and Gustavo Perez   Vilhena Paiva. 2020. "Further Study of the Electropolishing of Ti6Al4V Parts Made via Electron Beam Melting." In *Proceedings – 26th Annual International Solid Freeform Fabrication Symposium – An Additive Manufacturing Conference, SFF 2015*, 1730–1737.

Yi, Rong, Yi Zhang, Xinquan Zhang, Fengzhou Fang, and Hui Deng. 2020. "A Generic Approach of Polishing Metals via Isotropic Electrochemical Etching." *International Journal of Machine Tools and Manufacture* 150. Elsevier: 103517. doi:10.1016/j.ijmachtools.2020.103517.

Zhang, Baicheng, Xiaohua Lee, Jiaming Bai, Junfeng Guo, Pan Wang, Chen-nan Sun, Muiling Nai, Guojun Qi, and Jun Wei. 2017. "Study of Selective Laser Melting (SLM) Inconel 718 Part Surface Improvement by Electrochemical Polishing." *Materials and Design* 116. Elsevier: 531–537. doi:10.1016/j.matdes.2016.11.103.

Zhang, Yifei, Jianzhong Li, and Shuanghang Che. 2018. "Electropolishing Mechanism of Ti-6Al-4V Alloy Fabricated by Selective Laser Melting." *International Journal of Electrochemical Science* 13 (5). ESG BORIVOJA STEVANOVICA 25-7, BELGRADE, 11000, SERBIA: 4792–4807. doi:10.20964/2018.05.79.

8 Theoretical Modeling Considerations for Post-Processing

8.1 MODELING METHODOLOGY

Experimental studies are highly beneficial in understanding material removal processes as they give the most realistic representation of the expected outcome during practical application. However, in-depth experimental studies will require substantial testing with multiple variables where a large data set with multiple increments of each parameter would provide a fuller understanding of the quantitative research. Within each set of test parameters, tests would also be typically performed multiple times to evaluate the repeatability of the results.

The combination of these requirements results in the extensive use of test materials and machine-tool setup and operation time, which is often only realistic with the support of substantial funding for research and development (i.e., it is an expensive trial-and-error methodology). Therefore, theoretical frameworks are developed to construct more efficient methods designed to get relevant information to answer problem statements that arise during research. These theoretical frameworks also link assumptions used to describe experimentally observed phenomena with the corresponding fundamental science. Thus, theoretical modeling and simulation are vital toward any form of in-depth investigation alongside experimental evaluations.

Theoretical modeling for additive manufacturing generally covers several subtopics such as residual stress, microstructure evolution, powder trajectory, and melt pool boundary modeling (Gunasegaram et al. 2017). Post-process modeling, including that of material removal processes, is another important aspect of the manufacturing process that can be regarded as equally important. However, just as it is important to understand the material properties of as-built metal characteristics, it is crucial to develop an understanding of the available modeling and simulation capabilities for additive manufacturing that can be brought forward to post-process modeling. Hence, this chapter is structured to discuss the techniques to model as-built material states as the benchmark for further modeling of the post-processes, and to serve as the foundation to account for the complexities involved during post-processing of additively manufactured metals. The importance of this aspect in employing simulations and modeling to derive the material conditions prior to post-processing is not stressed enough. In this chapter, post-processing will be discussed in the context of mechanical material removal processes.

DOI: 10.1201/9781003272601-8

Modeling studies generally comprise analytical modeling and simulation modeling. Analytical studies quantitatively derive closed-form solutions to the problem statement based on idealized formulations of the deformed system. They require a precise definition of the purpose of the study, which typically makes predictions regarding the physical process. A majority of analytical models developed for additive manufacturing encompass heat exchanges during the laser irradiation process where latent heat and scanning strategies are critical for stabilizing the melt pools in the built material (Ning et al. 2020; Huang, Khamesee, and Toyserkani 2016; Steuben et al. 2019) and the heat transfer during cooling (Ning et al. 2019). Modeling of thermal effects are also employed in other analytical models describing the recrystallization and grain growth phenomenon during additive manufacturing (Ji et al. 2020). However, there are not many analytical studies designed for the material removal process of additively manufactured metals. Most analytical models employed for material removal revert to simplified models used to describe conventional materials. Material complexities developed from the additive manufacturing process will require a few more years of investigations for the derivation of analytical models specially designed for additively manufactured metals. Instead, most of the post-process modeling for additive manufacturing have been performed by iterative-based simulations.

8.2   FINITE ELEMENT METHOD MODELING

Numerical modeling represents a series of mathematical formulations to describe physical phenomena that occur during experimentation. Simulated results would typically be compared with the experimental outcomes for validation and serve as the basis for further parametric studies that largely rely on computational power. Numerical modeling of the cutting process is typically performed using finite element method (FEM) simulations, which incorporates the discretization of a spatial domain into a mesh of elements arbitrarily shaped and sized. Each element will have predefined nodes within and along each of its edges, which would be used to give a global interpolated map of the geometric domain comprising the assembly of elements that determine the final state of the element based on the numerically solved differential equations. The final state of each node holds a set of information that describes physical values, such as the new position, which will then determine the strain.

A stress–strain relationship can then be implemented in the model to dictate the kinematics involved in each elemental deformation, which would eventually determine the global deformation of the material under study. The allowable stress–strain relationship is governed by the material models that describe the viscoelastic or viscoplastic properties during deformation, which provide critical information in machining, such as forces, plastic strain, heat generation, and material failure, among others.

However, before jumping to the modeling of material removal for additively manufactured metals in accordance with the scope of this book, a critical aspect of the thermal history in the pre-machined surface should be acknowledged for additively manufactured materials. A full coverage of all theoretical modeling for additive manufacturing can be extensive and fitting as a topic on its own. This section aims to highlight the use of FEM to theoretically validate experimental observations and the importance of these considerations for post-processing (i.e., the use of modeling methods to derive the as-built material state, which would affect the material removal process).

Heat distribution during manufacturing affects the final geometry, residual stress state, and the resultant microstructure (Papazoglou, Karkalos, and Markopoulos 2020), which correspondingly influences the material removal process during post-processing. Thus, it is advisable to avoid the typical assumption of a completely defect-free initial material state prior to the modeling of machining for the additively manufactured metals. On this note, the thermo-mechanical aspects of the additive manufacturing process must be given equal emphasis for discussion in tandem with the considerations for mechanical processing simulations (Hajializadeh and Ince 2020).

Figure 8.1 gives an overview of thermal and mechanical modeling. Thermal analysis models the heat transfer during additive manufacturing, which involves the heat generation by irradiation and heat dissipation. These simulations are typically employed to study the effect of the heat energy influx on the temperature distribution of the powder bed. Thermal modeling also extends to thermal-fluid flow analysis (Chen and Yan 2020; Jamshidinia, Kong, and Kovacevic 2013), which is critical for simulation of melt pools in additively manufactured metals.

Mechanical modeling describes the material response to a defined loading condition, which is typically intended for structure integrity evaluation and material removal analysis. Both types of simulations meet in the middle where the final stages of the thermal analysis may be used as the starting point for mechanical evaluation. This should also be the prescribed procedure for modeling post-processing simulations where the derived as-built material state is subjected to the deformation involved during material removal.

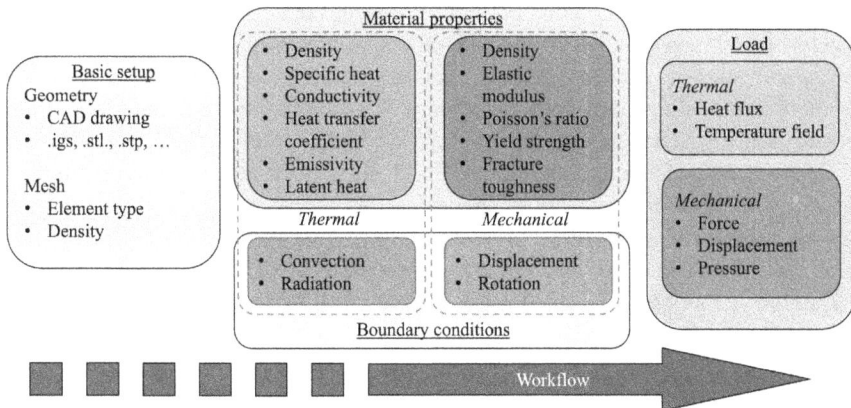

FIGURE 8.1 Overview of considerations for thermal and mechanical finite element method analysis.

Material parameters for both thermal and mechanical simulations are largely correlated when addressing the coupling problem. Coupled simulations, also known as thermomechanical simulations, may be applied in the weakly or strongly coupled mode. In a weakly coupled simulation, a thermal model is first implemented and the thermal history is then applied to subsequent modeling of the microstructure and mechanical deformation. Strongly coupled simulations yield more accurate results by simultaneously accounting for both thermal and mechanical effects in an idealized chain of events flowing from material addition to material removal. However, the latter would inevitably be more computationally expensive.

The workflow for numerical simulations typically begins by designing the geometry to analyze, which can be drawn in the simulation software or imported from other computer-aided design (CAD) software that are designed to construct more complex geometries in 2D or 3D file formats (e.g., .sat, .dxf, .igs, .iges, .stp, .step). With the given geometry, a mesh will be assigned to the geometry with the appropriate selection of element size. A finer mesh increases the accuracy of the model in terms of resolution but increases the computational cost for the increase in number of equations for each element. It follows that a decrease in mesh density lowers computational time but reduces the overall accuracy of the nodal output results describing deformation and heat information. Boundary conditions can be defined for the meshed geometry, which includes position constraints and ambient conditions (e.g., temperature, heat dissipation parameters, pressure, etc.). Subsequently, material properties will be defined to dictate the simulated reaction of the part under the applied loads. The last critical component for simulation setup is the assignment of loading conditions, which would differ between thermal and mechanical analysis with the input of either heat or kinetic energy. Other parameters in this subsection of the simulation workflow include determining the location, duration, and rates of these applied loads.

8.2.1 Thermal Modeling

Several aspects of thermal modeling will be covered in this section comprising heat transfer, metal deposition, stress development resulting in distortion and residual stress, and microstructure evolution. There are some essential material properties for thermal modeling. Thermal conductivity of not only the work material but also surrounding elements, such as the ambient air and supporting base, are necessary to determine the degree and rate of thermal flow. Specific heat capacity that defines the energy required for the increase in temperature is also necessary as it determines the transfer of heat energy from the source to the work material. With these critical parameters, heat addition may then be modelled accordingly.

Input heat energy is described using the classical method to model heat transfer with Equation 8.1:

$$\rho c \frac{\partial T}{\partial t} + \ldots \nabla \cdot \left(\kappa \nabla T \right) = q \tag{8.1}$$

where c and κ are the heat capacity and heat conductivity, respectively, T is the temperature, ρ is the density, and q is the heat source. The heat source is quantified by the parameters used during additive manufacturing such as the laser beam diameter, laser power, and power distribution. This input heat energy for a laser beam is often described based on the Gaussian heat source model as:

$$q(r) = \frac{2P}{\pi r_0} \exp\left(-\frac{2r^2}{r_0^2}\right) \tag{8.2}$$

where P is the laser power, r_0 is the radius of the Gaussian laser, and r is the radial position of evaluation from the center of the laser beam.

Additionally, heat transfer by convection and radiation should also be accounted for based on the respective Newton's law of cooling and Stefan-Boltzmann law as:

$$q = h\left(T - T_0\right) \tag{8.3}$$

$$q = \sigma\epsilon\left(T^4 - T_0^4\right) \tag{8.4}$$

where h is the convection heat transfer coefficient, T_0 is the ambient temperature, σ is the Stefan Boltzmann constant, and ϵ is the emissivity.

The corresponding temperature field derived from the conservation of energy (Equation 8.1) then defines the corresponding strain and stress as:

$$\varepsilon^{th} = \alpha_e\left(T - T_0\right) \tag{8.5}$$

$$\nabla \cdot \sigma = 0 \tag{8.6}$$

where α_e is the thermal expansion coefficient.

8.2.2 SIMULATING THE AM PROCESS

The simulation setup for thermal modeling incorporates suitable boundary conditions and a defined heat source to represent the absorption of heat energy transferred during the additive manufacturing process. In many ways, the modeling of additive manufacturing is largely like simulations developed for multi-pass welding procedures. The simplest method for analyzing the influence of heat addition to the material is by constructing an as-built geometry and applying the thermal loads to the part. There are, however, more advanced and closer representations of the actual additive manufacturing process where the layer-by-layer material addition occurs in the simulation.

This model setup would comprise the substrate material and the powder bed where the former would be physically constrained and assumed to be rigid, and the latter initially consists of inactivated elements for the subsequent deposition step just as how it occurs during the additive manufacturing process. This concept goes by different names such as element birth (Li et al. 2021; Cao, Gharghouri, and Nash 2016), element birth-and-death (Ha et al. 2020; Lu et al. 2021), thermal-mechanical macroscopic workflow (Stender et al. 2018), and progressive element activation (Sun et al. 2021). Regardless of the naming convention, the concept of element activation as a means for material deposition remains the same, with the typical workflow illustrated in Figure 8.2. Upon the supply of heat energy to the elements to meet the appropriate conditions for deposition (i.e., melting the powder), the elements may then be activated to be considered as part of the computational domain for subsequent thermal analysis, while the inactivated elements are excluded from the calculations. Activated elements are assigned with near melt material properties and the joining element material state (e.g., position, stress, heat, etc.) are transferred to the activated elements for subsequent thermo-mechanical analysis. Then the deformed state is mapped back to the reference mesh for iteration and the final state of the step is recognized as the initial configuration for the next calculation step.

As element activation is dependent on the input of heat energy, efficient computation is dependent on the mesh density, time step, and heat input parameters. Stender et al. (2018) assessed the relationship between mesh configuration and the laser parameters for element activation and identified the ideal conditions for an efficient mix of resolution and computational efficacy as illustrated in Figure 8.3. Figure 8.3(a) shows a case where the heat source is too small to affect the nodes of the elements and, therefore, results in no elements being activated. Figure 8.3(b) illustrates the heat source overlapping at least two elements, but with the laser spots controlled to be adjacent but not overlapping each other. This scenario is the most efficient for

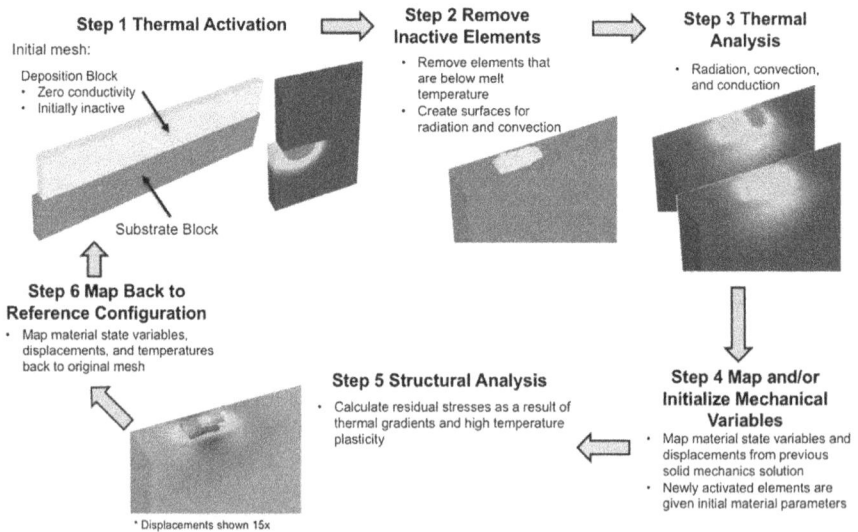

FIGURE 8.2 Element activation workflow. Reprinted with permission from Elsevier (Stender et al. 2018).

Initial Laser and Mesh Configuration	Spatial Activation Process For 3 Time Steps	Activated Volume
$\vec{v_l}\Delta t > l_e$ $\eta < l_e$ $\vec{v_l}$ η	$\vec{v_l}\Delta t$	No activated Elements
$\vec{v_l}\Delta t = l_e$ $\eta = l_e$ $\vec{v_l}$ η	$\vec{v_l}\Delta t$	
$\vec{v_l}\Delta t < l_e$ $\eta > l_e$ $\vec{v_l}$ η	$\vec{v_l}\Delta t$	

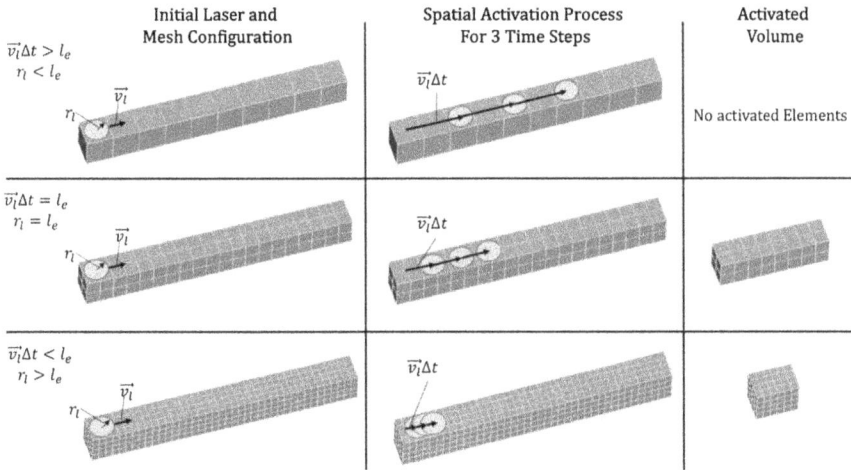

FIGURE 8.3 Simulation laser parameter considerations relative to element density and the corresponding results: (a) inadequate information; (b) high efficiency; (c) high accuracy. Reprinted with permission from Elsevier (Stender et al. 2018).

element activation relative to computational power. The last scenario is where laser spots overlap each other with reduced time steps and affecting multiple elements giving the most accurate solution.

An alternative solution is the "quiet element" method where the block of elements representing the areas of material to be deposited are physically present but are assigned properties that do not affect the analysis such as scaling back the thermal conductivity and specific heat in Equations 8.7 and 8.8 to mitigate heat conduction into these elements.

$$k_{quiet} = \alpha_k k \qquad (8.7)$$

$$C_{p_{quiet}} = \alpha_{C_p} C_p \qquad (8.8)$$

where α_k and α_{C_p} are scaling factors given a typically small value of 1e⁻⁶ to represent inactive elements. The method is also applicable for the Young's modulus and stiffness when assessing mechanical properties (Yang et al. 2016). Michaleris (2014) listed the advantages and disadvantages of the two metal deposition methods, discussed in Table 8.1.

Fluid flow during melting powders in additive manufacturing is also expected and can be modelled in FEM but these calculations tend to be computationally expensive with assessments of single or dual scan hatches. Such an occurrence is also more likely to have impactful consequences when the input energy is higher. Evaporation and fluid flow effects are commonly neglected when these conditions do not arise during FEM considerations (Promoppatum and Yao 2020).

TABLE 8.1
Advantages and Disadvantages of Material Deposition Methods (Michaleris 2014)

Material deposition method	Advantages	Disadvantages
Inactive element method	• Error avoidance with the use of scaling factors • Jacobian calculations performed only for active elements • Nodal degree of freedom only considered for active elements	• Relatively more challenging to implement in commercial finite element method software • Repeated renumbering of equations upon the activation of new elements increasing computational costs • Introduction of artificial energy by interpolation when nodes shared with active elements do not possess the same temperature
Quiet element method	• Ease of implementation in commercial finite element method software • Constant number of equations for a constant number of elements without the need for equation renumbering	• Possibility of thermal energy errors due to inappropriately selected scaling factors that result in thermal conduction • Ill-conditioning of the Jacobian resulting from inappropriate scaling factors • Computationally expensive with inappropriate model setup comprising majority of quiet elements

The importance of considering thermal analysis for post-processing is not so much for the heat generated during material building but the dissipation of heat to surrounding elements. Depending on the thermal gradient, structural distortion from designed geometries and residual stresses are the most common results of the thermal addition process. The current state of modeling has not reached the stage where structural distortion issues are majorly considered during post-processing. However, these deformities in actual additive manufacturing are bound to affect post-processing tool paths and could potentially create more adverse events during the fabrication of multi-material components. Hence, it is needful to include a brief discussion on the use of FEM to model these deformities and the potential solutions that can be numerically tested to mitigate the problems.

8.2.3 Distortion after Additive Manufacturing

Figure 8.4 shows an example of the deformation that occurs in overhanging structures constructed through additive manufacturing. The occurrence is attributed to the large

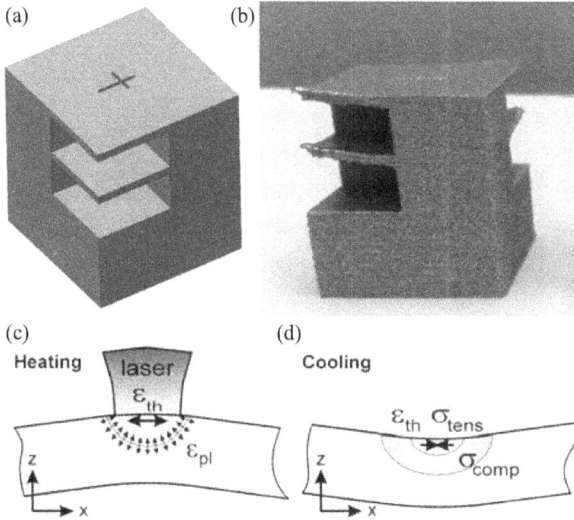

FIGURE 8.4 Deformation of overhanging structures: (a) designed structure; (b) deformed structure post-additive manufacturing; (c) deformation of the material during heating; (d) deformation of the material during cooling. Reprinted with permission from Elsevier (Cheng and Chou 2015; Mercelis and Kruth 2006).

temperature gradient between the surface and the bulk material (Mercelis and Kruth 2006). As the heat source heats the upper surface of the powder to melt the material, heat is allowed to conduct into the bulk material. However, due to slower speeds of heat conduction, a temperature gradient is formed. In the meantime, the heated layer of material on the surface undergoes thermal expansion but is restricted by the bulk material resulting in compressive stresses being developed at the regions with lower temperatures. This causes a bending moment acting from the underlying material and causes the part to bend outwards like Figure 8.4(c). As the heat source leaves the location and cooling takes place, the thermal gradient causes a sudden change in temperature and thermal contraction, which results in a reversal of the bending moment and the material to experience a tensile stress on the top layer and an inward bending motion like Figure 8.4(d). In normal cases of a thicker bulk material, the upper layer that experiences bending forces is restricted by the bulk material to result in residual stresses. However, overhanging structures like Figure 8.4(a) are allowed to deform freely and, therefore, result in warping (Figure 8.4(b)). The magnitude of the temperature gradient mechanism is affected by the temperature-dependent mechanical and thermal properties in that the underlying material retains its mechanical strength along with lower temperatures due to the ineffective transfer of heat by conduction. The underlying material then is sufficiently strong enough to resist expansion occurring in the upper layer of material, which is concurrently experiencing thermal softening effects where the material yields more easily under higher temperatures.

Understanding the thermal distribution and stress developed during additive manufacturing further advances laser scanning strategies designed to reduce these

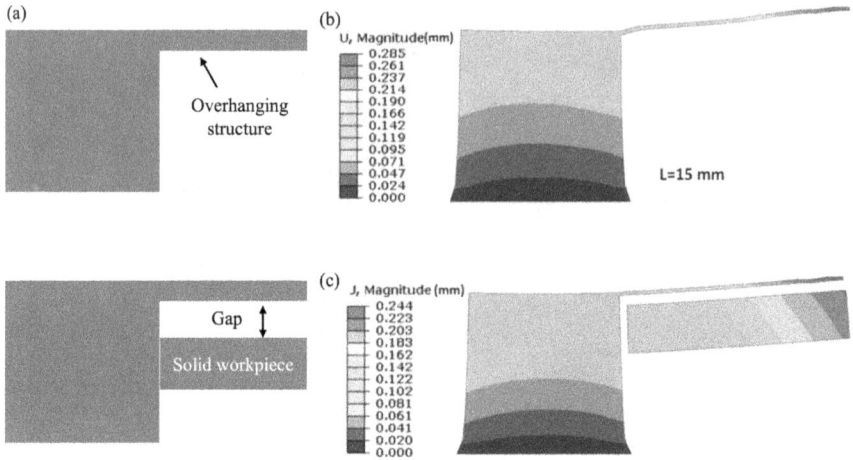

FIGURE 8.5 Simulation of overhanging structure deformation: (a) illustration of the initial configurations; (b) simulated distortion of the overhanging structure with solid workpiece beneath; (c) simulated distortion of the overhanging structure under normal conditions. Reprinted with permission from Elsevier (Cheng and Chou 2015).

phenomena such as the modifying the heating track vectors (Ramos, Belblidia, and Sienz 2019; Denlinger and Michaleris 2015), or the inclusion of additional material to reconfigure the heat dissipation process as demonstrated by Cheng and Chou (2015) in the use of an additional solid workpiece to enhance heat dissipation (Figure 8.5).

8.2.4 DEVELOPMENT OF RESIDUAL STRESSES FROM ADDITIVE MANUFACTURING

Thermal gradients will also occur across different thickness of material and result in the generation of residual stresses that were not dissipated through the deformation observed in overhanging structures. The process of generating these stresses is similar to those that result in distortion with the only difference being the stresses remaining in the material in the absence of external loads after processing. The phenomenon generally occurs across three regions. Melting occurs when the heat source is applied to the metallic powders during which heat is also transferred to the material ahead of the irradiation path. As the melted material experiences a substantial reduction in yield strength, the semi-heated material ahead of the irradiation path is allowed to expand and results in compressive yielding. Progressive solidification occurs as the heat source leaves the locality and the solidified material undergoes tension. This fundamental concept of residual stress generation excludes material specific effects such as microstructural reconfiguration and precipitation (Fang et al. 2020). It is important to account for these stresses as post-processing loads will be coupled with these residual stresses and affect the way the material subsequently reacts during deformation.

Residual stresses are assessed by observing the stress distribution of the part, which can be directional to evaluate the stress along different structures, as shown in the analysis of walls constructed on Ti-6Al-4V by LENS (Figures 8.6(a) and 8.6(b)), or the overall stressed state of the material using the von Mises stress distribution (Equation 8.9) in Figure 8.6(c).

$$\sigma_{VM} = \sqrt{\frac{1}{2}\left[\left(\sigma_{11}-\sigma_{22}\right)^2 + \left(\sigma_{22}-\sigma_{33}\right)^2 + \left(\sigma_{33}-\sigma_{11}\right)^2\right] + 3\left(\sigma_{12}^2 + \sigma_{23}^2 + \sigma_{31}^2\right)} \quad (8.9)$$

where σ_{ij} describes the orthogonal components that make up the stress tensor. Residual stresses are not only isolated to the surface but can also extend into the bulk material beneath the heat affected zone. FEM simulations allow cross-sectioning of the model to observe the stress development from the surface, as shown in Figure 8.7.

Validation of residual stresses in simulations may be performed by systematically substantiating the modelled input heat energy and the surface temperature with experimental results, followed by determining the residual stress in the material. While it may seem intuitive to go straight to comparing the theoretical and experimental residual stresses, the successful modeling of the first component (i.e., heat addition) is essential for future adoptions of the modeling approach to predict the residual stress. For instance, irradiation and scanning parameters are often varied and the effects of these changes on the residual stresses are common themes in additive manufacturing studies. This procedure of experimentally validating the surface temperature; however, is not an easy task (Vastola et al. 2016) and requires unique solutions with several assumptions to give an approximated experimental value such as observing the brightness of the surface with a CCD camera during the additive manufacturing process (Yadroitsev, Krakhmalev, and Yadroitsava 2014).

8.2.5 AM MICROSTRUCTURE MODELING

Microstructural evolution during additive manufacturing is another hot topic for analytical and simulation modeling. The importance of this characterization from the perspective of post-processing lies in the micrometric considerations of the grain size effect induced when the grains and the material removal length scales are of similar magnitude (Simoneau, Ng, and Elbestawi 2007; Lauro et al. 2015; Rahman et al. 2018; Venkatachalam et al. 2015). Although microstructure appearances and characteristics differ between materials, the theoretical analysis for its evolution during additive manufacturing most often begins from assessing the thermal distribution of the built material. The subsequent process then involves microstructure calculations, which cover a wide spectrum of methods specific to each material, to determine the phase and morphology of the final microstructure and the corresponding mechanical properties.

One aspect of microstructure modeling is in the prediction of phase volume fraction. For instance, Nie, Ojo, and Li (2014) predict the microstructure of Inconel 718 using stochastic analysis for grain growth, while Lindgren et al. (2016) chose

(a) σ_x longitudinal stress

(b) σ_y longitudinal stress

(c) σ_{VM} von Mises stress

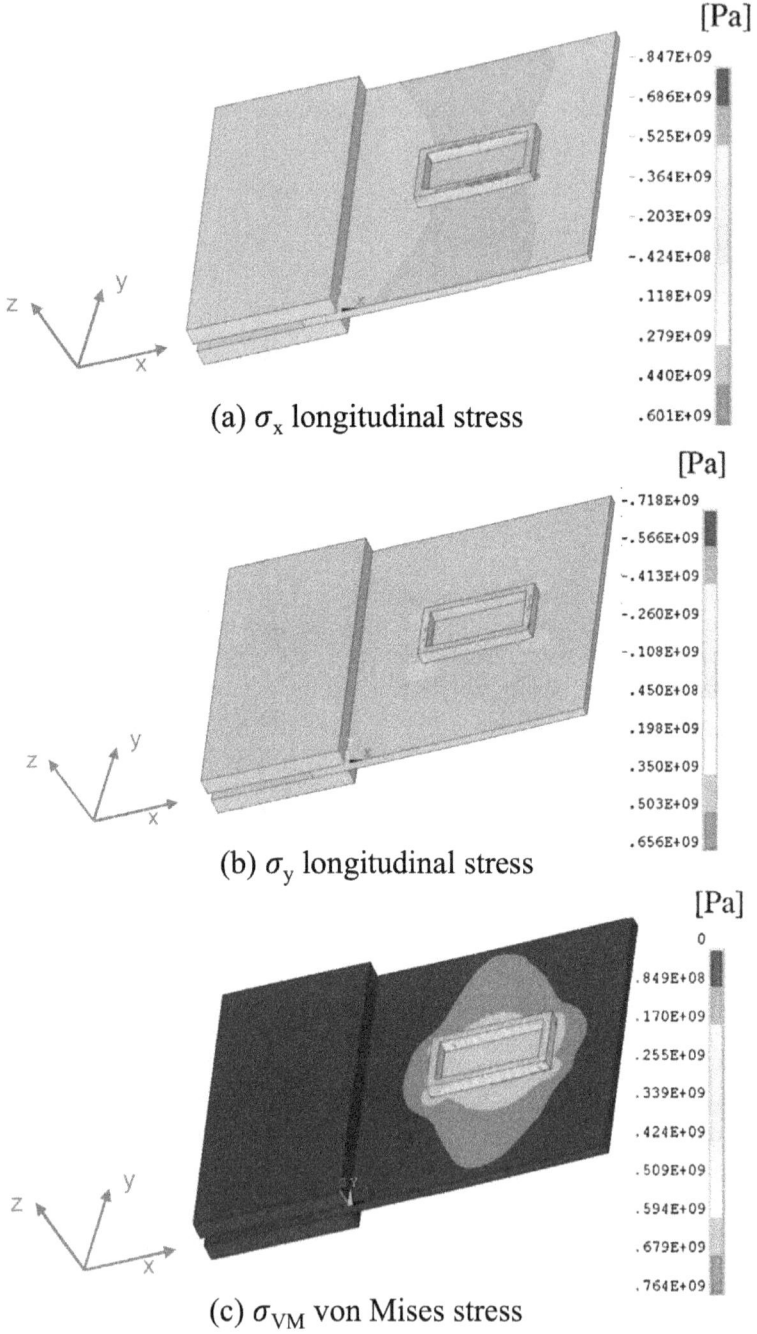

FIGURE 8.6 Simulated stress distribution: (a) longitudinal stress σ_x; longitudinal stress σ_y; (c) von Mises stress σ_{VM}. Reprinted with permission from Elsevier (Yang et al. 2016).

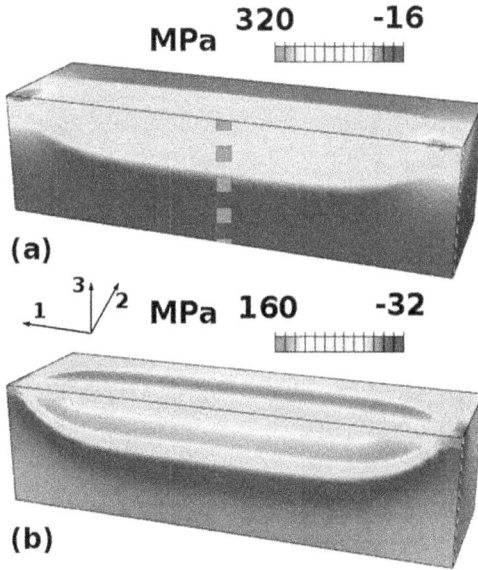

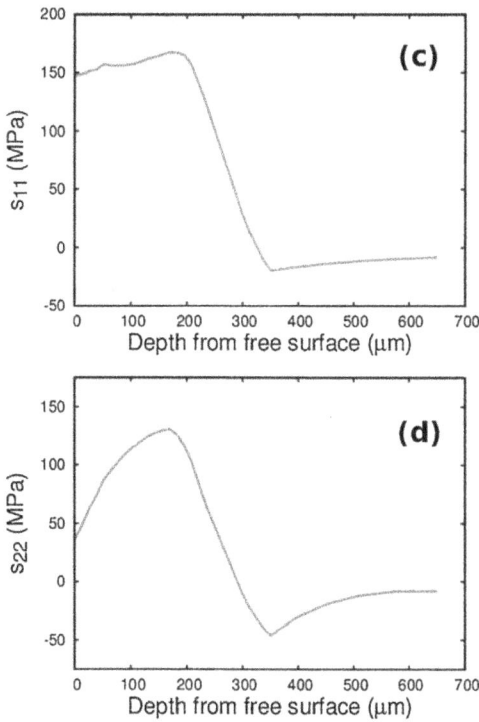

FIGURE 8.7 Cross-sectional subsurface stress distribution from simulations: (a) longitudinal stress σ_{11}; longitudinal stress σ_{22}; (c) and (d) stress development for σ_{11} and σ_{22} through a vertical line as exemplified in (a). Reprinted with permission from Elsevier (Vastola et al. 2016).

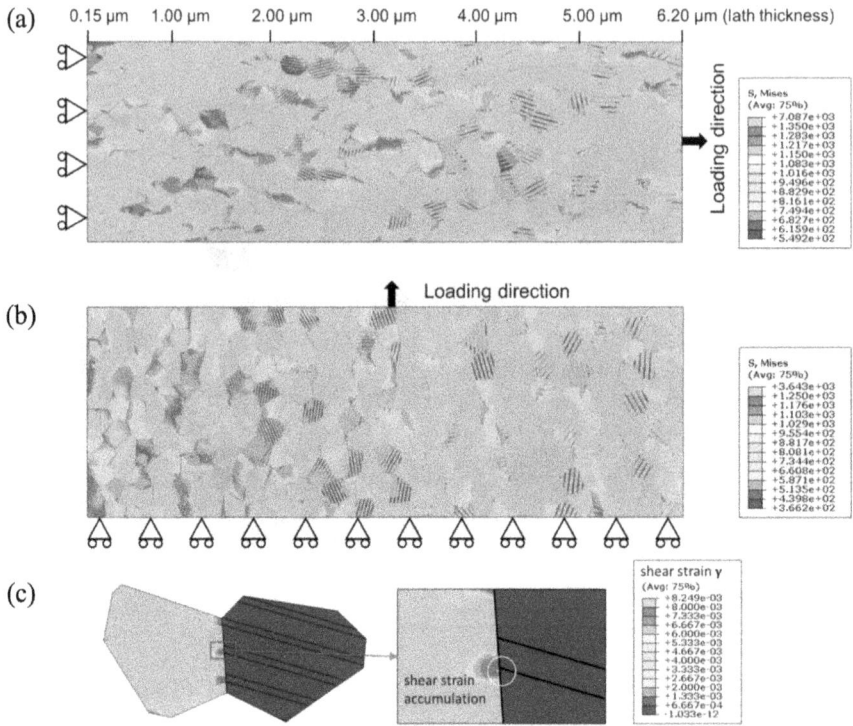

FIGURE 8.8   Simulated stress during loading of Ti-6Al-4V: (a) applied load is parallel to lath gradient direction; (b) applied load perpendicular to lath gradient direction; (c) accumulated shear strain distribution and location of stress concentration. Reprinted with permission from Elsevier (Geng and Harrison 2020).

to model phase transformations and precipitation with the appropriate temperature conditions to determine the final microstructure of the alloy. A similar method of studying the phase transformation volume fractions using the JMAK theory was also done for Ti-6Al-4V (Chen, Xu, and Jiao 2018).

The other hand, microstructure modeling is in simulating the mechanical properties of the material resulting from the grain orientation, crystallographic orientation, and distribution of phases within the microstructure. This was demonstrated in the inverse relationship between lath thickness and yield strength of Ti-6Al-4V (Geng and Harrison 2020). FEM further enabled the plotting of stress distributions with respect to the loading direction and grain orientations (Figures 8.8(a) and 8.8(b)) while also identifying stress concentration at intersections of three grain boundaries (Figure 8.8(c)).

Aside FEM modeling, there are several other notable methods to theoretically model the microstructure such as phase field modeling, cellular automata, kinetic Monte Carlo, and computational fluid dynamics (Tan, Sing, and Yeong 2020). Yet, present studies on the microstructural evolution stop at this stage of validating the simulated microstructure with as-built material characterization. The theoretically generated microstructure, regardless of methodology, should be put back into the simulation setup for effective analysis on the modeling for post-processing.

Melt pool boundary modeling is another important factor to consider for post-process modeling. Notwithstanding stress concentrations stemming from grain boundaries, melt pool boundaries serve as another source of major influence to the chip formation process during machining (Oliveira, Jardini, and Del Conte 2020). Melt pools comprise the modeling of fluid behavior as the heat source melts powder particles to set in the new layer of the fabricated part. Hence, simulation of melt pools typically involves computational fluid dynamics (CFD) for mesh-based simulations and other less mature mesh-free techniques – e.g., Lattice-Boltzmann method (LBM) and smoothed particle hydrodynamics (SPH) (Cook and Murphy 2020).

Thus, effective modeling for post-processing of additively manufactured materials requires additional considerations that have been studied in the modeling of the additive manufacturing process particularly in the areas of post-built distortion, residual stress, and microstructural evolution.

8.2.6 Mechanical Modeling

Modeling for material removal in post-processes is to study the mechanical response of a material, which is to determine the changes in nodal position of the elements during deformation. The motion is governed by mechanical properties that would dictate the input energy required to displace each element. In the same way, the change in position would also provide the information necessary to determine the stress state of the system. The purpose of analyzing the mechanical model is to assess the machining process of additively manufactured parts that would be associated with machined surface quality. This is typically achieved by evaluating forces, deformation morphology (e.g., cutting chips), and stress distributions in FEM simulations.

The manner and degree of deformation during mechanical simulation depend on the constitutive equations governing the mechanical behavior of the material. Many of these models are included in commercial simulation software and named after material types (e.g., cast iron plasticity, ductile metals, low density foams, etc.) and their deformation behavior on the stress–strain curve (e.g., elastic-plastic, viscoelastic, anisotropic elasticity, etc.). Material models may also be named after the researchers who developed the plasticity models – e.g., Johnson Cook (JC) (Johnson and Cook 1983), Drucker-Prager (Drucker and Prager 1952), etc.

Additively manufactured metals are effectively treated as another class of metals no different from the differences understood when approaching annealed, work-hardened, and casted metals. Thus, these material models are often applied for the modeling of additively manufactured metals during deformation, which not only evaluate the elastic–plastic behavior but also up to the failure stages. In most cases, the governing equations are guided by the progression of the stress–strain curve derived from standardized experimental methods on universal testing machines such as quasi-static tensile and compression testing (Wang et al. 2019; Kayacan et al. 2019).

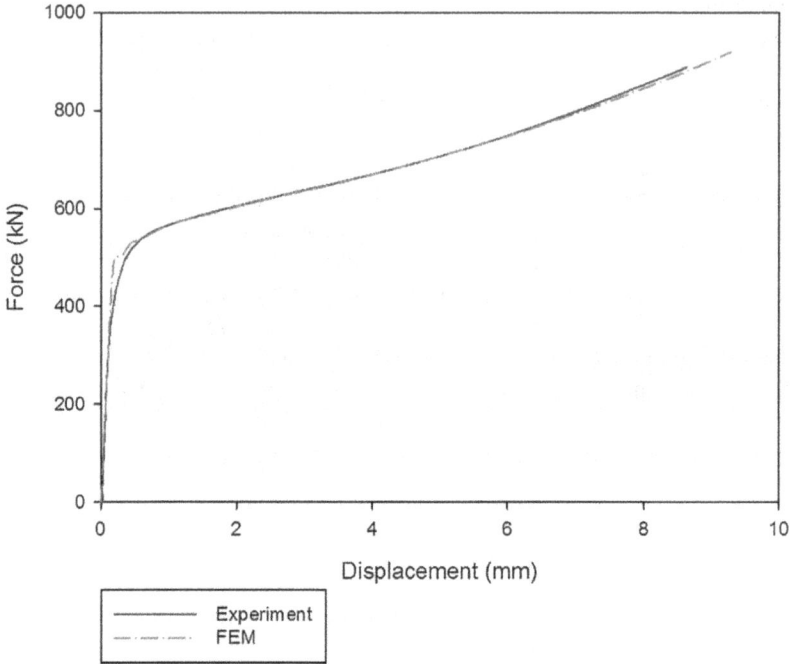

FIGURE 8.9 Force-displacement comparison between finite element method (FEM) modeling and experimental results for compression tests on DMLS-MS1. Reprinted with permission from Elsevier (Ebrahimi and Mohammadi 2018).

Basic material parameters such as the yield strength, ultimate strength, Young's modulus, and fracture strain can be derived from these quasi-static tests and sufficiently predict the isotropic elastic–plastic deformation of additively manufactured metals. Ebrahimi and Mohammadi (2018) demonstrated this coherency with tensile and compression test data on maraging steel (MS1) and hybridized MS1-H13 (H13) where force-displacement data (Figure 8.9) and the simulated strain distribution (Figure 8.10) were relatively consistent with the experimental results. An exception was given for the left face of the digital imaging correlation (DIC) cube obtained from experiments, which was attributed to experimental setup loading conditions (Ebrahimi and Mohammadi 2018).

Plastic energy is the largest product of a deformation process, but there are other participating components of energy conversion process such as thermal and acoustic energy. Where thermal effects are concerned, temperature-dependent mechanical properties may also be employed to model the physical response of the specimen under study. This set of information may be critical to certain materials such as Ti6Al4V and can be fed to the simulation software (Denlinger, Irwin, and Michaleris 2014). Table 8.2 shows a series of mechanical properties for Ti6Al4V over different temperatures to enable simulations factoring in thermal softening effects during deformation and even the thermo-mechanical deformation resulting from heat

(a) Experimental strain distribution

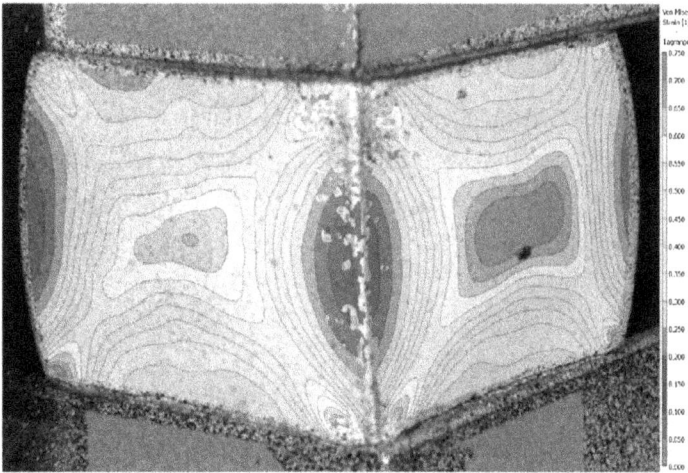

(b) Simulated strain distribution

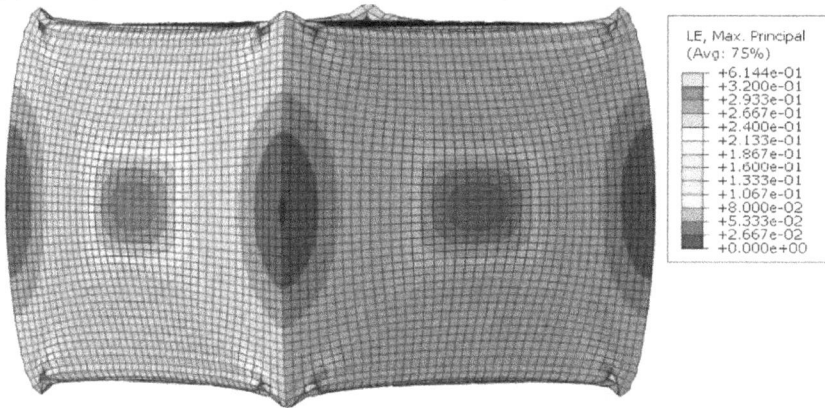

FIGURE 8.10   Comparison of strain distribution in a DMLS-MS1 cube during (a) experimental and (b) simulated compression testing. Reprinted with permission from Elsevier (Ebrahimi and Mohammadi 2018).

addition such as beam deflection during cooling (Heigel, Michaleris, and Reutzel 2015; Denlinger and Michaleris 2015) and overhanging support structures (Cheng and Chou 2015; Ramos, Belblidia, and Sienz 2019).

It is, however, important to consider the type of deformation process, which would require slight modifications to the material parameters. This is given by the different material parameters used for compression and tensile tests to accommodate the presence of pores in the additively manufactured metals that will affect the plastic flow during deformation. As an example, Figure 8.11 illustrates a set of mechanical properties experimentally derived for additively manufactured maraging steels (Cyr, Lloyd, and Mohammadi 2018; Cyr et al. 2018), which were successfully implemented

TABLE 8.2

Temperature-Dependent Material Properties of Ti-6Al-4V (Anca et al. 2011; Goldak, Chakravarti, and Bibby 1984; Heigel, Michaleris, and Reutzel 2015)

T (°C)	K (W/m/°C)	C_p (J/kg/°C)	B (GPa)	σ_y (MPa)	α_{cte} (μm/m/°C)
20	6.6	565	103.95	768.15	8.64
93	7.3	565	100.10	735.30	8.82
205	9.1	574	94.19	684.90	9.09
250	9.7	586	91.81	664.65	9.20
315	10.6	603	88.38	635.40	9.33
424	12.6	649	82.58	585.90	9.55
500	13.9	682	78.63	552.15	9.70
540	14.6	699	76.52	534.15	9.70
650	17.5	770	70.72	484.65	9.70
760	17.5	858	64.91	435.15	9.70
800	17.5	895	62.80	417.15	9.70
870	17.5	959	62.80	417.15	9.70

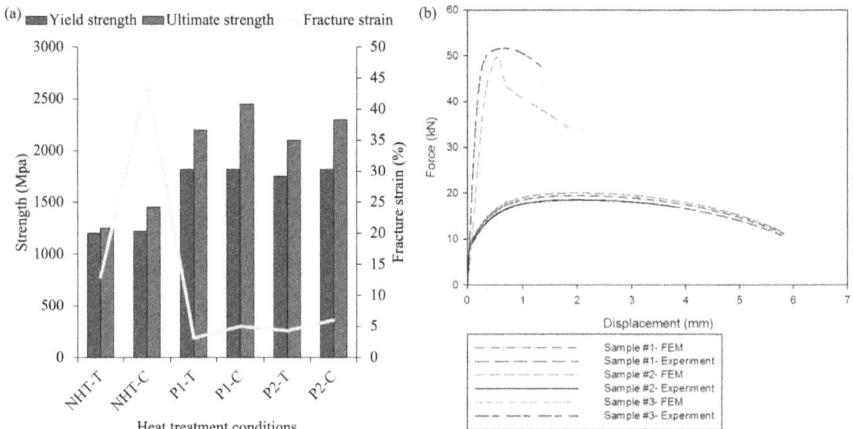

FIGURE 8.11 Implementation of (a) experimentally derived mechanical properties for numerical tensile (T) and compression (C) simulation tests (NHT: Non-heat treated; P1: aged at 490 °C for 6 h followed by air cooling; P2: preheating to 815 °C then rapidly heated to 982 °C for 1 h followed by air quenching) (Edward Cyr, Lloyd, and Mohammadi 2018; E Cyr et al. 2018); (b) comparison of forces between experiments and FEM during tensile testing. Reprinted with permission from Elsevier (Ebrahimi and Mohammadi 2018).

in numerical simulations of tensile and compression testing – see Figure 8.11(b) (Ebrahimi and Mohammadi 2018).

Quasi-static deformation can also be used for cost-effective computation of cutting forces. The coupling of tool kinematics with FEM simulated contact

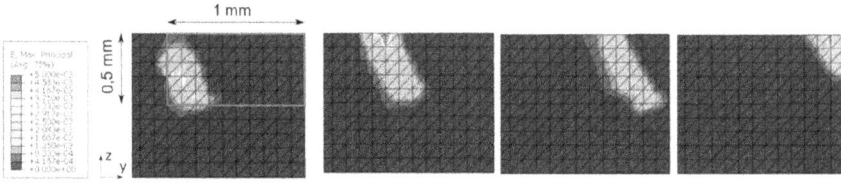

FIGURE 8.12 Tool-workpiece pressure exerted on the top surface of the workpiece over time where the leftmost image represents the partial contact of the tool at the initial engagement followed by two depictions of full engagement and lastly the exit of the tool out of the material. Reprinted with permission from Springer Nature (Didier et al. 2022).

pressures can determine the cutting forces (Didier et al. 2022). The tool-workpiece contact area determines the elements that will undergo displacements matched to the tool path trajectory where the displacement would give feedback of the pressure applied on the workpiece. This will give simulated results resembling Figure 8.12, which depicts the tool-workpiece contact of one cutter in a milling tool moving through the workpiece.

8.2.7 MATERIAL MODEL FOR LARGE DEFORMATION OF METALS

A commonly employed material model for illustrating large deformation and the progressive process of degradation till failure for most metals is the Johnson Cook (JC) material model (Johnson and Cook 1983), which describes the flow stress as:

$$\sigma_{eq} = \left[A + B\varepsilon_{pl}^{\ n} \right] \left[1 + C \ln\left(\frac{\dot{\varepsilon}}{\dot{\varepsilon}_0} \right) \right] \left[1 - \left(\frac{T - T_R}{T_M - T_R} \right)^m \right] \tag{8.10}$$

where A is the yield strength, B is the strain hardening coefficient, n is the strain hardening exponent, ε_{pl} is the equivalent plastic strain, $\dot{\varepsilon}$ is the strain rate, $\dot{\varepsilon}_0$ is the reference strain rate, T is the ambient temperature, T_R is the reference temperature often taken as the room temperature, T_M is the melting temperature, and m is a thermal softening parameter.

As intentionally partitioned, Equation 8.10 was split into three segments where the first term describes the stress–strain progression during elastic and plastic deformation with the reference strain rate. Hence, data points obtained on the stress–strain curve in experiments performed at the reference strain rate and reference temperature can be fitted with Equation 8.11 to determine the initial yield strength A.

$$\sigma_{eq} = \left[A + B\varepsilon_{pl}^{\ n} \right] \tag{8.11}$$

Rearrangement of Equation 8.11 into Equation 8.12 and plotting the natural loga-
rithmic of the equation gives:

$$\ln\left(\sigma_{eq} - A\right) = \ln B + n \ln \varepsilon_{pl} \tag{8.12}$$

where the intercept of the plot along the Y-axis gives the strain hardening coefficient
B and the gradient of the plot is the exponent n.

Subsequently, the second term comes into play when strain rates vary from the
reference strain rate, which gives the equivalent flow stress as:

$$\sigma_{eq} = \left[A + B\varepsilon_{pl}^{\,n}\right]\left[1 + C \ln\left(\frac{\dot{\varepsilon}}{\dot{\varepsilon}_0}\right)\right] \tag{8.13}$$

Equation 8.13 can be rearranged into:

$$\frac{\sigma_{eq}}{\left[A + B\varepsilon_{pl}^{\,n}\right]} = C \ln \dot{\varepsilon}^* + 1 \tag{8.14}$$

where $\dot{\varepsilon}^* = \dot{\varepsilon}/\dot{\varepsilon}_0$, $\left[\sigma_{eq}/\left(A + B\varepsilon_{pl}^{\,n}\right)\right]$ against $\ln \dot{\varepsilon}^*$ may be plotted and the slope may
give the constant C for various strain rates $\dot{\varepsilon}$.

Temperature effects are considered in the third term during material parameter
derivation at high temperatures, which would account for thermal softening effects.
For simplicity, determining the thermal softening parameter m is often done at the
reference strain rate, such that Equation 8.10 would only take the form as:

$$\sigma_{eq} = \left[A + B\varepsilon_{pl}^{\,n}\right]\left[1 - \left(\frac{T - T_R}{T_M - T_R}\right)^m\right] \tag{8.15}$$

which can be rearranged into:

$$\ln\left(1 - \frac{\sigma_{eq}}{A + B\varepsilon_{pl}^{\,n}}\right) = m \ln T^* \tag{8.16}$$

where $T^* = (T - T_R)/(T_M - T_R)$. The slope of the graph of $\ln\left(1 - \sigma_{eq}/A + B\varepsilon_{pl}^{\,n}\right)$
against $\ln T^*$ will give the material constant m.

With these equations, the equivalent flow stress with respect to a strain must be
given through experiments performed on a Split Hopkinson Pressure Bar (SHPB),

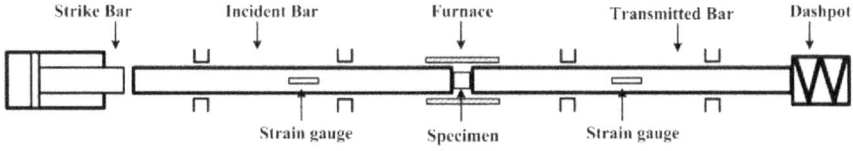

FIGURE 8.13 Schematic of the SPHB compression test rig. Reprinted with permission from Elsevier (Zhao et al. 2017).

which is essentially a compression testing setup shown in Figure 8.13. On one end of the setup, there is an incident bar, which transmits the axial impact delivered by the strike bar toward the sample positioned in the middle of the setup. During impact, part of the compression load is reflected into the incident bar, while the remaining travels through the specimen to the transmitted bar. Strain gauges positioned on both incident and transmitted bars record the energy transmitted in the form of strain, which would then be evaluated using elastic wave propagation theory to determine the strain rate, strain, and flow stress in the specimen as:

$$\dot{\varepsilon} = \frac{2C_0 \varepsilon_R}{L} \tag{8.17}$$

$$\varepsilon(t) = \frac{2C_0}{L} \int_0^t \varepsilon_R(t)\, dt \tag{8.18}$$

$$\sigma_s(t) = \frac{A_b}{A_s} E \varepsilon_T(t) \tag{8.19}$$

where ε_R is the measured strain reflected in the incident bar, C_0 is the elastic wave velocity, L is the initial length of the sample, A_b and A_s are the corresponding cross-sectional areas of the compression bar and the sample, E is the elastic modulus of the bars, and ε_T is the strain measured in the transmitted bar.

In most cases, the JC model has shown promise in delivering close approximations of the experimental work for various metals such as titanium alloys (Zheng, Ni, et al. 2021; Rypina et al. 2020), aluminum alloys (Liu and Melkote 2007; G. Chen et al. 2013; Zhang, Lee, and Wang 2021), and steels (Yao et al. 2020; Bergs, Hardt, and Schraknepper 2020; Zheng, Lee, et al. 2021). The most graphical representation of these close approximations is in cutting chips.

There are, however, also modifications made to the original constitutive equation to enhance the accurate representation of the experimental deformation. Common modifications include the Johnson-Cook Zerilli-Armstrong (JC-ZA) model, which accounts for the coupled effects of strain rate and heat generation when evaluating yielding and strain hardening (Lin and Chen 2010; Lin and Chen 2011). Strain softening effects were also unaccounted for in the JC model, which gave inaccurate representations of the adiabatic shear band formation during the cutting of Ti-6Al-4V. (Calamaz, Coupard, and Girot 2008). These strain softening effects coupled with strain hardening and thermal softening effects were modelled with a hyperbolic tangent (TANH) model that is another commonly employed modification of the JC material model for metal cutting simulations (Wang et al. 2018; Yameogo et al. 2017; Sima and Özel 2010; Ducobu, Rivière-Lorphèvre, and Filippi 2016). This chapter will not dive into assessing the merits of each modified JC model, but instead make it known to the reader that there are limitations of the original JC model that require certain modifications to suit the work material and deformation process under study.

8.2.8 POST-PROCESS RESIDUAL STRESSES

Residual stresses before and after post-processing are typical subjects of analysis in these simulations. Figure 8.14 is an example of determining the residual stresses developed after the additive process (Figure 8.14(a)) and the changes in stress distribution during the material removal process. Based on simulations, the final residual state of the post-processed part is relatively constant regardless of the magnitude of stress in the as-built state. Following the example in Figure 8.14, the residual stress after additive manufacturing can span across a wide range of 234–396 MPa, but post-processing brought the residual stresses down to a range of 145–205 MPa (Ren et al. 2021).

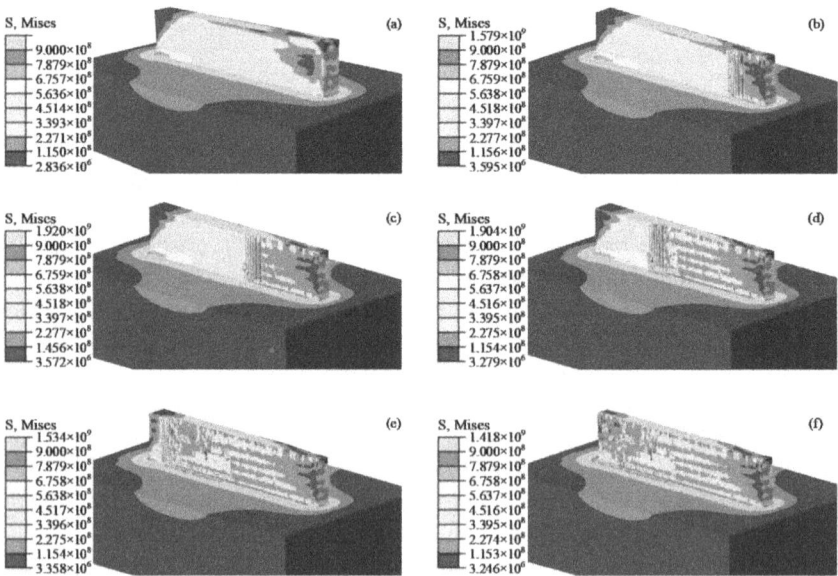

FIGURE 8.14 Simulated residual stresses at different time intervals during milling of Ti-6Al-4V: (a) 0s, (b) 0.037s, (c) 0.074s, (d) 0.111s, (e) 0.148s, and (f) 0.185s. Reprinted with permission from MDPI (Ren et al. 2021).

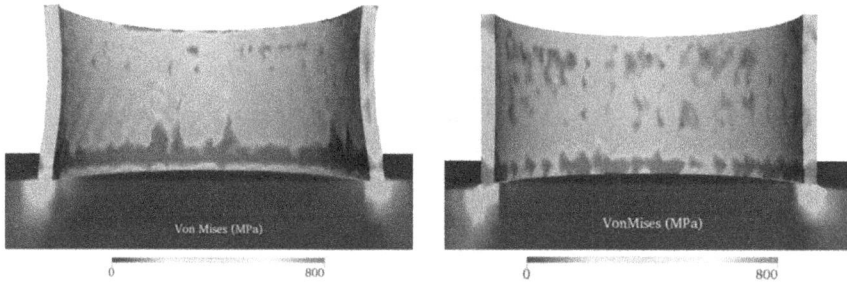

FIGURE 8.15 Simulated residual stresses and distortion of steel after additive manufacturing (left) and after simulated material removal along the walls (right). Reprinted with permission from Springer Nature (Salonitis et al. 2016).

Residual stress of additively manufactured parts is critical for post-processing analysis where distortions in the material will affect tool path planning. This is a mistake that most simulations on post-processes makes where they model the workpiece as stress-free and perform the machining process to determine the residual stresses. However, tool path designations conforming to the distorted geometry of the additively manufactured part in simulations can be computationally expensive. Fortunately, there have been innovative methods to work around this issue with the assumption that machining thin layers of material induces negligible mechanical and thermal loads on the part (Salonitis et al. 2016). An additively manufactured part is first simulated showing heavy distortions due to thermal gradients, as shown in Figure 8.15(a). Subsequently, a layer of material is subtracted from the model to simulate the post-process, which then allows the relaxation of the residual stresses in the walls and the restoration of the part geometry as illustrated in Figure 8.15(b). This method simply capitalizes on the relaxation of residual stresses generated from the additive manufacturing process and saves on the computational cost to simulate the chip formation process.

8.2.9 POST-PROCESSING OF SUPPORT STRUCTURES

Modeling of material removal also extends beyond analyzing the actual additively manufactured part to the analysis on the removal of support structures. This area is often neglected although it is critical to optimizing the all-rounded surface finish of the manufactured component. These structures are designed and constructed to provide a uniaxial support to the fabricated component, which would suggest that loads oriented perpendicular to these support axes by cutting tools would potentially result in undesirable deformation and failure. An example is given in the deformation of 316L stainless steel support structures that started to bend instead of the expected shearing during cutting (Figure 8.16). By identifying these overlooked occurrences, the flexibility of FEM simulations allows further modeling of different support structure geometries to provide quick evaluations and planning for the structural integrity and removal characteristics (Cao et al. 2020).

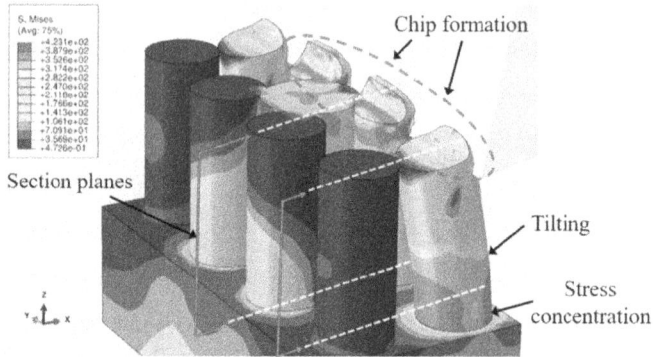

FIGURE 8.16 Von Mises stress distribution and the resultant deformation of support structures. Reprinted with permission from Elsevier (Cao et al. 2020).

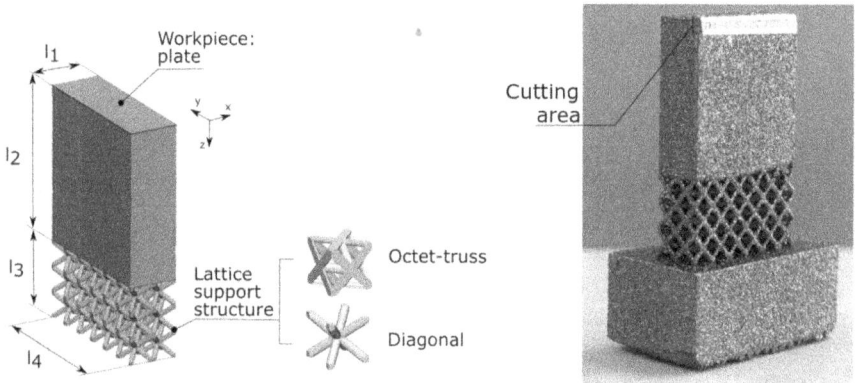

FIGURE 8.17 Model and fabricated workpiece supported and "clamped" by a lattice support structure for additive manufacturing and post-processing. Reprinted with permission from Springer Nature (Didier et al. 2022).

Support structures are typically designed to support the part during additive manufacturing with minimal material usage, which would mean thinner structures and hollow gaps between them. In some cases, support structures may also be used as the "clamping" device that upholds the parts as mechanical post-processing is performed on the part. An example is illustrated in Figure 8.17 where the additively manufactured workpiece on top is to be processed by machining while it is being supported by the lattice support structure. Successful machining then heavily relies on the stiffness of the support structures to attenuate vibrations involved during mechanical cutting. Thin walls (Thevenot et al. 2006; Seguy, Dessein, and Arnaud 2008) and hollow gaps are known to lower the stiffness of the "clamp" and affect the machining process on the workpiece. Hence, it is important to account for the stiffness of these "clamps" where applicable.

The modeling of the complex lattice structure, however, can be computationally expensive, which is where simplified homogenization of the support structure can ease off the computational loads. Instead of accurately modeling the intricacies of the lattice structure, a block structure can be modelled with a modified stiffness matrix C_{ijkl} (Didier et al. 2022). For different lattices, the 9 stiffness constants (reduced from 81 due to the orthogonal symmetry of the cubic structure) can be derived from the stress tensors measured for a given strain condition following Hooke's law ($\sigma_{ij} = C_{ijkl} \times \varepsilon_{kl}$).

While modeling of the additive manufacturing and post-processing in FEM gives a close representation of the actual occurrences in view of the appropriate material models designed for the deformation length scale, there are still areas that FEM is not quite feasible to simulate. This is particularly in the microstructural features that narrow down to the interatomic interactions. Multi-scale modeling is required for this, where atomistic events may be derived through molecular dynamics simulations (MDS) and fitted back into FEM simulations on the appropriate length scale.

8.3 MOLECULAR DYNAMICS SIMULATIONS

MDS is a solution that models the reaction of materials under physical fields on the atomistic level. One of the earliest employments of MDS in nanometric machining modeling of copper, silver, and silicon was in the 1990s (Belak, Boercker, and Stowers 1993; Belak and Stowers 1990). With technological improvements in computational power, MDS has gained significant advancement to model larger supercells with more atoms for larger scale deformation simulations to provide good insights on dislocation activities. MDS also includes the applicability of simulating the building process and the final configuration of atoms that form different phases and defect structures during additive manufacturing. Thus, it is critical to construct supercells of metal atoms based on the deposition process to understand the default configurations and structures, before advancing further into simulating the material removal processes.

Fundamentally, the purpose of MDS was to determine the motion and equilibrium status between atoms based on solid–liquid interfacial free energy and kinetic coefficients. At the most basic level, MDS describes the interatomic displacement between two individual atoms held at equilibrium or experiencing an attractive or repulsive force, such that the loss of attractive forces would be indicative of debonding. It is the agglomeration of these atoms that make up the physical mass that can be physically touched, and the severance of interatomic bonds results in plastic deformation, brittle failure, phase transformations, and so forth. Hence, MDS utilizes interatomic potentials to construct a system of atoms with defined lattice constants, cohesive energies, and elastic constants for the simulation of events that occur under externally applied forces (Antwi, Liu, and Wang 2018). In each time step, the pairwise forces, velocities, and displacement positions of each atom are then recalculated to reconstruct the equilibrized system (Komanduri and Raff 2001).

Alike numerical modeling, MDS for additive manufacturing largely focused on the prediction of thermal history encompassing the heating, cooling, and cyclical heating procedures (Etesami, Laradji, and Asadi 2020). During each of these processes, MDS interatomic potentials are required to account for the temperature-dependent liquid and solid parameters such as the density and enthalpies, among others. Additionally, the phase transition temperature and energy requirements need to be accounted for the modeling of two-phase coexistences during the cooling stages. Typical MDS procedures involve the selection of the appropriate interatomic potential energy function, model initialization, model relaxation at equilibrium, simulation run, and finding back the equilibrium state for a subsequent simulation run. There are several interatomic potential energy functions available for different materials where each of them are defined as the sum of N-bodied potentials (Tersoff 1988a):

$$E = \sum_i U_1\left(r_i\right) + \sum_{i<j} U_2\left(r_i,r_j\right) + \sum_{i<j<k} U_3\left(r_i,r_j,r_k\right) + \ldots + \sum_{i<j<k<\ldots<N} U_N\left(r_i,r_j,r_k,\ldots,r_N\right)$$

(8.20)

where U_1, U_2, ..., U_N are the m-body potentials. Each class of materials demands a suitable interatomic potential to accurately describe the way the material deforms and undergoes microstructural changes. Some of the more common interatomic potentials are Lennard-Jones (Edward and Sydney 1925) for rare gases, Morse potential (Morse 1929) for cubic metals, Tersoff potential (Tersoff 1988b; Tersoff 1988a) for covalent bonded materials, and Born-Mayer potential (Born and Mayer 1932) for metals and ceramics. The most common interatomic potential employed for metal modeling is the embedded-atom method (EAM) described as:

$$E_{tot} = \sum_i G_i\left(\rho_{h,i}\right) + \frac{1}{2}\sum_{i,j} U_{ij}\left(r_{ij}\right)$$

(8.21)

where ρ_{ij} is the electron density at atom i and G_i is the embedding energy to place an atom within the electrons (Foiles, Baskes, and Daw 1986; Foiles 1985).

The present progress in MD simulations has yet to reach the material removal processing of AM metals but is still largely focused on the building process and predictions of AM metals, which include defect formations during solidification and grain nucleation and growth.

8.3.1 Melting and Solidification

As MDS model the motion of atoms within the defined boundaries of the simulation supercell, the deposition process of the additive manufacturing process can be modelled by layering amorphous randomly distributed atoms representing the liquified metal atoms onto its solid counterpart. Figure 8.18 illustrates this process where the solid phase of the metal is first constructed to equilibrium followed by the deposition of the liquid phase atop the solid phase, which would be run to achieve an equilibrium configuration as the liquid phase cools to room temperature. Through this process of simulating the atomistic deposition process, the structure of defect formations can be easily observed as a function of the building parameter such as temperature and layer thickness.

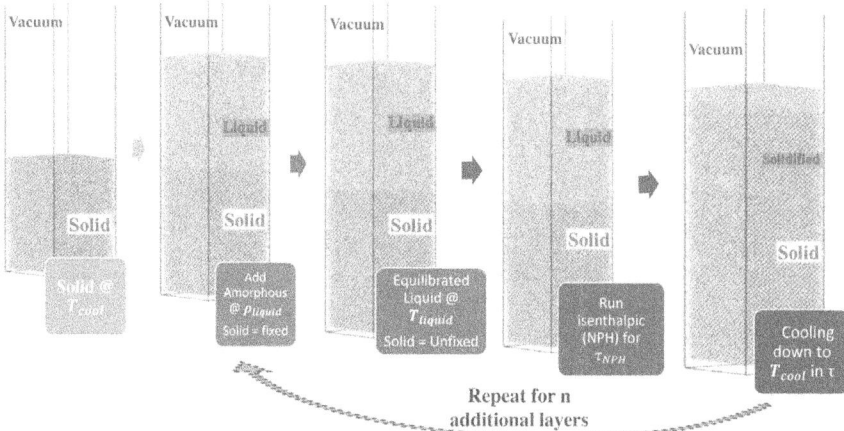

FIGURE 8.18 Simulation procedure for the additive manufacturing process. Reprinted with permission from Elsevier (Singh, Waas, and Sundararaghavan 2021).

The brief description of the simulation process deserves a more detailed elaboration. The solid phase must be first prepared by finding the equilibrium for the lattice structure of the metal at room temperature for the defined volume of the supercell. The solid is then heated to the melting temperature (T_{liquid}) in an NPT simulation (i.e., constant number of atoms, pressure, and temperature) where the density of the randomly distributed liquid phase (ρ_{liquid}) would be derived. With the known density, the deposition process can be simulated by depositing the liquid phase at T_{liquid} onto the solid of identical volume. At this juncture, the solid atoms remain fixed during the NVT simulation (i.e., constant number of atoms, volume, and temperature) while the atoms in the liquid phase equilibrate. Interaction between the two phases follows through a constant enthalpic step where the restrain on the solid atom positions will be removed and the two phases may begin interaction. Cooling from T_{liquid} to room temperature is the final step of the deposition process where a fully solidified structure will be produced. Subsequent melt pools can be simulated by repeating the procedure of depositing another layer of the liquid phase.

Melting and solidification simulations may also be performed between particles, which is particularly useful for direct metal laser sintering modeling and bimetallic material integration. Bimetallic material integration is physically possible by adopting a core–shell approach of two different materials to be melted and form a cohesive multi-material with multi-functions. Multi-material simulations require additional interatomic potentials between the two different atoms for instance the Al-Ti interatomic potentials to calculate results shown in Figure 8.19. The potential energy (PE) curves (Figure 8.19(a)) describe the phase changes and melting sequences by the corresponding large increase and plateauing of the PE (Rahmani et al. 2018). The atomic distribution (AD) in Figure 8.19(b) describes the ideal case of a multi-material being evenly distributed, while the structural analysis (SA) describes the transition from a purely crystalline material into an amorphous structure that supports the even distribution of atoms within the particle.

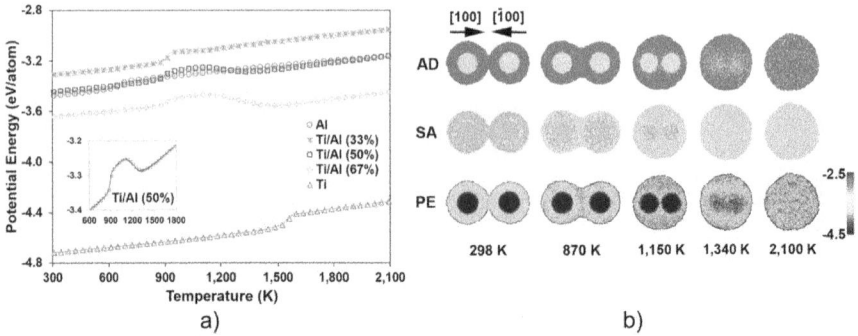

FIGURE 8.19 MDS of the melting process for Ti-Al multi-material: (a) potential energy per atom with different core volume fractions of Ti; (b) atomic distribution (AD), structural analysis (SA), and potential energy (PE) for a Ti/Al 50% core/shell combination. Reprinted with permission from Springer Nature (Rahmani et al. 2018).

DMLS simulation involves a relatively similar procedure of modeling two particles of the same material, where external heat will be applied to the system to give a sintering temperature just below the melting temperature (Figure 8.18). Contact between the two particles first occurs, followed by necking and diffusion before the two particles are completely merged. It is important to recall that DMLS does not involve the melting of the particles, hence diffusion is the main mode of these atomistic movements. Aside from analyzing the thermodynamic requirements for diffusion of these particles, defect formations may also be observed through MDS such as the clustering of Mg and Si particles circled in Figure 8.20(e). Such particles could serve to be hindrances as sources of high stress concentration and challenges in post-processing. The subsequent section will discuss further on the use of MDS to identify defect formations.

8.3.2 DEFECT MODELING

Defect modeling in MDS provides critical information on the final material characteristics such as phase formations (Etesami, Laradji, and Asadi 2020), lattice defects (Zhang et al. 2018), and structural density defects (Singh, Waas, and Sundararaghavan 2021).

Lattice defects such as Shockley partials, dislocation twinning, and grain boundaries can be modelled during the solidification process. The solidification process becomes increasingly complex with the involvement of preferential crystallographic orientation grain growth. Thus, the location for grain nucleation and the growth rates depends on the initial configuration of the powders. In the meantime, the highest cool rate is experienced at the center of the melt pool, which promotes the nucleation of new grains, as shown in Figure 8.21. The competitive growth of different grain types eventually meet together at grain boundaries to find equilibrium in the system. The misalignment of atoms is inevitable during this process resulting in twin boundaries and stacking faults, typically observed by the formation of another lattice structure (e.g., hexagonally closed packed).

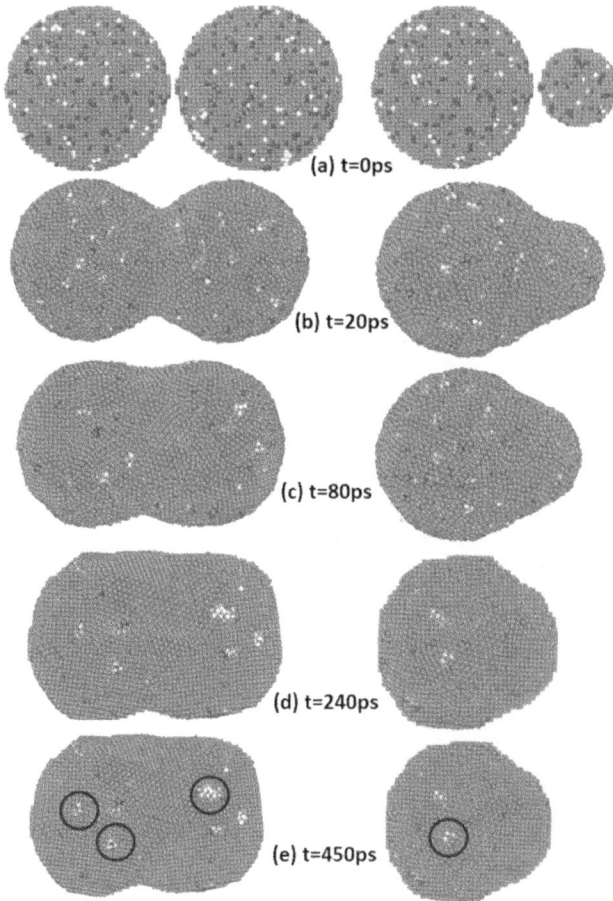

FIGURE 8.20 MDS of AlSi10Mg during DMLS: (a) initial configuration of the two particles; (b) contact between particles; (c) diffusion of atoms; (d) complete merging; (e) final stable configuration. Reprinted with permission from Elsevier (Nandy et al. 2019).

The importance of studying the formation of lattice structure defects is in their influence on the strength of the material (Mahata and Asle Zaeem 2019). With the existence of dislocation twins within the lattice structure, detwinning, formation of new twins, dislocation stretching can occur under applied stress and consequently affect the yield strength, ultimate tensile strength, and final failure. The direction of movement for these partial dislocations relative to other key features such as grain boundary growth orientations and crystallographic orientations can make it inconvenient for deformation and consequently increase material strengths.

Another critically important defect is the formation of pores in the material resulting from voids found between particles of the initial powder bed. Figure 8.22 illustrates the progressive formation of these defects in an adjacent melt pool formation where a laser is applied to melt the initial configuration of powder particles and

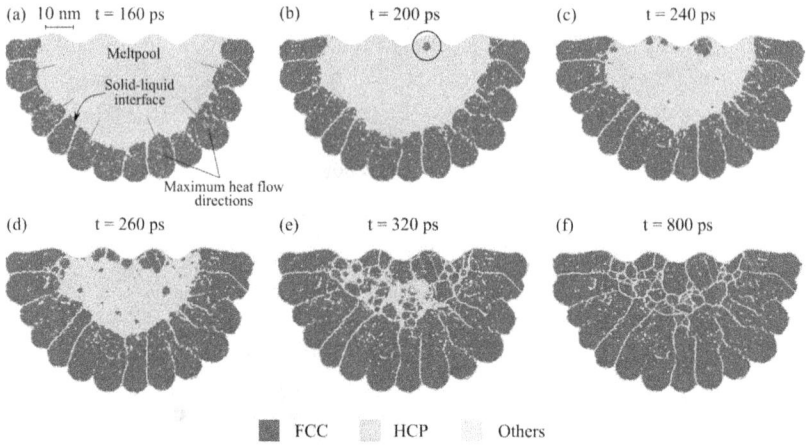

FIGURE 8.21 Simulated solidification process of an Al melt pool: (a) initial state of a liquid melt pool in the center comprising grey atoms; (b) onset nucleation of grains; (c) expansion and further nucleation of new grains; (d) formation of grain boundaries between early nucleated grains; (e) epitaxial grain growth hindered by initially nucleated grains; (f) final solidified microstructure. Reprinted with permission from Elsevier (Kurian and Mirzaeifar 2020).

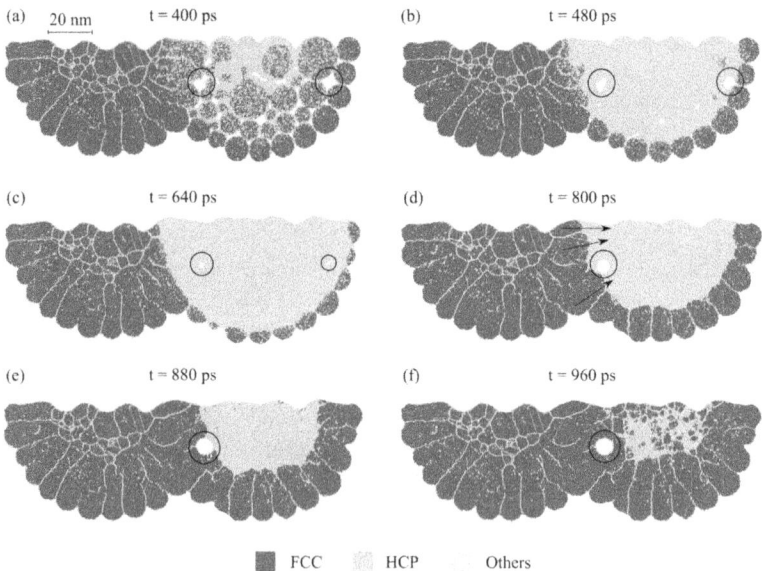

FIGURE 8.22 Simulated melting and solidification process of a subsequent melt pool in Al: (a) initial formation of a liquid meltpool demarcated by grey atoms; (b) collapse of smaller voids while larger voids remain (circled locations); (c) end of heating; (d) epitaxial growth of grains from previous melt track; (e) grain growth from previous melt track circumventing void; (f) remnants of the void as solidification continues. Reprinted with permission from Elsevier (Kurian and Mirzaeifar 2020).

gradually removes most of the smaller inter-particle voids. However, larger voids remained throughout the solidification process due to the hastened heating and cool rates during the irradiation process. Voids alone are critical engineering problems to be solved due to the discontinuity in load transfer during structural application. More specifically, MDS enables the observation of dislocation formation and flow around these voids to support the reduction in mechanical strength (Cui et al. 2022).

Although modeling the additively manufactured process with MDS shows promise, the current application of this modeling method in this area remains premature. As of now, there has yet to be a continuity on the modeling of the mechanical material removal process of these additively manufactured material models in MDS, which is necessary to achieve the goal of multi-scale modeling for post-processing.

8.4   CONCLUDING REMARKS

It is evident that the development of theoretical models for post-processing technologies is relatively limited despite a wealth of information to model the additive process. This fact serves as a good indicator of the depth of unexplored knowledge in post-processing. The chapter emphasized the importance of building the correct material setup that simulates the actual material conditions after the additive manufacturing process, where residual stresses would have been stored or dissipated through geometrical distortions. This extends to both finite element method simulations and molecular dynamic simulations that were covered in this chapter. Accurate theoretical studies would be achievable with a disciplined adherence to the sequential multi-scale modeling.

REFERENCES

Anca, Andrés, Víctor D Fachinotti, Gustavo Escobar-Palafox, and Alberto Cardona. 2011. "Computational Modelling of Shaped Metal Deposition." *International Journal for Numerical Methods in Engineering* 85 (1). John Wiley & Sons, Ltd: 84–106. doi:10.1002/nme.2959.

Antwi, Elijah Kwabena, Kui Liu, and Hao Wang. 2018. "A Review on Ductile Mode Cutting of Brittle Materials." *Frontiers of Mechanical Engineering* 13 (2). Springer: 251–263. doi:10.1007/s11465-018-0504-z.

Belak, J F, and I F Stowers. 1990. "A Molecular Dynamics Model of the Orthogonal Cutting Process." In *Proceedings of American Society of Photoptical Engineers Annual Conference. Rochester: Lawrence Livermore National Lab.* United States.

Belak, James, David B Boercker, and Irving F Stowers. 1993. "Simulation of Nanometer-Scale Deformation of Metallic and Ceramic Surfaces." *MRS Bulletin* 18 (5). Cambridge University Press: 55–60. doi:10.1557/S088376940004714X.

Bergs, Thomas, Marvin Hardt, and Daniel Schraknepper. 2020. "Determination of Johnson-Cook Material Model Parameters for AISI 1045 from Orthogonal Cutting Tests Using the Downhill-Simplex Algorithm." *Procedia Manufacturing* 48: 541–552. doi:10.1016/j.promfg.2020.05.081.

Born, Max, and Joseph E Mayer. 1932. "Zur Gittertheorie Der Ionenkristalle." *Zeitschrift Für Physik* 75 (1): 1–18. doi:10.1007/BF01340511.

Calamaz, Madalina, Dominique Coupard, and Franck Girot. 2008. "A New Material Model for 2D Numerical Simulation of Serrated Chip Formation When Machining Titanium

Alloy Ti–6Al–4V." *International Journal of Machine Tools and Manufacture* 48 (3–4). Elsevier: 275–288. doi:10.1016/j.ijmachtools.2007.10.014.

Cao, Jun, Michael A Gharghouri, and Philip Nash. 2016. "Finite-Element Analysis and Experimental Validation of Thermal Residual Stress and Distortion in Electron Beam Additive Manufactured Ti-6Al-4V Build Plates." *Journal of Materials Processing Technology* 237: 409–419. doi:10.1016/j.jmatprotec.2016.06.032.

Cao, Qiqiang, Yuchao Bai, Jiong Zhang, Zhuoqi Shi, Jerry Ying Hsi Fuh, and Hao Wang. 2020. "Removability of 316L Stainless Steel Cone and Block Support Structures Fabricated by Selective Laser Melting (SLM)." *Materials and Design* 191. Elsevier: 108691. doi:10.1016/j.matdes.2020.108691.

Chen, Fan, and Wentao Yan. 2020. "High-Fidelity Modelling of Thermal Stress for Additive Manufacturing by Linking Thermal-Fluid and Mechanical Models." *Materials & Design* 196: 109185. doi:10.1016/j.matdes.2020.109185.

Chen, Guang, Chengzu Ren, Pan Zhang, Kuihu Cui, and Yuanchen Li. 2013. "Measurement and Finite Element Simulation of Micro-Cutting Temperatures of Tool Tip and Workpiece." *International Journal of Machine Tools and Manufacture* 75: 16–26. doi:10.1016/j.ijmachtools.2013.08.005.

Chen, Shaohua, Yaopengxiao Xu, and Yang Jiao. 2018. "A Hybrid Finite-Element and Cellular-Automaton Framework for Modeling 3D Microstructure of Ti–6Al–4V Alloy during Solid–Solid Phase Transformation in Additive Manufacturing." *Modelling and Simulation in Materials Science and Engineering* 26 (4). IOP Publishing: 45011. doi:10.1088/1361-651x/aabcad.

Cheng, Bo, and Kevin Chou. 2015. "Geometric Consideration of Support Structures in Part Overhang Fabrications by Electron Beam Additive Manufacturing." *Computer-Aided Design* 69: 102–111. doi:10.1016/j.cad.2015.06.007.

Cook, Peter S, and Anthony B Murphy. 2020. "Simulation of Melt Pool Behaviour during Additive Manufacturing: Underlying Physics and Progress." *Additive Manufacturing* 31: 100909. doi:10.1016/j.addma.2019.100909.

Cui, Can, Xiaoguo Gong, Lijia Chen, Weiwei Xu, and Lijie Chen. 2022. "Atomic-Scale Investigations on Dislocation-Precipitate Interactions Influenced by Voids in Ni-Based Superalloys." *International Journal of Mechanical Sciences* 216: 106945. doi:10.1016/j.ijmecsci.2021.106945.

Cyr, E, H Asgari, S Shamsdini, M Purdy, K Hosseinkhani, and M Mohammadi. 2018. "Fracture Behaviour of Additively Manufactured MS1-H13 Hybrid Hard Steels." *Materials Letters* 212: 174–177. doi:10.1016/j.matlet.2017.10.097.

Cyr, Edward, Alan Lloyd, and Mohsen Mohammadi. 2018. "Tension-Compression Asymmetry of Additively Manufactured Maraging Steel." *Journal of Manufacturing Processes* 35: 289–294. doi:10.1016/j.jmapro.2018.08.015.

Denlinger, Erik R, Jeff Irwin, and Pan Michaleris. 2014. "Thermomechanical Modeling of Additive Manufacturing Large Parts." *Journal of Manufacturing Science and Engineering* 136 (6). doi:10.1115/1.4028669.

Denlinger, Erik R, and Pan Michaleris. 2015. "Mitigation of Distortion in Large Additive Manufacturing Parts." *Proceedings of the Institution of Mechanical Engineers, Part B: Journal of Engineering Manufacture* 231 (6). IMECHE: 983–993. doi:10.1177/0954405415578580.

Didier, P, G Le Coz, G Robin, P Lohmuller, B Piotrowski, A Moufki, and P Laheurte. 2022. "Consideration of Additive Manufacturing Supports for Post-Processing by End Milling: A Hybrid Analytical–Numerical Model and Experimental Validation." *Progress in Additive Manufacturing* 7 (1): 15–27. doi:10.1007/s40964-021-00211-4.

Drucker, D. C., and W. Prager. 1952. "Soil Mechanics and Plastic Analysis or Limit Design." *Quarterly of Applied Mathematics* 10 (2). Brown University: 157–165. doi:10.1090/qam/48291.

Ducobu, F, E Rivière-Lorphèvre, and E Filippi. 2016. "Material Constitutive Model and Chip Separation Criterion Influence on the Modeling of Ti6Al4V Machining with Experimental Validation in Strictly Orthogonal Cutting Condition." *International Journal of Mechanical Sciences* 107: 136–149. doi:10.1016/j.ijmecsci.2016.01.008.

Ebrahimi, Alireza, and Mohsen Mohammadi. 2018. "Numerical Tools to Investigate Mechanical and Fatigue Properties of Additively Manufactured MS1-H13 Hybrid Steels." *Additive Manufacturing* 23: 381–393. doi:10.1016/j.addma.2018.07.009.

Edward, Lennard-Jones John, and Chapman Sydney. 1925. "On the Forces between Atoms and Ions." *Proceedings of the Royal Society of London. Series A, Containing Papers of a Mathematical and Physical Character* 109 (752). Royal Society: 584–597. doi:10.1098/rspa.1925.0147.

Etesami, S Alireza, Mohamed Laradji, and Ebrahim Asadi. 2020. "Reliability of Molecular Dynamics Interatomic Potentials for Modeling of Titanium in Additive Manufacturing Processes." *Computational Materials Science* 184: 109883. doi:10.1016/j.commatsci.2020.109883.

Fang, Ze-Chen, Zhi-Lin Wu, Chen-Guang Huang, and Chen-Wu Wu. 2020. "Review on Residual Stress in Selective Laser Melting Additive Manufacturing of Alloy Parts." *Optics & Laser Technology* 129: 106283. doi:10.1016/j.optlastec.2020.106283.

Foiles, S M. 1985. "Application of the Embedded-Atom Method to Liquid Transition Metals." *Physical Review B* 32 (6). American Physical Society: 3409–3415. doi:10.1103/PhysRevB.32.3409.

Foiles, S M, M I Baskes, and M S Daw. 1986. "Embedded-Atom-Method Functions for the Fcc Metals Cu, Ag, Au, Ni, Pd, Pt, and Their Alloys." *Physical Review B* 33 (12). American Physical Society: 7983–7991. doi:10.1103/PhysRevB.33.7983.

Geng, Yaoyi, and Noel Harrison. 2020. "Functionally Graded Bimodal Ti6Al4V Fabricated by Powder Bed Fusion Additive Manufacturing: Crystal Plasticity Finite Element Modelling." *Materials Science and Engineering: A* 773: 138736. doi:10.1016/j.msea.2019.138736.

Goldak, John, Aditya Chakravarti, and Malcolm Bibby. 1984. "A New Finite Element Model for Welding Heat Sources." *Metallurgical Transactions B* 15 (2): 299–305. doi:10.1007/BF02667333.

Gunasegaram, Dayalan R, Anthony B Murphy, Sharen J Cummins, Vincent Lemiale, Gary W Delaney, Vu Nguyen, and Yuqing Feng. 2017. "Aiming for Modeling-Assisted Tailored Designs for Additive Manufacturing." In *TMS 2017 146th Annual Meeting & Exhibition Supplemental Proceedings. The Minerals, Metals & Materials Series*, edited by The Minerals TMS Metals & Materials Society, 91–102. Cham: Springer International Publishing. doi:10.1007/978-3-319-51493-2_10.

Ha, Kyeongsik, Taehwan Kim, Gyeong Yun Baek, Jong Bae Jeon, Do-sik Shim, Young Hoon Moon, and Wookjin Lee. 2020. "Numerical Study of the Effect of Progressive Solidification on Residual Stress in Single-Bead-on-Plate Additive Manufacturing." *Additive Manufacturing* 34: 101245. doi:10.1016/j.addma.2020.101245.

Hajializadeh, Farshid, and Ayhan Ince. 2020. "Short Review on Modeling Approaches for Metal Additive Manufacturing Process." *Material Design & Processing Communications* 2 (2). John Wiley & Sons, Ltd: e56. doi:10.1002/mdp2.56.

Heigel, J C, P Michaleris, and E W Reutzel. 2015. "Thermo-Mechanical Model Development and Validation of Directed Energy Deposition Additive Manufacturing of Ti–6Al–4V." *Additive Manufacturing* 5: 9–19. doi:10.1016/j.addma.2014.10.003.

Huang, Yuze, Mir Behrad Khamesee, and Ehsan Toyserkani. 2016. "A Comprehensive Analytical Model for Laser Powder-Fed Additive Manufacturing." *Additive Manufacturing* 12: 90–99. doi:10.1016/j.addma.2016.07.001.

Jamshidinia, Mahdi, Fanrong Kong, and Radovan Kovacevic. 2013. "Numerical Modeling of Heat Distribution in the Electron Beam Melting of Ti-6Al-4V." *Journal of Manufacturing Science and Engineering* 135 (6). doi:10.1115/1.4025746.

Ji, Xia, Elham Mirkoohi, Jinqiang Ning, and Steven Y Liang. 2020. "Analytical Modeling of Post-Printing Grain Size in Metal Additive Manufacturing." *Optics and Lasers in Engineering* 124: 105805. doi:10.1016/j.optlaseng.2019.105805.

Johnson, Gordon R, and William H Cook. 1983. "A Computational Constitutive Model and Data for Metals Subjected to Large Strain, High Strain Rates and High Pressures." In *The Seventh International Symposium on Ballistics*, 541–547.

Kayacan, Mevlüt Yunus, Koray Özsoy, Burhan Duman, Nihat Yilmaz, and Mehmet Cengiz Kayacan. 2019. "A Study on Elimination of Failures Resulting from Layering and Internal Stresses in Powder Bed Fusion (PBF) Additive Manufacturing." *Materials and Manufacturing Processes* 34 (13). Taylor & Francis: 1467–1475. doi:10.1080/10426914.2019.1655151.

Komanduri, R., and L. M. Raff. 2001. "A Review on the Molecular Dynamics Simulation of Machining at the Atomic Scale." *Proceedings of the Institution of Mechanical Engineers, Part B: Journal of Engineering Manufacture* 215 (12): 1639–1672. doi:10.1177/095440540121501201.

Kurian, Sachin, and Reza Mirzaeifar. 2020. "Selective Laser Melting of Aluminum Nano-Powder Particles, a Molecular Dynamics Study." *Additive Manufacturing* 35: 101272. doi:10.1016/j.addma.2020.101272.

Lauro, C H, L C Brandão, D Carou, and J P Davim. 2015. "Specific Cutting Energy Employed to Study the Influence of the Grain Size in the Micro-Milling of the Hardened AISI H13 Steel." *The International Journal of Advanced Manufacturing Technology* 81 (9): 1591–1599. doi:10.1007/s00170-015-7321-x.

Li, Runsheng, Guilan Wang, Xushan Zhao, Fusheng Dai, Cheng Huang, Mingbo Zhang, Xi Chen, Hao Song, and Haiou Zhang. 2021. "Effect of Path Strategy on Residual Stress and Distortion in Laser and Cold Metal Transfer Hybrid Additive Manufacturing." *Additive Manufacturing* 46: 102203. doi:10.1016/j.addma.2021.102203.

Lin, Y C, and Xiao-Min Chen. 2010. "A Combined Johnson–Cook and Zerilli–Armstrong Model for Hot Compressed Typical High-Strength Alloy Steel." *Computational Materials Science* 49 (3): 628–633. doi:10.1016/j.commatsci.2010.06.004.

Lin, Y C, and Xiao-Min Chen. 2011. "Erratum to: 'A Combined Johnson–Cook and Zerilli–Armstrong Model for Hot Compressed Typical High-Strength Alloy Steel' [Comput. Mater. Sci. 49 (2010) 628–633]." *Computational Materials Science* 50 (10): 3073. doi:10.1016/j.commatsci.2010.09.001.

Lindgren, Lars-Erik, Andreas Lundbäck, Martin Fisk, Robert Pederson, and Joel Andersson. 2016. "Simulation of Additive Manufacturing Using Coupled Constitutive and Microstructure Models." *Additive Manufacturing* 12: 144–158. doi:10.1016/j.addma.2016.05.005.

Liu, Kai, and Shreyes N. Melkote. 2007. "Finite Element Analysis of the Influence of Tool Edge Radius on Size Effect in Orthogonal Micro-Cutting Process." *International Journal of Mechanical Sciences* 49 (5): 650–660. doi:10.1016/j.ijmecsci.2006.09.012.

Lu, Xufei, Miguel Cervera, Michele Chiumenti, and Xin Lin. 2021. "Residual Stresses Control in Additive Manufacturing." *Journal of Manufacturing and Materials Processing* 5 (4). doi:10.3390/jmmp5040138.

Mahata, Avik, and Mohsen Asle Zaeem. 2019. "Effects of Solidification Defects on Nanoscale Mechanical Properties of Rapid Directionally Solidified Al-Cu Alloy: A Large Scale Molecular Dynamics Study." *Journal of Crystal Growth* 527: 125255. doi:10.1016/j.jcrysgro.2019.125255.

Mercelis, Peter, and Jean-Pierre Kruth. 2006. "Residual Stresses in Selective Laser Sintering and Selective Laser Melting." *Rapid Prototyping Journal* 12 (5). Emerald Group Publishing Limited: 254–265. doi:10.1108/13552540610707013.

Michaleris, Panagiotis. 2014. "Modeling Metal Deposition in Heat Transfer Analyses of Additive Manufacturing Processes." *Finite Elements in Analysis and Design* 86: 51–60. doi:10.1016/j.finel.2014.04.003.

Morse, Philip M. 1929. "Diatomic Molecules According to the Wave Mechanics. II. Vibrational Levels." *Physical Review* 34 (1). American Physical Society: 57–64. doi:10.1103/PhysRev.34.57.

Nandy, Jyotirmoy, Natraj Yedla, Pradeep Gupta, Hrushikesh Sarangi, and Seshadev Sahoo. 2019. "Sintering of AlSi10Mg Particles in Direct Metal Laser Sintering Process: A Molecular Dynamics Simulation Study." *Materials Chemistry and Physics* 236: 121803. doi:10.1016/j.matchemphys.2019.121803.

Nie, Pulin, O A Ojo, and Zhuguo Li. 2014. "Numerical Modeling of Microstructure Evolution during Laser Additive Manufacturing of a Nickel-Based Superalloy." *Acta Materialia* 77: 85–95. doi:10.1016/j.actamat.2014.05.039.

Ning, Jinqiang, Daniel E Sievers, Hamid Garmestani, and Steven Y Liang. 2019. "Analytical Modeling of Transient Temperature in Powder Feed Metal Additive Manufacturing during Heating and Cooling Stages." *Applied Physics A* 125 (8): 496. doi:10.1007/s00339-019-2782-7.

Ning, Jinqiang, Daniel E Sievers, Hamid Garmestani, and Steven Y Liang. 2020. "Analytical Modeling of In-Process Temperature in Powder Feed Metal Additive Manufacturing Considering Heat Transfer Boundary Condition." *International Journal of Precision Engineering and Manufacturing-Green Technology* 7 (3): 585–593. doi:10.1007/s40684-019-00164-8.

Oliveira, A R, A L Jardini, and E G Del Conte. 2020. "Effects of Cutting Parameters on Roughness and Residual Stress of Maraging Steel Specimens Produced by Additive Manufacturing." *The International Journal of Advanced Manufacturing Technology* 111 (9): 2449–2459. doi:10.1007/s00170-020-06309-3.

Papazoglou, E L, N E Karkalos, and A P Markopoulos. 2020. "A Comprehensive Study on Thermal Modeling of SLM Process under Conduction Mode Using FEM." *The International Journal of Advanced Manufacturing Technology* 111 (9): 2939–2955. doi:10.1007/s00170-020-06294-7.

Promoppatum, Patcharapit, and Shi-Chune Yao. 2020. "Influence of Scanning Length and Energy Input on Residual Stress Reduction in Metal Additive Manufacturing: Numerical and Experimental Studies." *Journal of Manufacturing Processes* 49: 247–259. doi:10.1016/j.jmapro.2019.11.020.

Rahman, M A, K S Woon, V C Venkatesh, and M Rahman. 2018. "Modelling of the Combined Microstructural and Cutting Edge Effects in Ultraprecision Machining." *CIRP Annals* 67 (1): 129–132. doi:10.1016/j.cirp.2018.03.019.

Rahmani, Farzin, Jungmin Jeon, Shan Jiang, and Sasan Nouranian. 2018. "Melting and Solidification Behavior of Cu/Al and Ti/Al Bimetallic Core/Shell Nanoparticles during Additive Manufacturing by Molecular Dynamics Simulation." *Journal of Nanoparticle Research* 20 (5): 133. doi:10.1007/s11051-018-4237-z.

Ramos, Davi, Fawzi Belblidia, and Johann Sienz. 2019. "New Scanning Strategy to Reduce Warpage in Additive Manufacturing." *Additive Manufacturing* 28: 554–564. doi:10.1016/j.addma.2019.05.016.

Ren, Zhaohui, Xingwen Zhang, Yunhe Wang, Zhuhong Li, and Zhen Liu. 2021. "Finite Element Analysis of the Milling of Ti6Al4V Titanium Alloy Laser Additive Manufacturing Parts." *Applied Sciences* 11 (11). doi:10.3390/app11114813.

Rypina, Łukasz, Dariusz Lipiński, Błażej Bałasz, Wojciech Kacalak, and Tomasz Szatkiewicz. 2020. "Analysis and Modeling of the Micro-Cutting Process of Ti-6Al-4V Titanium Alloy with Single Abrasive Grain." *Materials* 13 (24). doi:10.3390/ma13245835.

Salonitis, Konstantinos, Laurent D'Alvise, Babis Schoinochoritis, and Dimitrios Chantzis. 2016. "Additive Manufacturing and Post-Processing Simulation: Laser Cladding Followed by High Speed Machining." *The International Journal of Advanced Manufacturing Technology* 85 (9): 2401–2411. doi:10.1007/s00170-015-7989-y.

Seguy, Sébastien, Gilles Dessein, and Lionel Arnaud. 2008. "Surface Roughness Variation of Thin Wall Milling, Related to Modal Interactions." *International Journal of Machine Tools and Manufacture* 48 (3–4): 261–274. doi:10.1016/j.ijmachtools.2007.09.005.

Sima, Mohammad, and Tuğrul Özel. 2010. "Modified Material Constitutive Models for Serrated Chip Formation Simulations and Experimental Validation in Machining of Titanium Alloy Ti–6Al–4V." *International Journal of Machine Tools and Manufacture* 50 (11): 943–960. doi:10.1016/j.ijmachtools.2010.08.004.

Simoneau, A., E. Ng, and M. A. Elbestawi. 2007. "Grain Size and Orientation Effects When Microcutting AISI 1045 Steel." *CIRP Annals* 56 (1): 57–60. doi:10.1016/j.cirp.2007.05.016.

Singh, Gurmeet, Anthony M Waas, and Veera Sundararaghavan. 2021. "Understanding Defect Structures in Nanoscale Metal Additive Manufacturing via Molecular Dynamics." *Computational Materials Science* 200: 110807. doi:10.1016/j.commatsci.2021.110807.

Stender, Michael E, Lauren L Beghini, Joshua D Sugar, Michael G Veilleux, Samuel R Subia, Thale R Smith, Christopher W San Marchi, Arthur A Brown, and Daryl J Dagel. 2018. "A Thermal-Mechanical Finite Element Workflow for Directed Energy Deposition Additive Manufacturing Process Modeling." *Additive Manufacturing* 21: 556–566. doi:10.1016/j.addma.2018.04.012.

Steuben, John C, Andrew J Birnbaum, John G Michopoulos, and Athanasios P Iliopoulos. 2019. "Enriched Analytical Solutions for Additive Manufacturing Modeling and Simulation." *Additive Manufacturing* 25: 437–447. doi:10.1016/j.addma.2018.10.017.

Sun, Li, Xiaobo Ren, Jianying He, and Zhiliang Zhang. 2021. "Numerical Investigation of a Novel Pattern for Reducing Residual Stress in Metal Additive Manufacturing." *Journal of Materials Science & Technology* 67: 11–22. doi:10.1016/j.jmst.2020.05.080.

Tan, Joel Heang Kuan, Swee Leong Sing, and Wai Yee Yeong. 2020. "Microstructure Modelling for Metallic Additive Manufacturing: A Review." *Virtual and Physical Prototyping* 15 (1). Taylor & Francis: 87–105. doi:10.1080/17452759.2019.1677345.

Tersoff, J. 1988a. "New Empirical Approach for the Structure and Energy of Covalent Systems." *Physical Review B* 37 (12). American Physical Society: 6991–7000. doi:10.1103/PhysRevB.37.6991.

Tersoff, J. 1988b. "Empirical Interatomic Potential for Silicon with Improved Elastic Properties." *Physical Review B* 38 (14). American Physical Society: 9902–9905. doi:10.1103/PhysRevB.38.9902.

Thevenot, Vincent, Lionel Arnaud, Gilles Dessein, and Gilles Cazenave-Larroche. 2006. "Integration of Dynamic Behaviour Variations in the Stability Lobes Method: 3D Lobes Construction and Application to Thin-Walled Structure Milling." *The International Journal of Advanced Manufacturing Technology* 27 (7): 638–644. doi:10.1007/s00170-004-2241-1.

Vastola, G, G Zhang, Q X Pei, and Y.-W. Zhang. 2016. "Controlling of Residual Stress in Additive Manufacturing of Ti6Al4V by Finite Element Modeling." *Additive Manufacturing* 12: 231–239. doi:10.1016/j.addma.2016.05.010.

Venkatachalam, Siva, Omar Fergani, Xiaoping Li, Jiang Guo Yang, Kuo-Ning Chiang, and Steven Y. Liang. 2015. "Microstructure Effects on Cutting Forces and Flow Stress in Ultra-Precision Machining of Polycrystalline Brittle Materials." *Journal of Manufacturing Science and Engineering* 137 (2): 021020-1–8. doi:10.1115/1.4029648.

Wang, Bing, Zhanqiang Liu, Qinghua Song, Yi Wan, and Xiaoping Ren. 2018. "A Modified Johnson–Cook Constitutive Model and Its Application to High Speed Machining of 7050-T7451 Aluminum Alloy." *Journal of Manufacturing Science and Engineering* 141 (1). doi:10.1115/1.4041915.

Wang, Panding, Hongshuai Lei, Xiaolei Zhu, Haosen Chen, and Daining Fang. 2019. "Influence of Manufacturing Geometric Defects on the Mechanical Properties of AlSi10Mg Alloy Fabricated by Selective Laser Melting." *Journal of Alloys and Compounds* 789: 852–859. doi:10.1016/j.jallcom.2019.03.135.

Yadroitsev, I, P Krakhmalev, and I Yadroitsava. 2014. "Selective Laser Melting of Ti6Al4V Alloy for Biomedical Applications: Temperature Monitoring and Microstructural Evolution." *Journal of Alloys and Compounds* 583: 404–409. doi:10.1016/j.jallcom.2013.08.183.

Yameogo, D, B Haddag, H Makich, and M Nouari. 2017. "Prediction of the Cutting Forces and Chip Morphology When Machining the Ti6Al4V Alloy Using a Microstructural Coupled Model." *Procedia CIRP* 58: 335–340. doi:10.1016/j.procir.2017.03.233.

Yang, Qingcheng, Pu Zhang, Lin Cheng, Zheng Min, Minking Chyu, and Albert C To. 2016. "Finite Element Modeling and Validation of Thermomechanical Behavior of Ti-6Al-4V in Directed Energy Deposition Additive Manufacturing." *Additive Manufacturing* 12: 169–177. doi:10.1016/j.addma.2016.06.012.

Yao, Yang, Hongtao Zhu, Chuanzhen Huang, Jun Wang, Pu Zhang, Peng Yao, and Xiaodan Wang. 2020. "Determination of the Minimum Chip Thickness and the Effect of the Plowing Depth on the Residual Stress Field in Micro-Cutting of 18 Ni Maraging Steel." *The International Journal of Advanced Manufacturing Technology* 106 (1): 345–355. doi:10.1007/s00170-019-04439-x.

Zhang, Jiayi, Yan Jin Lee, and Hao Wang. 2021. "Surface Texture Transformation in Micro-Cutting of AA6061-T6 with the Rehbinder Effect." *International Journal of Precision Engineering and Manufacturing – Green Technology* 8 (4). Springer: 1151–1162. doi:10.1007/s40684-020-00260-0.

Zhang, Jing, Yi Zhang, Weng Hoh Lee, Linmin Wu, Hyun-Hee Choi, and Yeon-Gil Jung. 2018. "A Multi-Scale Multi-Physics Modeling Framework of Laser Powder Bed Fusion Additive Manufacturing Process." *Metal Powder Report* 73 (3): 151–157. doi:10.1016/j.mprp.2018.01.003.

Zhao, Yanhua, Jie Sun, Jianfeng Li, Yuqin Yan, and Ping Wang. 2017. "A Comparative Study on Johnson-Cook and Modified Johnson-Cook Constitutive Material Model to Predict the Dynamic Behavior Laser Additive Manufacturing FeCr Alloy." *Journal of Alloys and Compounds* 723: 179–187. doi:10.1016/j.jallcom.2017.06.251.

Zheng, Zhongpeng, Yan Jin Lee, Jiayi Zhang, Xin Jin, and Hao Wang. 2021. "Ultra-Precision Micro-Cutting of Maraging Steel 3J33C under the Influence of a Surface-Active Medium." *Journal of Materials Processing Technology* 292 (June). Elsevier: 117054. doi:10.1016/j.jmatprotec.2021.117054.

Zheng, Zhongpeng, Chenbing Ni, Yun Yang, Yuchao Bai, and Xin Jin. 2021. "Numerical Analysis of Serrated Chip Formation Mechanism with Johnson-Cook Parameters in Micro-Cutting of Ti6Al4V." *Metals* 11 (1). doi:10.3390/met11010102.

9 Hybrid Additive Manufacturing

9.1 SUCCESSFUL INTEGRATIONS

There are three types of hybrid integrations in additive manufacturing (Pragana et al. 2021; Merklein et al. 2016): (1) Hybrid additive/subtractive manufacturing; (2) multi-energy field coupled hybrid additive manufacturing; and (3) hybrid additive/semi-forming manufacturing. Hybrid additive/subtractive manufacturing is used to improve the surface quality and dimensional accuracy of metal parts with intricate geometries and desirable surface finishing. Multi-energy field coupled hybrid additive manufacturing integrates multiple energy sources to simultaneously improve the part quality, such as reducing porosity, refining microstructure, and improving dimensional stability. Hybrid additive/semi-forming manufacturing adds part features by additive manufacturing onto semi-finished parts fabricated by traditional processes with the intention to improve production efficiency and lower costs.

9.1.1 Hybrid Additive/Subtractive Manufacturing

As shown in the early chapters of this book, metal additive manufacturing technology is a promising way to fabricate functional metal components with complex geometries both externally and internally. However, the complex metallurgy involved during the layer-by-layer fabrication results in inferior surface quality that is usually far from desirable in precision fields and requires post-processing. Therefore, the combination of additive and subtractive manufacturing technologies is the solution for rapid fabrication of functional metal parts with high surface quality and dimensional accuracies. This manufacturing process flow is termed as hybrid additive/subtractive manufacturing. Thus far, the two most employed additive manufacturing techniques for hybrid manufacturing are laser powder bed fusion (LPBF) and directed energy deposition (DED), coupled with CNC milling as the post-process method.

These integrations can be categorized into offline and online processes. Offline hybrid processes are where post-processing is performed separately and after the additive manufacturing procedure. The post-processing can be effectively viewed as a standard machining process of a pre-formed workpiece, which is the perspective of most research works on the post-processing for additive manufacturing topic.

DOI: 10.1201/9781003272601-9

However, it is important to view the processes to fabricate a component collectively where the hybridization of both additive and subtractive technologies is essential.

The near-net shapes of components are first formed by an additive manufacturing process, such as gas metal arc welding (GMAW) deposition for molds and dies shown in Figure 9.1. Post-process milling then subsequently processes the surface to obtain the desirable finishing and dimensional accuracies. Milling will substantially reduce the surface roughness of the additively manufactured part from > 20 μm Ra to < 1 μm Ra as reported for A131 steel (Bai, Chaudhari, and Wang 2020) fabricated by DED (Figure 9.2) and from 13 μm Ra to 0.6 μm Ra for Ti6Al4V (Ni et al. 2020) manufactured by SLM.

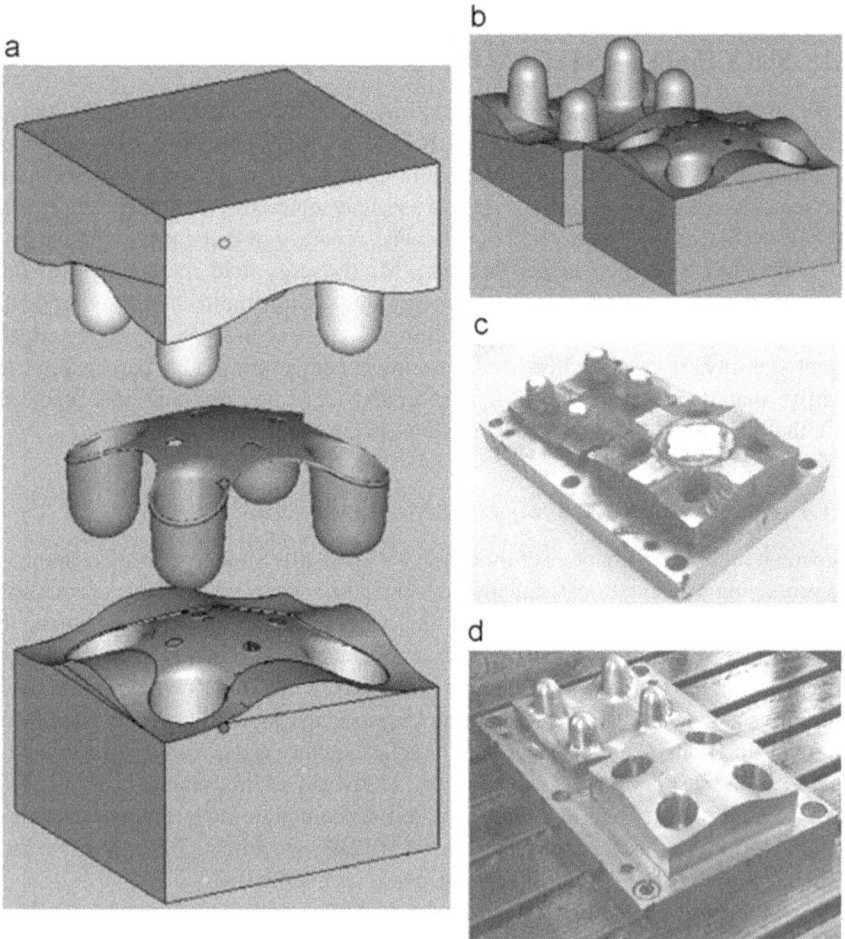

FIGURE 9.1 Injection mold and die fabrication using hybrid additive/subtractive manufacturing. (a) Massager and its dies, (b) both dies arranged for hybrid additive/subtractive manufacturing, (c) additively manufactured near-net shape of the die pair, and (d) finished die pair. Reprinted with permission from Elsevier (Karunakaran et al. 2010).

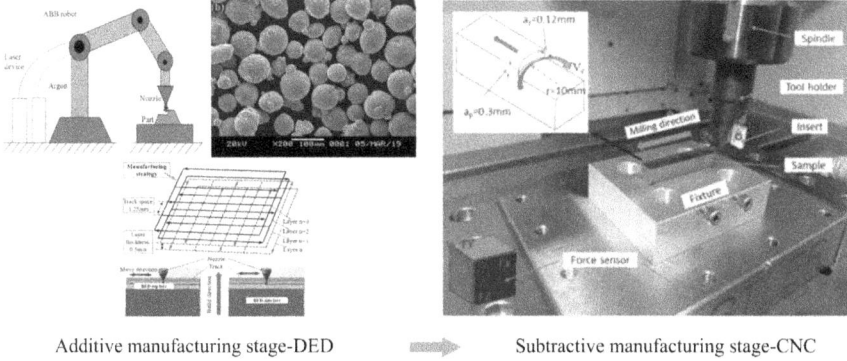

Additive manufacturing stage-DED   Subtractive manufacturing stage-CNC

FIGURE 9.2 Manufacturing A131 steel using offline hybrid additive/subtractive manufacturing. Reprinted with permission from Elsevier (Bai, Chaudhari, and Wang 2020).

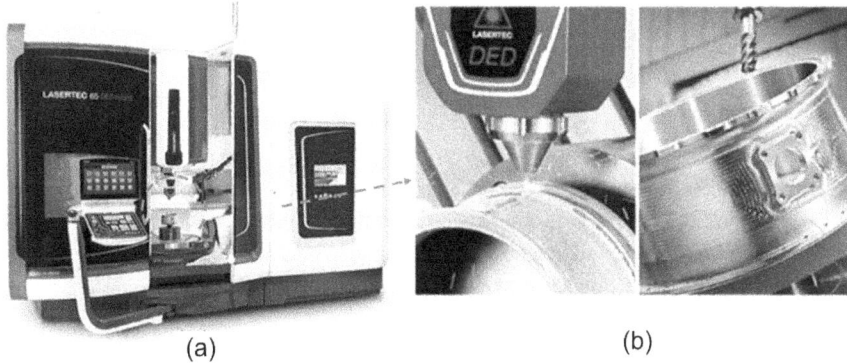

(a)   (b)

FIGURE 9.3 Laser deposition setup installed onto a CNC milling machine for offline hybrid additive/subtractive manufacturing: (a) DMG Mori Lasertec 65 hybrid manufacturing machine, (b) laser head with five-axis deposition coaxial nozzle (left) and five-axis milling machine (right). Reprinted with permission from Elsevier (Sefene, Hailu, and Tsegaw 2022).

Offline hybrid additive/subtractive manufacturing also includes systems that integrate both additive and subtractive processes into one machine tool. The caveat for this classification is that the subtractive process is performed after additive manufacturing is complete. Modern CNC machines are highly versatile and can easily integrate additive manufacturing tools into its workspace. An example is given in Figure 9.3 where a powder-flown laser deposition additive manufacturing unit is installed onto a CNC milling machine with relative low-costs for integration (Sefene, Hailu, and Tsegaw 2022).

The process workflow for offline hybrid manufacturing in a machine tool, as shown in Figure 9.3, has a critical flaw where cutting tools may not be able to access all surfaces of the fabricated part. A simplified illustration of this problem is given in Figure 9.4(a) where imminent collision may occur between the cutting tool and

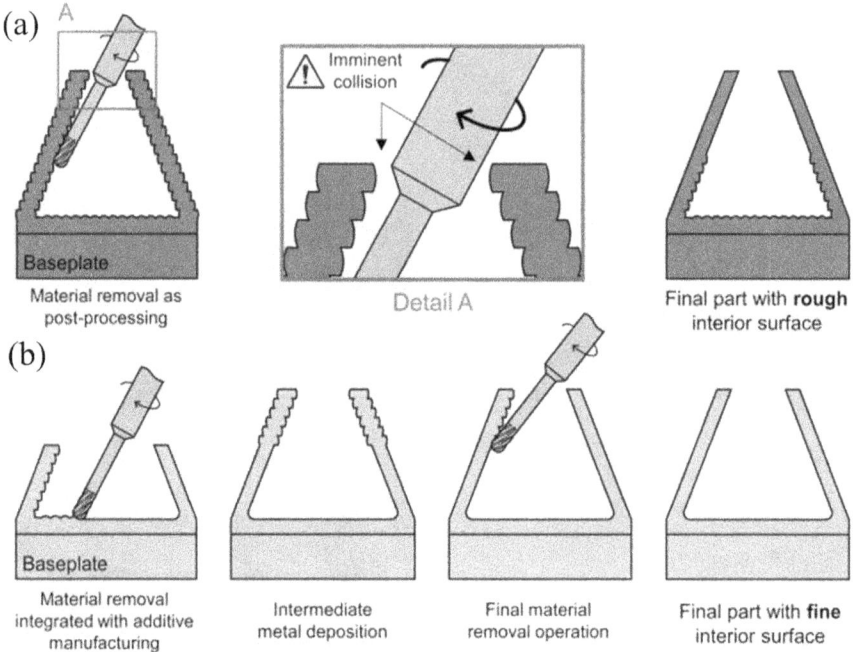

FIGURE 9.4 Illustration of the challenge in post-processing of internal surfaces: (a) Demonstrating imminent collision between cutting tool and workpiece in offline hybrid manufacturing; and (b) the solution provided by online hybrid manufacturing. Reprinted with permission from Elsevier (Pragana et al. 2021).

the workpiece in the attempt to machine internal surfaces. As such, internal surfaces may only be partially machined and the desirable surface quality for internal surfaces will be unsatisfactory. This is where online hybrid additive/subtractive manufacturing (Figure 9.4(b)) presents the capabilities to overcome the issue.

Online hybrid additive/subtractive manufacturing integrates subtractive manufacturing into the real-time additive manufacturing process for in-situ processing of a wide variety of complex surfaces that would be inaccessible upon completion of additive manufacturing. As displayed in Figure 9.4(b), subtractive machining is performed on the internal surfaces of a part that is partially constructed by additively manufacturing. Collision can then be avoided, and the desirable internal surface quality can be achieved. This revolutionizes manufacturing capabilities for complex structures due to the increase in degrees of freedom for hard-to-reach surfaces.

Online hybrid additive/subtractive manufacturing first came in the form of powder-blown (DED)/milling. One of the earlier innovations of this technology was the integration of a laser-based DED process with conventional a-axis milling (Klocke, Wirtz, and Meiners 1996). Powder feed additive manufacturing systems were also configured with a 5-axis CNC milling machine with the capabilities of manufacturing stainless steel, bronze, titanium, aluminum, and INVAR alloy (Fessler et al. 1996).

FIGURE 9.5   Hybrid additive/subtractive manufacturing machine (Sodick OPM250L) with LPBF and CNC milling capabilities. Reprinted with permission from Elsevier (Du, Bai, and Zhang 2016).

Moving into the 21st century, online hybrid systems are integrating laser cladding technology (Kerschbaumer and Ernst 2004), gas metal arc welding (Song et al. 2005), and plasma deposition (Xiong, Zhang, and Wang 2009) with milling processes. Modern technology digitalizes the subtractive process with additive manufacturing to further enhance manufacturing quality (Manogharan et al. 2015).

Online hybrid additive/subtractive manufacturing became commercial with hybrid LPBF/CNC milling machines, such as the OPM250L by Sodick (Figure 9.5), the Lumex Avance-25 and Lumex Avance-60 by Matsuura, INTEGREX i-400 AM by Mazak Corporation, LASERTEC 3000 DED hybrid by DMG MORI company, HYBRID HSTM 1000 by Hamuel Reichenbacher Ltd, and CybaCAST Hybrid by Cybaman Technologies, among others. These machines are claimed to be able to achieve final surface roughness values of 0.19 μm Ra and dimensional accuracy of within ± 1/100 mm, which are adequate for some mold and die applications (Ahn 2011). However, most of these hybrid machines are used for manufacturing small metal parts and can be costly due to the efforts taken to meet the high demand for mechanical stability in the manufacturing environment.

The process workflow of an online hybrid additive/subtractive manufacturing machine for the manufacturing of a component with internal channels is illustrated in Figure 9.6. A layer of material is first constructed like a normal additive manufacturing

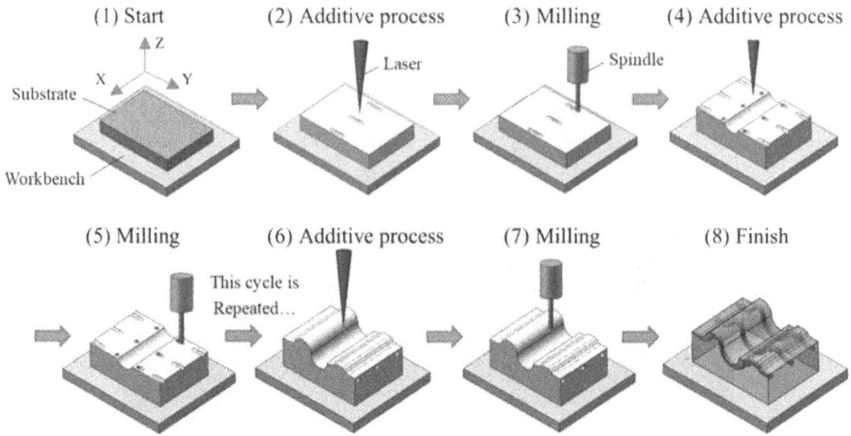

FIGURE 9.6 Schematic of the online hybrid additive/subtractive process workflow in manufacturing of a part with internal channels. Reprinted with permission from (Du, Bai, and Zhang 2016).

process. Figure 9.6 illustrates the LPBF process with a laser beam scanning a bed of powder particles to melt and solidify the metal. The build platform is lowered after each layer has been scanned, followed by the spreading of a new layer of powder particles for a repeat of this additive process. After a designated number of layers where the internal surfaces of some features are formed, the milling tool is activated to process the surface of the partially built part. Upon completion of the surface processing, the milling tool is retracted, and the additive manufacturing process resumes to deposit more layers of material. This process repeats until the full component is constructed. The swap between additive and subtractive processes will be dictated by the accessibility of the cutting tool, as illustrated in Figure 9.4. This requires careful consideration as tool swapping procedures in machine tools can significantly delay the time to complete the part fabrication. Another upside of the online system is the ability to ensure dimensional and geometrical accuracy during manufacturing where errors can be corrected in real-time before the final part is complete.

While Figure 9.6 illustrates an LPBF process, DED is the most common process used in hybrid additive/subtractive manufacturing. DED has more flexibility in the additive manufacturing process showing the feasibility to be mounted on robot arms with multiple degrees of motion (Zhang et al. 2019), and it can be easily integrated with machining machines without considering the high stability and high manufacturing resolution required in the LPBF process. DED is also able to fabricate large-scale structural parts despite the downside of significantly poor quality on as-built surfaces (measuring roughly 200 µm Rz for plasma-based DED). The coupling with subtractive machining enables superior surface finishing down to 2.32 µm Rz alongside an improvement in dimensional accuracy to ±0.05% (Xiong, Zhang, and Wang 2009). Part sizes will typically range up to 210 mm in height and 190 mm in diameter when employing a robot equipped with wire arc additive/subtractive manufacturing (Zhang et al. 2019). There are also unique cases of constructing thin-walled aluminum structures measuring ~3 m in length with hybrid wire arc additive/sub-tractive manufacturing (Ma et al. 2019).

9.1.2 MULTI-ENERGY FIELD COUPLED HYBRID ADDITIVE MANUFACTURING

Energy fields in additive manufacturing are essentially the thermal and mechanical in nature. Multi-energy would indicate the employment of more than one type of thermal and/or mechanical energy in the additive manufacturing process. This will exclude certain combinations such as laser-based additive manufacturing with laser-based eroding/remelting, which employs the same heat source for different processing (Yasa, Kruth, and Deckers 2011; Yasa, Deckers, and Kruth 2011).

Multi-thermal energy fields are designed to improve process stability during additive manufacturing for better quality products. For instance, the combination of laser and plasma in the additive manufacturing process will utilize the plasma arc compression effect induced by the laser beam for higher shape precision and better surface state of the deposited layer (Zhang et al. 2006). This is because the diameter of the plasma arc beam becomes smaller, and the plasma arc beam becomes more stable under the compression effect. Additionally, in terms of the nickel-based superalloys, the increase in melt pool depths, the reduction in melt pool widths, and the more uniform distribution of elements in the microstructure led to improvements in tensile strength and elongation (Zhang, Qian, and Wang 2006). Other benefits of the laser and plasma combination include improvements to energy density for the production of low-porosity and well-bonded layers of material (Qian et al. 2008).

Low-power pulsed-lasers can also be coupled with metal inert gas (MIG) arc welding in the additive manufacturing process (an improvement based on the work of Zhang et al. 2006), otherwise referred to as hybrid laser-arc additive manufacturing. This also benefits from the plasma arc compression effect induced by the laser beam to reduce arc diameter and stabilize the arc. Here, through the continuous action of pulsed laser (without need for thermal action, reducing the energy consumption), the liquid metal formed during the welding process will not flow down the previously solidified layer. Therefore, more uniformly thin walls can be fabricated with a 50% reduction in fluctuations of width and height and 15% to 91.12% improvements in material utilization (Zhang et al. 2018). Refined grains and smaller heat-affected regions caused by the higher cooling rate in a small melt pool result in the increase in micro-hardness and tensile strength, which are also advantages of the laser-arc technique (Liu et al. 2020). Wu et al. (2020) also reported an increase in tensile strength in Al-Cu alloy fabricated by this hybrid process. Reductions in distortion and residual stresses can be achieved with laser-cold metal transfer (a new welding technology without spatters) additive manufacturing with the appropriate scanning strategy (Li et al. 2020; Li et al. 2021).

Coupling of thermal and mechanical energies in conventional manufacturing resembles forging where the principle is to induce hardening of the workpiece. Hybridization of metal inert gas welding-based additive manufacturing with in-situ rolling results in grain refinement, avoidance of hot cracking, and improvements in mechanical properties comparable to forged counterparts (Zhang et al. 2013). Additionally, overall energy consumption can be reduced to 35% when compared with conventional forging, and the energy efficiency of the deforming process is a 150-fold higher than the forging process (Zhang et al. 2021). Hybrid additive/rolling manufacturing was also reported to avoid residual stress, distortion and the formation of large-sized grains (Colegrove et al. 2013). Figure 9.7(a) illustrates the hybrid process where rolling is performed on each layer of material that is newly deposited by wire and arc additive manufacturing (WAAM). Such a process helps to lower the surface roughness and promotes dynamic recrystallization with the heavy plastic deformation. Additionally, the rolling operation can eliminate anisotropic mechanical properties introduced during WAAM and also increase the ultimate tensile strength and yield strength of the metal (Ti6Al4V in this case) (Colegrove et al. 2017). A combination of additive and hot rolling/hot forging can further reduce the porosity of 316L stainless steel and Ti6Al4V alloy, and refine the microstructure of 316L stainless steel (Sokolov et al. 2020; Duarte et al. 2020).

Another variation of coupling mechanical and thermal energies is with shot peening in the additive manufacturing process. Figure 9.8 illustrates the different types of shot peening processes that can be integrated with additive manufacturing, which includes ball shot peening (Figure 9.8(a)), ultrasonic peening (Figure 9.8(b)), and laser shot peening (Figure 9.8(c)). Shot peening differs from surface rolling in that the substantially lower loads are applied to the workpiece. Ball shot peening involves the repeated impact of small hard balls onto the surface to induce plastic deformation and compressive stresses. This operation releases residual tensile stresses and reduces distortion caused by the high-temperature gradient during additive manufacturing (Fritz and Lee 1993). Micro-strain and grain refinement from shot peening also result in the increase in micro-hardness, compressive yield stress, and wear resistance of direct metal laser sintered 17-4 stainless steel (AlMangour and Yang 2016). Improved fatigue resistance and fatigue limits can also be obtained on AlSi10Mg coupled with both SLM (Uzan et al. 2018) and DMLS (Book and Sangid 2016).

Ball shot peening of additively manufactured metals has attracted various corporations such as General Electric Co., United Technologies Corp., Bae Systems, Lawrence

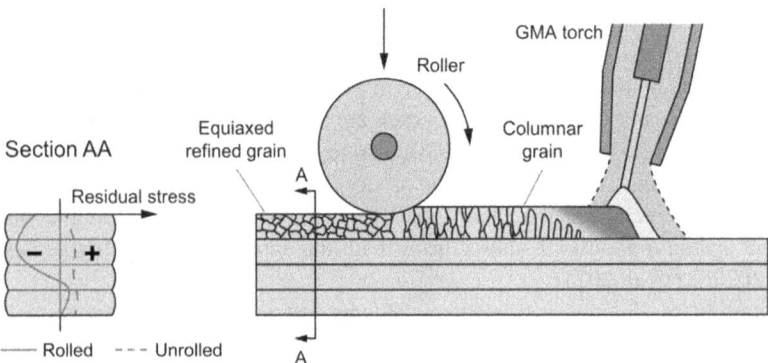

FIGURE 9.7 Schematic of the hybridization of wire arc additive manufacturing (WAAM) with surface rolling and the impact on metallurgical properties (i.e., reduction of grain size and its effect on residual stresses). Reprinted with permission from Elsevier (Pragana et al. 2021).

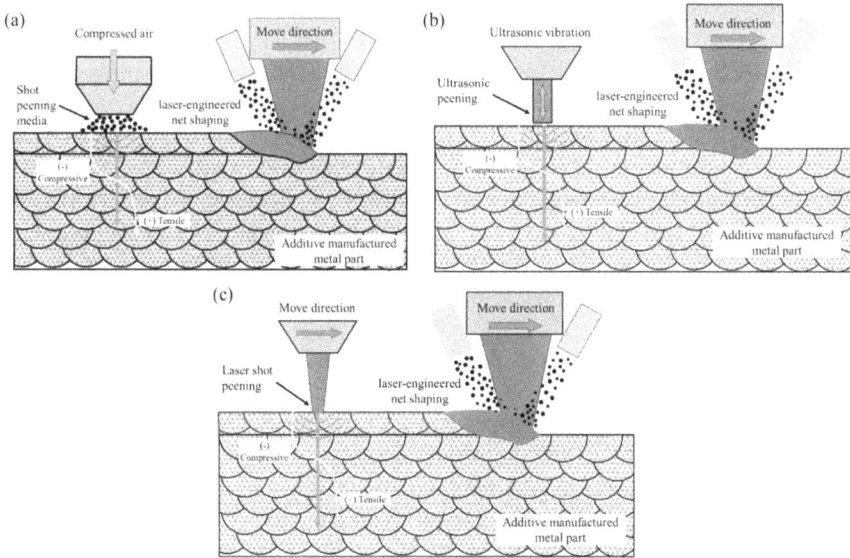

FIGURE 9.8 Schematic of hybrid metal additive manufacturing process with (a) ball shot peening, (b) ultrasonic peening, and (c) laser shock peening.

Livermore National Security, École Polytechnique Fédérale de Lausanne in Lausanne, and others to find further innovation to improve the process leading to the application of multiple patents (El-Wardany, Lynch, Viens, and Grelotti 2014; Kalentics, Logé, and Boillat 2017). An example of a tweak to the process is the use of high pressure (Bamberg, Hess, Hessert, and Satzger 2012).

Ultrasonic peening uses an electro-mechanical transducer to apply ultrasonic mechanical impact energy to a workpiece. It is a low-cost and easy-to-integrate variation of shot peening with additive manufacturing processes. The application of ultrasonic peening resulted in grain refinement and reduced porosity in wire arc additive manufacturing of Al alloy (using a ER 4043 aluminum wire) (Tian et al. 2021). Reduction in residual stresses and solidification cracking can also be achieved with the combination of ultrasonic peening and DED (Wang et al. 2022).

Laser shock peening (LSP) use shock waves emitted from rapidly expanding plasma to induce plastic deformation on a workpiece, which serves as a type of mechanical action. Like the other shot peening variants, LSP will result in the increase in surface hardness, transition of surface stresses from tensile to compressive, and improvement of both tensile strength and elongation along the horizontal and vertical directions in the case of Ti6Al4V (Lu et al. 2020).

9.1.3 Hybrid Additive/Semi-Forming Manufacturing

Although additive manufacturing has various advantages especially in the fabrication of complex-geometry structures, it is not suitable for high-volume manufacturing considering the time and energy costs. Therefore, a better solution would be to combine additive manufacturing with a conventional process to manufacture metal parts that comprise both simple and complex structures. The simple segment of the part can be fabricated by a conventional process such as casting, forging, milling, and sheet metal processing while the complex structure of the part will be completed by the additive manufacturing process. This is defined as hybrid additive/semi-forming manufacturing, which will take advantage of two or more processes to lower manufacturing costs and integrate additive manufacturing in current production chains (Merklein et al. 2016). This hybrid process also allows for the joining of similar and/or dissimilar materials for the final part, for example, the simple segment constructed out of material A while material A or B is constructed for the complex segment.

Figure 9.9 shows an example of using additive/semi-forming manufacturing to fabricate a Ti6Al4V alloy part (Merklein et al. 2016). The lower segment was first manufactured using a deep drawing process while the upper segment was constructed on the lower segment by LPBF. Another example is the fabrication of a mold with internal cooling channels using the hybrid additive/semi-forming method (Tan et al. 2020). The mold comprises of a lower and upper segment. The lower segment was made of stainless-steel, which was processed by milling and drilling while the upper segment was made of maraging steel with a series of curved internal cooling channels fabricated by LPBF onto the lower stainless-steel segment.

The sequence of conventionally manufacturing the base of a part followed by additive manufacturing to add on complex features is a low-cost and efficient manufacturing workflow. However, the interfacial bond between the two materials is critical and will affect the final part quality. Yet, intermetallic bonds between different materials have thus far been promising. For instance, maraging steel and casted CrMn steel showed excellent bonding due to the Marangoni convection that facilitated the embedment of the two materials at the interface (Bai et al. 2021). Maraging steel by LPBF also showed robust interfacial bonds with hot work H13 tool steel (Shakerin

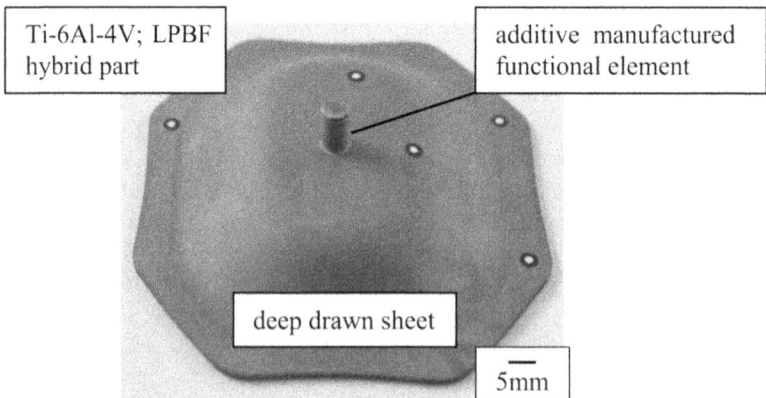

FIGURE 9.9 Ti6Al4V part manufactured by a combination of deep drawing and LPBF. Reprinted with permission from Elsevier (Merklein et al. 2016).

et al. 2019). Aside from interfacial bonding, other critical issues such as the mismatch in residual stresses and distortion problems must also be accounted for.

9.2 MULTI-MATERIAL, STRUCTURAL, AND FUNCTIONAL MANUFACTURING

Metal additive manufacturing has the capabilities for multi-material and multi-function design and manufacturing, which have growing demands in the aerospace, automotive, and energy production industries. The aspects of multi-material, multi-structural, and multi-function manufacturing may not necessarily be independent of each other. For instance, the material-structure-performance integrated additive manufacturing (MSPI-AM) concept was developed to combine multi-material and multi-structure manufacturing to achieve high-performance multi-function components (Gu et al. 2021). Nonetheless, each category will be individually discussed for clarity on their definitions, which will form the basis for further development of ideal part designs and manufacturing.

9.2.1 MULTI-MATERIAL MANUFACTURING

Multi-material additive manufacturing is the additive manufacturing of two or more materials in one part to enhance part performance, which can be in the form of mechanical properties, function and design freedom, traceability, and safety (Gibson et al. 2021). Multi-materials can be further categorized into bimetal (Bai et al. 2020), gradient (Ghanavati and Naffakh-Moosavy 2021), composite materials (Shi and Wang 2020) and their spatial combination (Tan, Chew, et al. 2021). In all cases of multi-material manufacturing, there are two main areas of focus: (1) the ability to control the position of the different materials, and (2) the metallurgical bond between dissimilar materials.

The LPBF and DED additive manufacturing techniques have been used to fabricate metallic multi-materials, as shown in Figure 9.10. LPBF is beneficial when

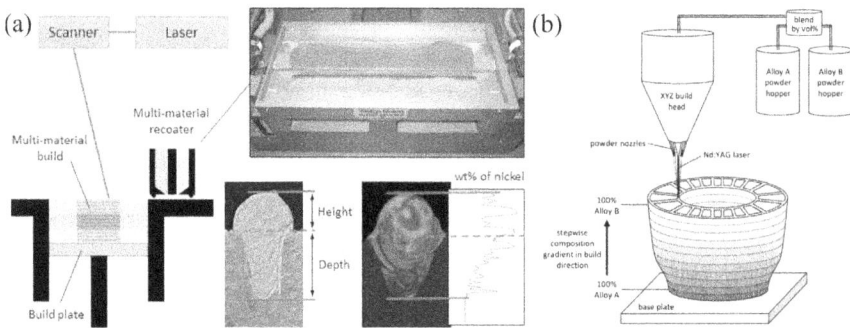

FIGURE 9.10 Multi-material manufacturing using (a) LPBF. Reprinted with permission from Elsevier (Wits and Amsterdam 2021); (b) DED. Reprinted with permission from Taylor & Francis (Reichardt et al. 2021).

it comes to changing materials along the build direction with each new layer being different powder materials. This bimetal sandwiching structure can be constructed by LPBF with high resolutions of alternating materials (Tey et al. 2020; Bai et al. 2021; Bai et al. 2020). However, this would also mean that the standard LPBF process is incapable of producing dissimilar materials along the horizontal direction. Special adaptations of the LPBF process will be required such as the unique selective powder deposition re-coater developed by Aerosint. This allows selectively depositing different materials of powder within a layer of the powder bed to enable multi-material manufacturing along both the horizontal and build directions. However, the efficiency of coating dissimilar powders still needs to be improved. Wang et al. (2022) reported the recent progress in multi-material manufacturing using LPBF systematically. They found that although lots of multi-material LPBF devices are successfully developed to print multi-material parts with complex geometries, low efficiency and powder cross-contamination remain critical problems to be solved. In addition, an efficient and high-precision powder delivery system for the flexible combination and precise space distribution of different materials is also lacking for industrial applications.

The DED process is much more suitable for multi-material manufacturing where it simply switches materials accordingly as the deposition position changes, which makes it a highly attractive process for large-scale multi-metal manufacturing (Feenstra et al. 2021). DED involves two raw material forms i.e., powders and wires, and a multi-material part can be fabricated with a combination of different metal wires (Hauser et al. 2021) or metal powders (Sahasrabudhe et al. 2015). The highly versatile DED process allows for rapid changes in material usage, which can rapidly fabricate new alloys, perform in-situ alloying (Klein, Birgmann, and Schnall 2020) and produce functionally graded materials (Feenstra, Molotnikov, and Birbilis 2020). Some of the alloys produced by DED include titanium alloys (Liu et al. 2018), nickel alloys (Savitha et al. 2020), copper (Zhang et al. 2020), steels (Kim et al. 2019) and aluminum alloys (Li et al. 2020). Despite the highly configurable capabilities of DED, it is inferior in terms of its construction resolution where it is unable to fabricate parts with fine structures due to the size of the powders, which are typically larger than those used in LPBF.

Although the ideas of multi-metal manufacturing are very attractive, the processes still face several limitations such as the low efficiency and precision of dissimilar powder supply device in the additive manufacturing equipment, limited variety of materials available, and ineffective bonding of dissimilar materials. Cross-contamination of different powder materials and recycling are also challenges in multi-material fabrication that greatly increase manufacturing costs.

9.2.2 Multi-Structural Manufacturing

Multi-structural additive manufacturing is characterized by the fabrication of complex shapes across multiple length scales (Thompson et al. 2016). Additive manufacturing has granted great freedom for revolutionary designs of parts that boast of material savings and light weight structures while still maintaining operational capabilities, such as the odd-shaped structure in Figure 9.11(a). Lattice structures are

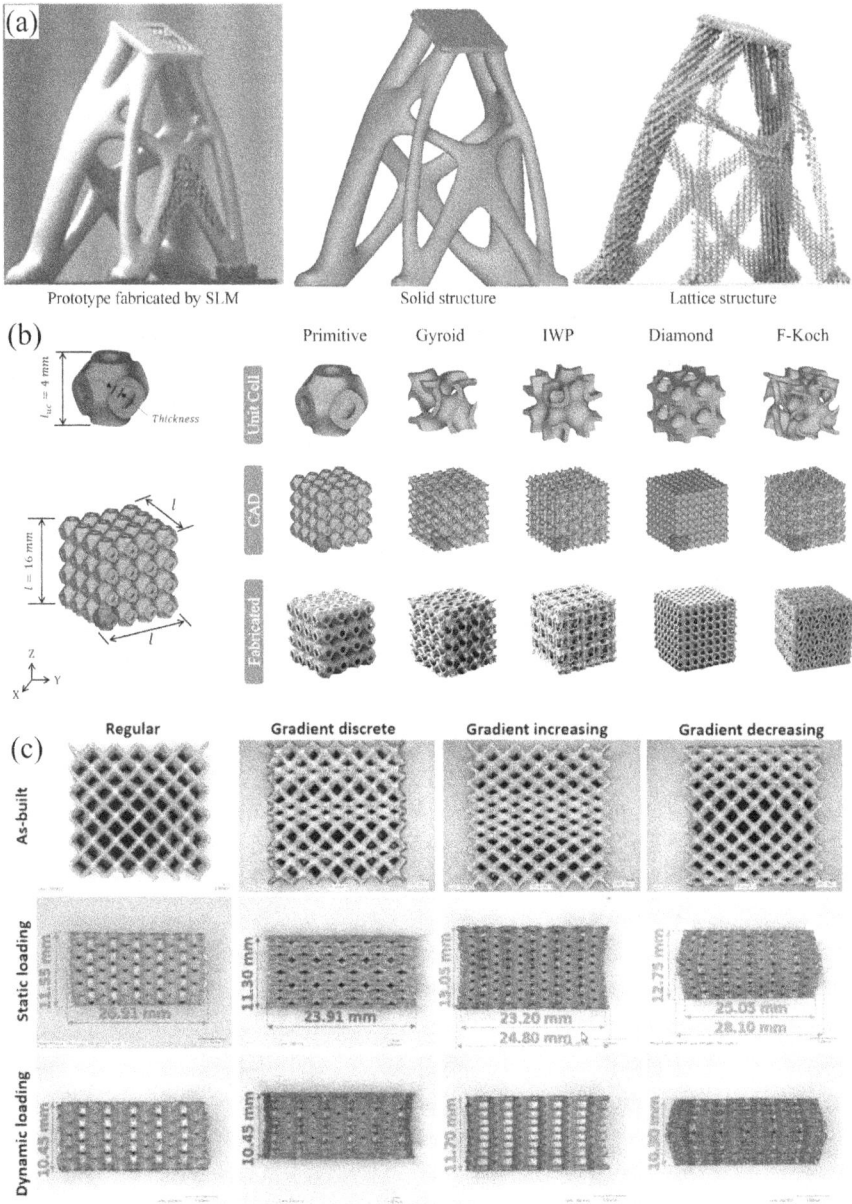

FIGURE 9.11 Various structures manufactured by metal additive manufacturing: (a) topology structure. Reprinted with permission from Elsevier (Zhu et al. 2021); (b) types of lattice structures. Reprinted with permission from Elsevier (AlMahri et al. 2021); (c) different gradient lattice structures. Reprinted with permission from MDPI (Sienkiewicz et al. 2020).

also one of the famous examples of complex shapes that can only be manufactured by additive manufacturing (Tilton et al. 2021). Different types of lattice structures, as shown in Figure 9.11(b), can be manufactured to give different mechanical properties (e.g., plateau stresses and specific energy absorption during high strain rate loading) (AlMahri et al. 2021) and deformation characteristics, as shown in Figure 9.11(c) (Sienkiewicz et al. 2020). These lattice structures can then be incorporated into the bodies of larger components (e.g., struts and beams), making multi-scale designs feasible, as illustrated in Figure 9.11(a).

Bioinspired structures are also attractive in additive manufacturing (Yang et al. 2018) where natural surface properties (e.g., mechanical, hydrodynamic, optical, and electrical) are replicated onto engineered surfaces. Some unique examples include layered shell textures of Chrysomallon squamiferum colloquially known as scaly-foot gastropod (Wang et al. 2022), bioinspired functionally graded lattice structures for enhanced biomechanical performance (Tan, Zou, et al. 2021), reticulated shell structures resembling the cross-link structure of spider silk (Wang et al. 2019), and sponge-like components resembling deep-sea glass sponges (Fernandes et al. 2021), and others.

The ability to manufacture these complex structures is subject to the constraints of structural forming limits in the additive manufacturing process. For instance, DED is typically unsuitable for such fabrication due to low forming resolution and LPBF is unable to fabricate large-scale parts. A combination of DED and LPBF may be feasible to build large-scale structures with fine features. Essentially, DED will be used to fabricate the base structure while the functional structures with intricate features will be deposited on top of the base structures by LPBF. The connection between these processes has to be resolved first. An innovative design of the existing LPBF equipment or innovative precision printing method is required to allow greater flexibility while maintaining high precision. However, the LPBF process is also needed to be further improved to extend the forming limits of fine structures.

9.2.3 MULTI-FUNCTIONAL MANUFACTURING

Multi-functional additive manufacturing is the result of successful multi-material manufacturing and multi-structural manufacturing. Multi-functional additive manufacturing is the fabrication of a device using a single additive manufacturing process that can serve multiple purposes by integrating multiple materials and/or multiple structures. For example, a multi-function sandwich panel that is light, able to sustain aerodynamic loading conditions and bird strikes. Moreover, anti-ice integrated functions can be manufactured by additive manufacturing of a panel with lattice structure and specially designed outer skins (Bici et al. 2018).

It should be noted that optimal multi-function integration often requires the integration of other materials aside from metals. One example is the additive manufacturing of a "smart cap" to seal and detect food deterioration with a LC-resonant circuit embedded into the cap during the additive process, as shown in Figure 9.12 (Wu et al. 2015).

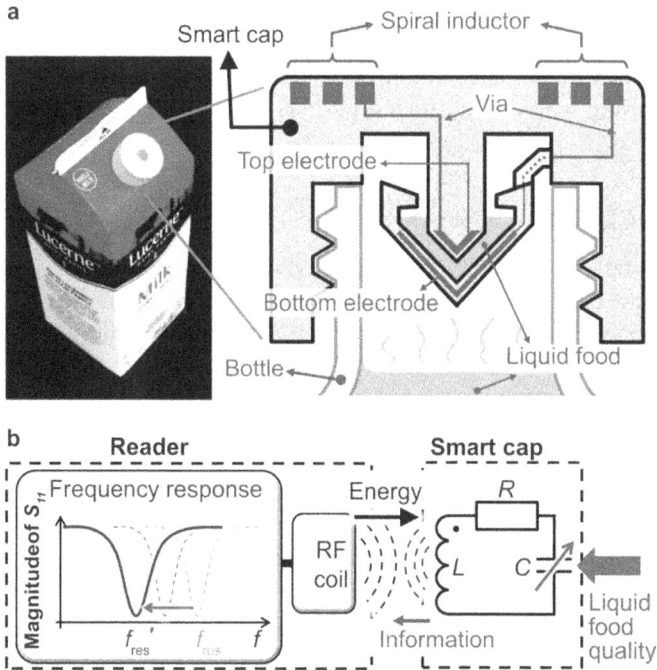

FIGURE 9.12 "Smart cap" for rapid detection of liquid food quality featuring wireless readout: (a) Smart cap with a half-gallon milk package and the cross-sectional diagram; (b) sensing principle and the circuit diagram. Reprinted with permission from Springer Nature (Wu et al. 2015).

Another example is the additive manufacturing of fluidic devices for electrochemical detection where electrode materials such as carbon, platinum, gold, and silver can be easily integrated into the device for neurotransmitter detection, NO detection, and oxygen tension measurements in blood (Erkal et al. 2014). Although the electrode materials were not fabricated by additive manufacturing technique, this study still successfully proved the feasibility of multi-functional manufacturing using additive manufacturing. In the future, with the improvement of technology and the development of new materials for additive manufacturing technology, it can be expected that above multi-functional components will be fabricated by additive manufacturing alone, where the successful manufacturing of assembly-free devices using additive manufacturing is a typical example. Moreover, Yee et al. (2019) also reported the feasibility of additive manufacturing multifunctional components. They presented a facile process for fabricating 3D-architected metal oxides via the use of an aqueous metal-ion-containing photoresin using stereolithography (SLA) additive manufacturing process. An emergent electromechanical response was observed when conducting a situ compression experiment on the printed nano-architected zinc oxide (ZnO) architectures. This research provides a pathway to create arbitrarily shaped 3D metal oxides with multiple functions such as piezoelectricity, superconductivity and semi-conductivity to be used in micro/nano-system device.

Hybrid manufacturing is a promising technology that aims to further revolutionize the manufacturing industry after significant breakthroughs in the additive manufacturing sector. The developments of this technology will further widen the perspective of manufacturing with the integration of more materials and functions within one component.

9.3 CONCLUDING REMARKS

Although additive manufacturing is good at fabricating components with complex geometries, the poor surface quality, low dimension accuracy, and low efficiency in fabricating regular structures continue to challenge the adoption of AM for many other engineering applications. This implies that despite the proclaimed advantage of additive manufacturing processes being able to produce near-net-shape products, the technology itself is not capable of operating independently. Therefore, hybrid additive manufacturing is the solution to overcome the challenges by combining the advantages of additive manufacturing processes and conventional manufacturing processes. This chapter evaluated different types of hybrid additive manufacturing processes such as hybrid additive/subtractive manufacturing, multi-energy field coupled hybrid additive manufacturing, and hybrid additive/semi-forming manufacturing. Hybrid additive/subtractive manufacturing can obtain complex components with high accuracy. Multi-energy field coupled hybrid additive manufacturing can improve printing stability and produce high-quality components comparable to forged counterparts. Hybrid additive/semi-forming manufacturing was also shown as an effective method to decrease the production costs for large-scale components with regular structures. The discussed benefits of hybridization make hybrid additive manufacturing an attractive and cost-efficient process for high-quality products.

REFERENCES

Ahn, Dong Gyu. 2011. "Applications of Laser Assisted Metal Rapid Tooling Process to Manufacture of Molding & Forming Tools – State of the Art." *International Journal of Precision Engineering and Manufacturing* 12 (5): 925–938. doi:10.1007/s12541-011-0125-5.

AlMahri, Sara, Rafael Santiago, Dong Wook Lee, Henrique Ramos, Haleimah Alabdouli, Mohamed Alteneiji, Zhongwei Guan, Wesley Cantwell, and Marcilio Alves. 2021. "Evaluation of the Dynamic Response of Triply Periodic Minimal Surfaces Subjected to High Strain-Rate Compression." *Additive Manufacturing* 46 (May). Elsevier B.V.: 102220. doi:10.1016/j.addma.2021.102220.

AlMangour, Bandar, and Jenn Ming Yang. 2016. "Improving the Surface Quality and Mechanical Properties by Shot-Peening of 17-4 Stainless Steel Fabricated by Additive Manufacturing." *Materials and Design* 110. Elsevier: 914–924. doi:10.1016/j.matdes.2016.08.037.

Bai, Yuchao, Akshay Chaudhari, and Hao Wang. 2020. "Investigation on the Microstructure and Machinability of ASTM A131 Steel Manufactured by Directed Energy Deposition." *Journal of Materials Processing Technology* 276. Elsevier: 116410. doi:10.1016/j.jmatprotec.2019.116410.

Bai, Yuchao, Jiayi Zhang, Cuiling Zhao, Chaojiang Li, and Hao Wang. 2020. "Dual Interfacial Characterization and Property in Multi-Material Selective Laser Melting of 316L Stainless Steel and C52400 Copper Alloy." *Materials Characterization* 167 (September): 110489. doi:10.1016/j.matchar.2020.110489.

Bai, Yuchao, Cuiling Zhao, Yu Zhang, and Hao Wang. 2021. "Microstructure and Mechanical Properties of Additively Manufactured Multi-Material Component with Maraging Steel on CrMn Steel." *Materials Science and Engineering A* 802. Elsevier: 140630. doi:10.1016/j.msea.2020.140630.

Bamberg, J., Hess, T., Hessert, R., and Satzger, W. 2012. "Method for the Production, Reparation or Replacement of a Component, Including a Compacting Step Using Pressure. Mtu Aero Engines Gmbh, Munich, Germany, Patent No. WO 2012152259 A1."

Bici, Michele, Salvatore Brischetto, Francesca Campana, Carlo Giovanni Ferro, Carlo Seclì, Sara Varetti, Paolo Maggiore, and Andrea Mazza. 2018. "Development of a Multifunctional Panel for Aerospace Use through SLM Additive Manufacturing." *Procedia CIRP* 67: 215–220. doi:10.1016/j.procir.2017.12.202.

Book, Todd A., and Michael D. Sangid. 2016. "Evaluation of Select Surface Processing Techniques for In Situ Application During the Additive Manufacturing Build Process." *Jom* 68 (7): 1780–1792. doi:10.1007/s11837-016-1897-y.

Colegrove, Paul A., Harry E. Coules, Julian Fairman, Filomeno Martina, Tariq Kashoob, Hani Mamash, and Luis D. Cozzolino. 2013. "Microstructure and Residual Stress Improvement in Wire and Arc Additively Manufactured Parts through High-Pressure Rolling." *Journal of Materials Processing Technology* 213 (10). Elsevier B.V.: 1782–1791. doi:10.1016/j.jmatprotec.2013.04.012.

Colegrove, Paul A., Jack Donoghue, Filomeno Martina, Jianglong Gu, Philip Prangnell, and Jan Hönnige. 2017. "Application of Bulk Deformation Methods for Microstructural and Material Property Improvement and Residual Stress and Distortion Control in Additively Manufactured Components." *Scripta Materialia* 135: 111–118. doi:10.1016/j.scriptamat.2016.10.031.

Du, Wei, Qian Bai, and Bi Zhang. 2016. "A Novel Method for Additive/Subtractive Hybrid Manufacturing of Metallic Parts." *Procedia Manufacturing* 5: 1018–1030. doi:10.1016/j.promfg.2016.08.067.

Duarte, Valdemar R., Tiago A. Rodrigues, N. Schell, R. M. Miranda, J. P. Oliveira, and Telmo G. Santos. 2020. "Hot Forging Wire and Arc Additive Manufacturing (HF-WAAM)." *Additive Manufacturing* 35 (February). Elsevier: 101193. doi:10.1016/j.addma.2020.101193.

El-Wardany, T. I., Lynch, M. E., Viens, D. V., and Grelotti, R. A. 2014. "Turbine Disk Fabrication With In Situ Material Property Variation, United Technologies Corporation, Farmington, CT, U.S. Patent No. US20140255198 A1."

Erkal, Jayda L., Asmira Selimovic, Bethany C. Gross, Sarah Y. Lockwood, Eric L. Walton, Stephen McNamara, R. Scott Martin, and Dana M. Spence. 2014. "3D Printed Microfluidic Devices with Integrated Versatile and Reusable Electrodes." *Lab on a Chip* 14 (12): 2023–2032. doi:10.1039/c4lc00171k.

Feenstra, D. R., R. Banerjee, H. L. Fraser, A. Huang, A. Molotnikov, and N. Birbilis. 2021. "Critical Review of the State of the Art in Multi-Material Fabrication via Directed Energy Deposition." *Current Opinion in Solid State and Materials Science* 25 (4). Elsevier: 100924. doi:10.1016/j.cossms.2021.100924.

Feenstra, D. R., A. Molotnikov, and N. Birbilis. 2020. "Effect of Energy Density on the Interface Evolution of Stainless Steel 316L Deposited upon INC 625 via Directed Energy Deposition." *Journal of Materials Science* 55 (27). Springer US: 13314–13328. doi:10.1007/s10853-020-04913-y.

Fernandes, Matheus C., Joanna Aizenberg, James C. Weaver, and Katia Bertoldi. 2021. "Mechanically Robust Lattices Inspired by Deep-Sea Glass Sponges." *Nature Materials* 20 (2). Springer US: 237–241. doi:10.1038/s41563-020-0798-1.

Fessler, Jr, R Merz, Ah Nickel, Fb Prinz, and Le Weiss. 1996. "Laser Deposition of Metals for Shape Deposition Manufacturing." *Solid Freeform Fabrication Symposium Proceedings, University of Texas at Austin*, 117–124. http://lov2ty04.free.fr/Projet/SMD.pdf.

Fritz, B. Prinz, and E.Weiss Lee. 1993. "Method and Apparatus for Fabrication of Three-Dimensional Metal Articles by Weld Deposition. United States Patent: US5207371."

Ghanavati, Reza, and Homam Naffakh-Moosavy. 2021. "Additive Manufacturing of Functionally Graded Metallic Materials: A Review of Experimental and Numerical Studies." *Journal of Materials Research and Technology* 13. Elsevier Ltd: 1628–1664. doi:10.1016/j.jmrt.2021.05.022.

Gibson, Ian, David Rosen, Brent Stucker, and Mahyar Khorasani. 2021. *Additive Manufacturing Technologies*. Vol. 17. Cham: Springer International Publishing. doi:10.1007/978-3-030-56127-7.

Gu, Dongdong, Xinyu Shi, Reinhart Poprawe, David L. Bourell, Rossitza Setchi, and Jihong Zhu. 2021. "Material-Structure-Performance Integrated Laser-Metal Additive Manufacturing." *Science* 372 (6545): eabg1487. doi:10.1126/science.abg1487.

Hauser, Tobias, Raven T. Reisch, Stefan Seebauer, Aashirwad Parasar, Tobias Kamps, Riccardo Casati, Joerg Volpp, and Alexander F.H. Kaplan. 2021. "Multi-Material Wire Arc Additive Manufacturing of Low and High Alloyed Aluminium Alloys with in-Situ Material Analysis." *Journal of Manufacturing Processes* 69 (July). Elsevier Ltd: 378–390. doi:10.1016/j.jmapro.2021.08.005.

Kalentics, N., Logé, R., and Boillat, E. 2017. "Method and Device for Implementing Laser Shock Peening or Warm Laser Shock Peening During Selective Laser Melting, École Polytechnique Fédérale de Lausanne, Lausanne, Switzerland, U.S. Patent No. US2017/0087670 A1."

Karunakaran, K. P., S. Suryakumar, Vishal Pushpa, and Sreenathbabu Akula. 2010. "Low Cost Integration of Additive and Subtractive Processes for Hybrid Layered Manufacturing." *Robotics and Computer-Integrated Manufacturing* 26 (5). Elsevier: 490–499. doi:10.1016/j.rcim.2010.03.008.

Kerschbaumer, Michael, and Georg Ernst. 2004. "Hybrid Manufacturing Process for Rapid High Performance Tooling Combining High Speed Milling and Laser Cladding." *ICALEO 2004 – 23rd International Congress on Applications of Laser and Electro-Optics, Congress Proceedings* 1710 (2004). doi:10.2351/1.5060234.

Kim, Dong Kyu, Wanchuck Woo, Eun Young Kim, and Shi Hoon Choi. 2019. "Microstructure and Mechanical Characteristics of Multi-Layered Materials Composed of 316L Stainless Steel and Ferritic Steel Produced by Direct Energy Deposition." *Journal of Alloys and Compounds* 774. Elsevier B.V: 896–907. doi:10.1016/j.jallcom.2018.09.390.

Klein, Thomas, Alois Birgmann, and Martin Schnall. 2020. "In Situ Alloying of Aluminium-Based Alloys by (Multi-)Wire-Arc Additive Manufacturing ." *MATEC Web of Conferences* 326: 01003. doi:10.1051/matecconf/202032601003.

Klocke, F, H Wirtz, and W Meiners. 1996. "Direct Manufacturing of Metal Prototypes and Prototype Tools." *Proceedings of the SFF Symposium*, 141–148.

Li, J. C., X. Lin, N. Kang, J. L. Lu, Q. Z. Wang, and W. D. Huang. 2020. "Microstructure, Tensile and Wear Properties of a Novel Graded Al Matrix Composite Prepared by Direct Energy Deposition." *Journal of Alloys and Compounds* 826. doi:10.1016/j.jallcom.2020.154077.

Li, Runsheng, Guilan Wang, Yaoyu Ding, Shangyong Tang, Xi Chen, Fusheng Dai, Rui Wang, Hao Song, and Haiou Zhang. 2020. "Optimization of the Geometry for the

End Lateral Extension Path Strategy to Fabricate Intersections Using Laser and Cold Metal Transfer Hybrid Additive Manufacturing." *Additive Manufacturing* 36 (August). Elsevier: 101546. doi:10.1016/j.addma.2020.101546.

Li, Runsheng, Guilan Wang, Xushan Zhao, Fusheng Dai, Cheng Huang, Mingbo Zhang, Xi Chen, Hao Song, and Haiou Zhang. 2021. "Effect of Path Strategy on Residual Stress and Distortion in Laser and Cold Metal Transfer Hybrid Additive Manufacturing." *Additive Manufacturing* 46: 102203. doi:10.1016/j.addma.2021.102203.

Liu, Miaoran, Guangyi Ma, Dehua Liu, Jingling Yu, Fangyong Niu, and Dongjiang Wu. 2020. "Microstructure and Mechanical Properties of Aluminum Alloy Prepared by Laser-Arc Hybrid Additive Manufacturing." *Journal of Laser Applications* 32 (2). Laser Institute of America: 022052. doi:10.2351/7.0000082.

Liu, Yang, Chao Liu, Wensheng Liu, Yunzhu Ma, Cheng Zhang, Qingshan Cai, and Bing Liu. 2018. "Microstructure and Properties of Ti/Al Lightweight Graded Material by Direct Laser Deposition." *Materials Science and Technology (United Kingdom)* 34 (8): 945–951. doi:10.1080/02670836.2017.1412042.

Lu, Jinzhong, Haifei Lu, Xiang Xu, Jianhua Yao, Jie Cai, and Kaiyu Luo. 2020. "High-Performance Integrated Additive Manufacturing with Laser Shock Peening –Induced Microstructural Evolution and Improvement in Mechanical Properties of Ti6Al4V Alloy Components." *International Journal of Machine Tools and Manufacture* 148 (October 2019). Elsevier Ltd: 103475. doi:10.1016/j.ijmachtools.2019.103475.

Ma, Guocai, Gang Zhao, Zhihao Li, Min Yang, and Wenlei Xiao. 2019. "Optimization Strategies for Robotic Additive and Subtractive Manufacturing of Large and High Thin-Walled Aluminum Structures." *International Journal of Advanced Manufacturing Technology* 101 (5–8). The International Journal of Advanced Manufacturing Technology: 1275–1292. doi:10.1007/s00170-018-3009-3.

Manogharan, Guha, Richard Wysk, Ola Harrysson, and Ronald Aman. 2015. "AIMS – A Metal Additive-Hybrid Manufacturing System: System Architecture and Attributes." *Procedia Manufacturing* 1. Elsevier B.V.: 273–286. doi:10.1016/j.promfg.2015.09.021.

Merklein, Marion, Daniel Junker, Adam Schaub, and Franziska Neubauer. 2016. "Hybrid Additive Manufacturing Technologies – An Analysis Regarding Potentials and Applications." *Physics Procedia* 83: 549–559. doi:10.1016/j.phpro.2016.08.057.

Ni, Chenbing, Lida Zhu, Zhongpeng Zheng, Jiayi Zhang, Yun Yang, Ruochen Hong, Yuchao Bai, Wen Feng Lu, and Hao Wang. 2020. "Effects of Machining Surface and Laser Beam Scanning Strategy on Machinability of Selective Laser Melted Ti6Al4V Alloy in Milling." *Materials & Design* 194 (September): 108880. doi:10.1016/j.matdes.2020.108880.

Pragana, J. P.M., R. F.V. Sampaio, I. M.F. Bragança, C. M.A. Silva, and P. A.F. Martins. 2021. "Hybrid Metal Additive Manufacturing: A State–of–the-Art Review." *Advances in Industrial and Manufacturing Engineering* 2 (October 2020). Elsevier Ltd: 100032. doi:10.1016/j.aime.2021.100032.

Qian, Ying Ping, Ju Hua Huang, Hai Ou Zhang, and Gui Lan Wang. 2008. "Direct Rapid High-Temperature Alloy Prototyping by Hybrid Plasma-Laser Technology." *Journal of Materials Processing Technology* 208 (1–3): 99–104. doi:10.1016/j.jmatprotec.2007.12.116.

Reichardt, Ashley, Andrew A. Shapiro, Richard Otis, R. Peter Dillon, John Paul Borgonia, Bryan W. McEnerney, Peter Hosemann, and Allison M. Beese. 2021. "Advances in Additive Manufacturing of Metal-Based Functionally Graded Materials." *International Materials Reviews* 66 (1). Taylor & Francis: 1–29. doi:10.1080/09506608.2019.1709354.

Sahasrabudhe, Himanshu, Ryan Harrison, Christian Carpenter, and Amit Bandyopadhyay. 2015. "Stainless Steel to Titanium Bimetallic Structure Using LENS™." *Additive Manufacturing* 5. Elsevier B.V.: 1–8. doi:10.1016/j.addma.2014.10.002.

Sefene, Eyob Messele, Yeabsra Mekdim Hailu, and Assefa Asmare Tsegaw. 2022. "Metal Hybrid Additive Manufacturing: State-of-the-Art." *Progress in Additive Manufacturing* 7 (4). Springer International Publishing: 737–749. doi:10.1007/s40964-022-00262-1.

Shakerin, Sajad, Amir Hadadzadeh, Babak Shalchi Amirkhiz, Seyedamirreza Shamsdini, Jian Li, and Mohsen Mohammadi. 2019. "Additive Manufacturing of Maraging Steel-H13 Bimetals Using Laser Powder Bed Fusion Technique." *Additive Manufacturing* 29 (June). Elsevier: 100797. doi:10.1016/j.addma.2019.100797.

Shi, Jing, and Yachao Wang. 2020. "Development of Metal Matrix Composites by Laser-Assisted Additive Manufacturing Technologies: A Review." *Journal of Materials Science* 55 (23). Springer US: 9883–9917. doi:10.1007/s10853-020-04730-3.

Sienkiewicz, Judyta, Paweł Płatek, Fengchun Jiang, Xiaojing Sun, and Alexis Rusinek. 2020. "Investigations on the Mechanical Response of Gradient Lattice Structures Manufactured via Slm." *Metals* 10 (2). doi:10.3390/met10020213.

Sokolov, P., A. Aleshchenko, A. Koshmin, V. Cheverikin, P. Petrovskiy, A. Travyanov, and A. Sova. 2020. "Effect of Hot Rolling on Structure and Mechanical Properties of Ti-6Al-4V Alloy Parts Produced by Direct Laser Deposition." *International Journal of Advanced Manufacturing Technology* 107 (3–4): 1595–1603. doi:10.1007/s00170-020-05132-0.

Song, Yong Ak, Sehyung Park, Doosun Choi, and Haesung Jee. 2005. "3D Welding and Milling: Part I-a Direct Approach for Freeform Fabrication of Metallic Prototypes." *International Journal of Machine Tools and Manufacture* 45 (9): 1057–1062. doi:10.1016/j.ijmachtools.2004.11.021.

Tan, Chaolin, Youxiang Chew, Fei Weng, Shang Sui, Fern Lan Ng, Tong Liu, and Guijun Bi. 2021. "Laser Aided Additive Manufacturing of Spatially Heterostructured Steels." *International Journal of Machine Tools and Manufacture* 172 (October 2021). Elsevier Ltd: 103817. doi:10.1016/j.ijmachtools.2021.103817.

Tan, Chaolin, Xinyue Zhang, Dongdong Dong, Bonnie Attard, Di Wang, Min Kuang, Wenyou Ma, and Kesong Zhou. 2020. "In-Situ Synthesised Interlayer Enhances Bonding Strength in Additively Manufactured Multi-Material Hybrid Tooling." *International Journal of Machine Tools and Manufacture* 155 (July). Elsevier Ltd: 103592. doi:10.1016/j.ijmachtools.2020.103592.

Tan, Chaolin, Ji Zou, Sheng Li, Parastoo Jamshidi, Alessandro Abena, Alex Forsey, Richard J. Moat, et al. 2021. "Additive Manufacturing of Bio-Inspired Multi-Scale Hierarchically Strengthened Lattice Structures." *International Journal of Machine Tools and Manufacture* 167 (February). Elsevier Ltd: 103764. doi:10.1016/j.ijmachtools.2021.103764.

Tey, Cher Fu, Xipeng Tan, Swee Leong Sing, and Wai Yee Yeong. 2020. "Additive Manufacturing of Multiple Materials by Selective Laser Melting: Ti-Alloy to Stainless Steel via a Cu-Alloy Interlayer." *Additive Manufacturing* 31 (November 2019). Elsevier: 100970. doi:10.1016/j.addma.2019.100970.

Thompson, Mary Kathryn, Giovanni Moroni, Tom Vaneker, Georges Fadel, R. Ian Campbell, Ian Gibson, Alain Bernard, et al. 2016. "Design for Additive Manufacturing: Trends, Opportunities, Considerations, and Constraints." *CIRP Annals – Manufacturing Technology* 65 (2). Elsevier: 737–760. doi:10.1016/j.cirp.2016.05.004.

Tian, Yinbao, Junqi Shen, Shengsun Hu, Jian Han, Qian Wang, and Yangchuan Cai. 2021. "Effects of Ultrasonic Peening Treatment Layer by Layer on Microstructure of Components Fabricated by Wire and Arc Additive Manufacturing." *Materials Letters* 284. Elsevier B.V.: 128917. doi:10.1016/j.matlet.2020.128917.

Tilton, Maryam, Alireza Borjali, Aaron Isaacson, Kartik Mangudi Varadarajan, and Guha P. Manogharan. 2021. "On Structure and Mechanics of Biomimetic Meta-Biomaterials Fabricated via Metal Additive Manufacturing." *Materials and Design* 201: 109498. doi:10.1016/j.matdes.2021.109498.

U., Savitha, G. Jagan Reddy, Vajinder Singh, A. A. Gokhale, and Sundararaman M. 2020. "Additive Laser Deposition of Compositionally Graded NiCrAlY-YSZ Multi-Materials on IN625-NiCrAlY Substrate." *Materials Characterization* 164 (November 2019). Elsevier: 110317. doi:10.1016/j.matchar.2020.110317.

Uzan, Naor Elad, Shlomo Ramati, Roni Shneck, Nachum Frage, and Ori Yeheskel. 2018. "On the Effect of Shot-Peening on Fatigue Resistance of AlSi10Mg Specimens Fabricated by Additive Manufacturing Using Selective Laser Melting (AM-SLM)." *Additive Manufacturing* 21 (September 2017). Elsevier: 458–464. doi:10.1016/j.addma.2018.03.030.

Wang, Di, Linqing Liu, Guowei Deng, Cheng Deng, Yuchao Bai, Yongqiang Yang, Weihui Wu, et al. 2022. "Recent Progress on Additive Manufacturing of Multi-Material Structures with Laser Powder Bed Fusion." *Virtual and Physical Prototyping* 17 (2): 329–365. doi:10.1080/17452759.2022.2028343.

Wang, Haoran, Dongdong Gu, Kaijie Lin, Lixia Xi, and Luhao Yuan. 2019. "Compressive Properties of Bio-Inspired Reticulated Shell Structures Processed by Selective Laser Melting." *Advanced Engineering Materials* 21 (4): 1–10. doi:10.1002/adem.201801168.

Wang, Rui, Dongdong Gu, Kaijie Lin, Caiyan Chen, Qing Ge, and Deli Li. 2022. "Multi-Material Additive Manufacturing of a Bio-Inspired Layered Ceramic/Metal Structure: Formation Mechanisms and Mechanical Properties." *International Journal of Machine Tools and Manufacture* 175 (February). Elsevier Ltd: 103872. doi:10.1016/j.ijmachtools.2022.103872.

Wang, Yachao, Sougata Roy, Hyunsuk Choi, and Touhid Rimon. 2022. "Cracking Suppression in Additive Manufacturing of Hard-to-Weld Nickel-Based Superalloy through Layer-Wise Ultrasonic Impact Peening." *Journal of Manufacturing Processes* 80 (May). Elsevier Ltd: 320–327. doi:10.1016/j.jmapro.2022.05.041.

Wits, Wessel W., and Emiel Amsterdam. 2021. "Graded Structures by Multi-Material Mixing in Laser Powder Bed Fusion." *CIRP Annals* 70 (1). Elsevier Ltd: 159–162. doi:10.1016/j.cirp.2021.03.005.

Wu, Dongjiang, Dehua Liu, Fangyong Niu, Qiuyu Miao, Kai Zhao, Bokai Tang, Guijun Bi, and Guangyi Ma. 2020. "Al–Cu Alloy Fabricated by Novel Laser-Tungsten Inert Gas Hybrid Additive Manufacturing." *Additive Manufacturing* 32 (October 2019). Elsevier: 100954. doi:10.1016/j.addma.2019.100954.

Wu, Sung Yueh, Chen Yang, Wensyang Hsu, and Liwei Lin. 2015. "3D-Printed Microelectronics for Integrated Circuitry and Passive Wireless Sensors." *Microsystems and Nanoengineering* 1 (April): 1–9. doi:10.1038/micronano.2015.13.

Xiong, Xinhong, Haiou Zhang, and Guilan Wang. 2009. "Metal Direct Prototyping by Using Hybrid Plasma Deposition and Milling." *Journal of Materials Processing Technology* 209 (1): 124–130. doi:10.1016/j.jmatprotec.2008.01.059.

Yang, Yang, Xuan Song, Xiangjia Li, Zeyu Chen, Chi Zhou, Qifa Zhou, and Yong Chen. 2018. "Recent Progress in Biomimetic Additive Manufacturing Technology: From Materials to Functional Structures." *Advanced Materials* 30 (36): 1–34. doi:10.1002/adma.201706539.

Yasa, E., J. P. Kruth, and J. Deckers. 2011. "Manufacturing by Combining Selective Laser Melting and Selective Laser Erosion/Laser Re-Melting." *CIRP Annals – Manufacturing Technology* 60 (1): 263–266. doi:10.1016/j.cirp.2011.03.063.

Yasa, Evren, Jan Deckers, and Jean Pierre Kruth. 2011. "The Investigation of the Influence of Laser Re-Melting on Density, Surface Quality and Microstructure of Selective Laser Melting Parts." *Rapid Prototyping Journal* 17 (5). Emerald Group Publishing Limited: 312–327. doi:10.1108/13552541111156450.

Yee, Daryl W., Max L. Lifson, Bryce W. Edwards, and Julia R. Greer. 2019. "Additive Manufacturing of 3D-Architected Multifunctional Metal Oxides." *Advanced Materials* 31 (33): 1–9. doi:10.1002/adma.201901345.

Zhang, Haiou, Yingping Qian, Guilan Wang, and Qiguang Zheng. 2006. "The Characteristics of Arc Beam Shaping in Hybrid Plasma and Laser Deposition Manufacturing." *Science in China, Series E: Technological Sciences* 49 (2): 238–247. doi:10.1007/s11431-006-0238-8.

Zhang, Haiou, Ying ping Qian, and Gui lan Wang. 2006. "Study of Rapid and Direct Thick Coating Deposition by Hybrid Plasma-Laser Manufacturing." *Surface and Coatings Technology* 201 (3–4): 1739–1744. doi:10.1016/j.surfcoat.2006.02.049.

Zhang, Haiou, Cheng Huang, Guilan Wang, Runsheng Li, and Gang Zhao. 2021. "Comparison of Energy Consumption between Hybrid Deposition & Micro-Rolling and Conventional Approach for Wrought Parts." *Journal of Cleaner Production* 279. Elsevier Ltd: 123307. doi:10.1016/j.jclepro.2020.123307.

Zhang, Haiou, Daoman Rui, Yang Xie, and Guilan Wang. 2013. "Study on Metamorphic Rolling Mechanism for Metal Hybrid Additive Manufacturing." *24th International SFF Symposium – An Additive Manufacturing Conference, SFF 2013*, 188–199.

Zhang, Shuai, Yazhou Zhang, Ming Gao, Fude Wang, Quan Li, and Xiaoyan Zeng. 2019. "Effects of Milling Thickness on Wire Deposition Accuracy of Hybrid Additive/Subtractive Manufacturing." *Science and Technology of Welding and Joining* 24 (5). Taylor & Francis: 375–381. doi:10.1080/13621718.2019.1595925.

Zhang, Xinchang, Cheng Sun, Tan Pan, Aaron Flood, Yunlu Zhang, Lan Li, and Frank Liou. 2020. "Additive Manufacturing of Copper – H13 Tool Steel Bi-Metallic Structures via Ni-Based Multi-Interlayer." *Additive Manufacturing* 36 (July). Elsevier: 101474. doi:10.1016/j.addma.2020.101474.

Zhang, Zhaodong, Chengshuai Sun, Xinkun Xu, and Liming Liu. 2018. "Surface Quality and Forming Characteristics of Thin-Wall Aluminium Alloy Parts Manufactured by Laser Assisted MIG Arc Additive Manufacturing." *International Journal of Lightweight Materials and Manufacture* 1 (2). Elsevier Taiwan LLC: 89–95. doi:10.1016/j.ijlmm.2018.03.005.

Zhu, Jihong, Han Zhou, Chuang Wang, Lu Zhou, Shangqin Yuan, and Weihong Zhang. 2021. "A Review of Topology Optimization for Additive Manufacturing: Status and Challenges." *Chinese Journal of Aeronautics* 34 (1). Chinese Society of Aeronautics and Astronautics: 91–110. doi:10.1016/j.cja.2020.09.020.

10 Case Studies

10.1 LIGHTWEIGHT OPTICAL SURFACE FABRICATION

10.1.1 APPLICATION

Optical surfaces are extremely smooth to allow light to be transmitted or reflected, which can be characterized by extremely low surface roughness. Roughness measurements can be performed based on several parameters, such as the most common arithmetic mean height (Ra), the average maximum peak to valley height (Rz), the root mean square or quadratic average (Rq), and the asymmetrical height of the mean line (Rsk). An optical surface is typically defined by surfaces with roughness values in the nanometric range (<10 nm).

These optical surfaces are used in almost all industries, with the most common products being lenses and reflectors. For instance, optical surfaces are necessary for windows, head lights, and mirrors in vehicles. These surfaces typically transmit light over the visible (380–700 nm) and infrared (780 nm–1 mm) light spectrums. More advanced optical surfaces involve capturing light at smaller wavelengths in the ultraviolet (<380 nm) spectrum. Instances of such advanced optics include the reflective surfaces used in space exploratory telescopes. Reflective mirrors on telescopes are designed to concentrate and magnify the light emitted from distant objects and channeled into a detector. The reverse of light being channeled through the optical lens is the working principal of light transmission, such as lasers. The transmission of light demands surfaces with geometric features in the range of 1/10–1/20 Rq of the desired wavelength. Additional requirements are necessary to reduce light scattering by further improving the optical surface quality in the range of 1/200–1/500 Rq of the desired wavelength. This explains the stringent surface finish requirements necessary for optical transmission. Aside from functionality, optical surfaces are also aesthetically more appealing with shinier surfaces for commercial products such as those used for mobile phones.

One of the goals during fabrication of optical lenses, especially for space applications, is to meet the need for lightweight solutions. A traditional optical mirror has a contoured back that is designed for easy mounting, but the mass of material used for this structure can be substantial and could result in costly payloads for space optics. To address this issue, lightweight core structures were developed for weight

DOI: 10.1201/9781003272601-10

reduction while still providing adequate support for mirror distortion prevention (Barnes 1972). Examples of these core structures include the honeycomb, periodic, and aperiodic structures. One side of these core structures is the mirror surface while the other side is left open or covered with a sandwich backing. The open back is lighter and uses less material, but the sandwich configuration is structurally stiffer.

Conventionally, lightweight mirrors with innovative core structures were manufactured by securing the mirror surface onto the core structures using adhesives, which tends to be problematic structurally and functionally when subjected to prolonged stress and when experiencing thermal distortion. The solution to these problems is through AM technology, which widens the type of internal structures that are not easily achieved using traditional cutting methodologies. Hence, the focus of this case study is on the procedures to produce optical mirrors on an additively manufactured surface. There are other notable considerations to account for in the additive manufacturing of optical mirrors, such as the morphology of the internal structure, but this section will focus on processing optical surfaces. The issue, as themed in this book, is the limited technological maturity for additive manufacturing as a sole process to produce products with optical-grade surface finishing. Hence, the manufacturing workflow involving traditional material removal processes. Specifications for the types of processes to produce a lightweight optical mirror will be discussed in further detail.

10.1.2 Material Selection

Materials are selected based on their optical performance, which include parameters such as the reflectance and thermal expansion coefficient. Reflectance varies over different wavelengths and is the primary optical property of a mirror. Thermal expansion coefficient determines the distortion of the mirror due to temperature changes, which is particularly important when capturing high resolution imaging and where micro-optics are involved. In this case study, the maturity of AM technology to process a material, the material density, and its machinability are important factors to consider and affect the overall cost in manufacturing.

For metallic mirrors, aluminum, silver, and beryllium are often considered for selection due to their optical reflectance. Aluminum is a relatively easier metal to machine, is light, and can exhibit strong mechanical properties with the right alloying elements. Beryllium has nearly no thermal expansion coefficient at cryogenic temperatures, which makes it highly suitable for space use. However, it has relatively poor machinability and exhibits high thermal expansion at high temperatures, which limits its usage in general as an optical material. Silver is an expensive precious metal and is generally employed as a coating material. Thus, aluminum-based alloys – AlSi12, AlSi40, and AlSi10Mg – are favorable metals for this case study on lightweight optical mirror manufacturing.

10.1.3 Process Flow

Optical surface fabrication on an additively manufactured material involves additive manufacturing followed by a material removal process to achieve the optical surface. But within this seemingly simple task lies multiple processes that were implemented to efficiently achieve the desired result of a surface with nanometric roughness.

The mirror will first be fabricated using AM with two faces, the optical face, and the back face where the latter would be constructed on support structures. The part will then be removed from the building chamber by sawing followed by milling of the support structure remnants. The back surface will then be lapped to achieve a flat surface, which will form the reference plane for further processing of the optical surface. The optical surface will then be diamond turned on an ultra-precision machine center where the mirror can be attached onto the spindle by positioning the back face on a vacuum chuck. Diamond turning of the optical surface will produce nanometric surface roughness, which would suffice for infrared wavelengths despite the presence of cutting tool feed marks on the surface. Further material removal processes can be subsequently employed to improve the surface quality, such as magnetorheological finishing (MRF) to smoothen the surface and chemical mechanical polishing (CMP) to lower the surface roughness.

Support structure removal will depend on the way the mirror is constructed. Hilpert et al. (2018) determined that an angle of 41° was optimal for the part to be tilted during construction to minimize overhanging features without support structures, particularly for the internal features. These overhanging features come from holes on the circumferential and internal surfaces to remove powder particles left within these internal structures of the sandwich mirror. With the mirror tilted at 41°, the main support structures – illustrated in Figure 10.1 – will require mechanical sawing and milling to remove the remaining features. The back surface is then milled and lapped. In the case of an optical mirror with a 76 mm diameter, up to 30 μm of material will be removed to achieve a flat surface (Hilpert et al. 2019).

10.1.4 Shape Generation by Diamond Turning

The mirror is inverted to machine the optical surface into the target shape (e.g., convex/concave) or micro-pattern formation. In cases where the initial surface roughness of the additively manufactured part is significantly high (~10 μm Ra), milling is the preferred choice to first lower the surface roughness (<1 μm Ra) before diamond turning is performed using an ultra-precision machine tool. The reason for such a procedure is to reduce undesirable mechanical stresses that lead to machine tool degradation on the ultra-precision machine when used on highly uneven surface asperities.

Machine tool parameters include the spindle speed, feed rate, and nominal cutting depth. The spindle speed, measured in revolutions per minute, determines the effective cutting speed at an instantaneous radius of the workpiece. Additionally, the spindle speed coupled with the feed rate (v_f) determines the tool feed (f), as shown in Figure 10.2 and determined by Equation 10.1. Single crystal diamond cutting tools are typically round-nosed for stronger mechanical strength during machining. The nose radius (r_n), nominal cutting depth, and feed determine the effective cutting area during diamond-turning, which have an impact on the machined surface quality when cutting brittle materials (Blackley and Scattergood 1991).

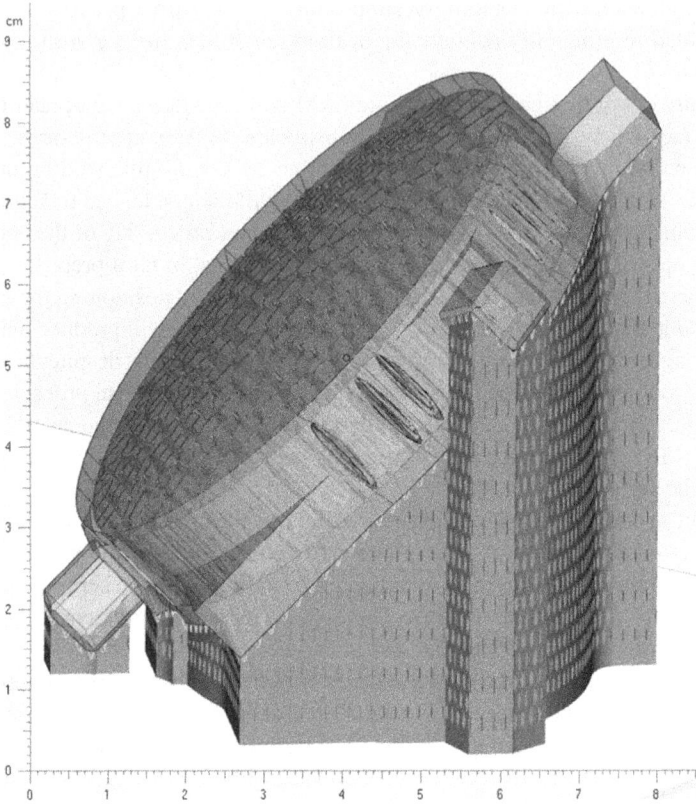

FIGURE 10.1   CAD drawing of an optical mirror (gray) with support structures (blue) constructed by selective laser melting. Reprinted with permission from SPIE (Hilpert et al. 2019).

$$f = \frac{v_f}{N} \qquad (10.1)$$

The cutting tool has other parameters such as rake angle, clearance angle, and tool edge radius (r_e) as illustrated in Figure 10.2. Cutting brittle materials typically employs negative rake angles to induce hydrostatic stresses for the suppression of crack propagation (Komanduri, Chandrasekaran, and Raff 1998), while ductile materials generally involve positive rake angled cutting tools to promote chip flow. The clearance angle is designed to maintain cutting tool integrity and reduce the obstruction of elastic recovery on the machined surface. The clearance angle must be sufficiently large for the machined surface to elastically recover but a larger clearance angle for a positive rake angled tool will cause the tool to be extremely sharp and susceptible to brittle failure during machining. The tool edge radius has particular significance when the cutting length scale is of similar magnitude with the edge radius. This is

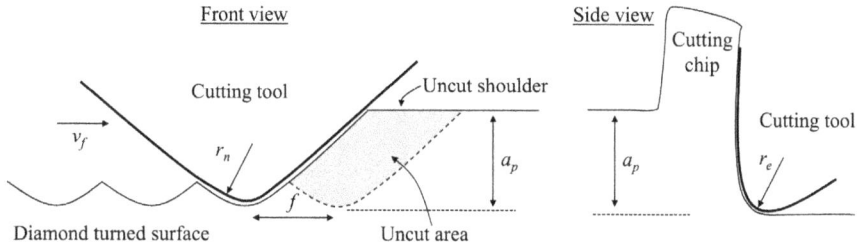

FIGURE 10.2   Illustration of the diamond turning process and machining parameters.

known as the tool-edge radius effect (Zhao et al. 2017) or the relative tool sharpness (Rahman et al. 2017) where cutting length scales smaller than the edge radius can result in change in material removal/deformation mechanism (Rahman, Rahman, and Kumar 2018).

With this in mind, optimal process parameters can be selected for the fabrication of the optical surfaces. Table 10.1 lists several combinations of parameters employed in diamond turning aluminum alloys and the achievable surface quality with nanometric roughness. It is important to clarify that some of the listed alloys were not additively manufactured and the table of parameters may only be used as a guide to parameter selection for diamond turning. As a guiding principle, the kinematic machined surface peak-to-valley height is determined using Equation 10.2 as a function of the feed and nose radius. This information is representative of the surface roughness, which is determined in the average mean height using Equation 10.3.

$$Rz = \frac{f^2}{8r_n} \tag{10.2}$$

$$Ra = 0.032 \frac{f^2}{r_n} \tag{10.3}$$

A surface with roughness of 1 nm Ra would require a feed of approximately 2.8 μm/rev with a 1 mm nose radius diamond tool, which would correspond to an example combination of 2.8 mm/min feed rate and 1,000 rpm spindle speed. The decrease in feed theoretically improves the surface roughness. However, drastically reducing the feed effectively sets the uncut chip thickness within the uncut shoulder (Figure 10.2) to be mostly smaller than the tool edge radius. This effects the relative tool sharpness where the material removal mechanism may no longer be shearing but ploughing, which results in a different type of machined surface formation.

A large feed will also result in a larger cutting area and consequently larger volume of material being removed, which may induce undesirable stress–strain conditions and tool-tip vibrations that cause degradation of machined surface. The geometrical parameters on the machine tool parameter selection are theoretically sufficient for optical surface generation and an ideal range of the feed will avoid tool edge radius effects and tool-tip vibrations.

TABLE 10.1

Diamond Turning Parameters for Aluminum Alloys

Material	Spindle speed (rpm)	Feed rate (mm/min)	Feed (µm/rev)	Cutting depth (µm)	Surface roughness	Reference
AlSi10Mg	1,200	–	2.0	2.0	8.0 nm Rq	(Han et al. 2019)
AlSi10Mg	1,500	2.0	–	2.0	2.9 nm Rq	(Wang et al. 2021)
AlSi10Mg	–	–	–	–	3.6 nm Sq	(Atkins et al. 2018)
Al6061	1,000	2.0	–	2.0	2.2 nm Ra	(Cheng et al. 2014)
Al6061-T6	1,500	3.0	–	–	3.0 nm Ra	(Chabot et al. 2019)
Al6061	2,500	–	1.0	5.0	2.6 nm Ra	(Cardenas et al. 2015)
Al6061	2,000	25.0	–	2.0	12.5 nm Ra	(Cheung and Lee 2000)
Al6061	3,000	5.0	–	5.0	3.8 nm Rq	(Gubbels et al. 2008)
Al6061	1,000	2.0	–	1.0	2.1 nm Ra	(Cheng et al. 2015)
RSA6061	1,000	2.0	–	1.0	1.5 nm Ra	(Cheng et al. 2015)
RSA6061	3,000	5.0	–	5.0	2.3 nm Rq	(Gubbels et al. 2008)
RSA6061	1,500	3.0	–	–	1.7 nm Ra	(Chabot et al. 2019)
RSA6061	1,000	2.0	–	2.0	1.7 nm Ra	(Cheng et al. 2014)
AlSi12					9.1 nm Sa	(Hilpert et al. 2018)

The next factor for consideration is the machining time, which can be shortened by increasing the feed rate. In the example above, the feed of 2.8 μm/rev may be assumed to be ideal to achieve a surface roughness of 1 nm Ra. An increase in feed rate to 5.6 mm/min would then require an increase in spindle speed to 2,000 rpm as determined in Equation 10.1. However, the increase in spindle speed indicates the increase in effective cutting speed, which would lead to other metal deformation events that may affect the machined surface quality. In the case of aluminum alloys, the increase in cutting speed promotes the production of low roughness surfaces (Revel, Khanfir, and Fillit 2006). However, this trend of metal deformation events may not hold true for other metals where an increase in cutting speed promotes formation of machined surface defects and increases surface roughness from the theoretical value. An increase in cutting speed often leads to an increase in heat generation, which can beneficially soften the material to ease machinability but can also result in thermal instabilities for both the cutting chip and the cutting tool, leading to undesirable events such as tool-tip vibrations (Wang et al. 2011).

The described causes for the machined surface deterioration while employing different cutting speeds is largely correlated to the microstructural properties of the work material, which is an important aspect to understand as part of the postprocessing workflow for the fabrication of an optical surface.

10.1.5 MICROSTRUCTURE MODIFICATIONS

Heat treatment is a post-processing procedure intended to relieve residual stresses developed during solidification of particles in the AM process. However, heat treatment also results in the alteration of microstructural features, which largely affect the material properties and its machinability.

For instance, a comparison of AlSiMg0.75 microstructures (Figure 10.3) present the micro-segregation of the Si phase along cell boundaries of the Al matrix in the as-built (AB) sample (Bai et al. 2020). After heat treatment at 300 °C for 3 h followed by furnace cooling, the dissolution and scattering of Si particles occurred throughout the heat treated (HT) microstructure. Dendrite structure widths will increase together with the formation of Si particles after stress relieving via heat treatment of AlSi12 at 200 °C (Siddique et al. 2015).

It is apparent the precipitation of silicon particles within the aluminum matrix is a common occurrence. Machining performance would be affected by the presence of these precipitations (Kikuchi et al. 2003). Cutting forces are known to decrease with the increase in silicon precipitation due to decrease in tool-chip interfacial adhesion (Zhang et al. 2020) and the easy breakage of chips (Kamiya et al. 2008). Silicon precipitation also affect the accumulation of material and dislodging of hard particles to severely affect the machined surface (Santos et al. 2016). In the case of Al6061, heat generated during machining itself has also been reported to be sufficient for the precipitation of Mg_2Si particles to subsequently affect the machined surface quality (Wang et al. 2015). Figure 10.4 presents optical images of machined surfaces on AlSiMg0.75 and illustrates the impact of precipitates on the machined surface quality. The hard precipitates embedded within the ductile aluminum matrix were

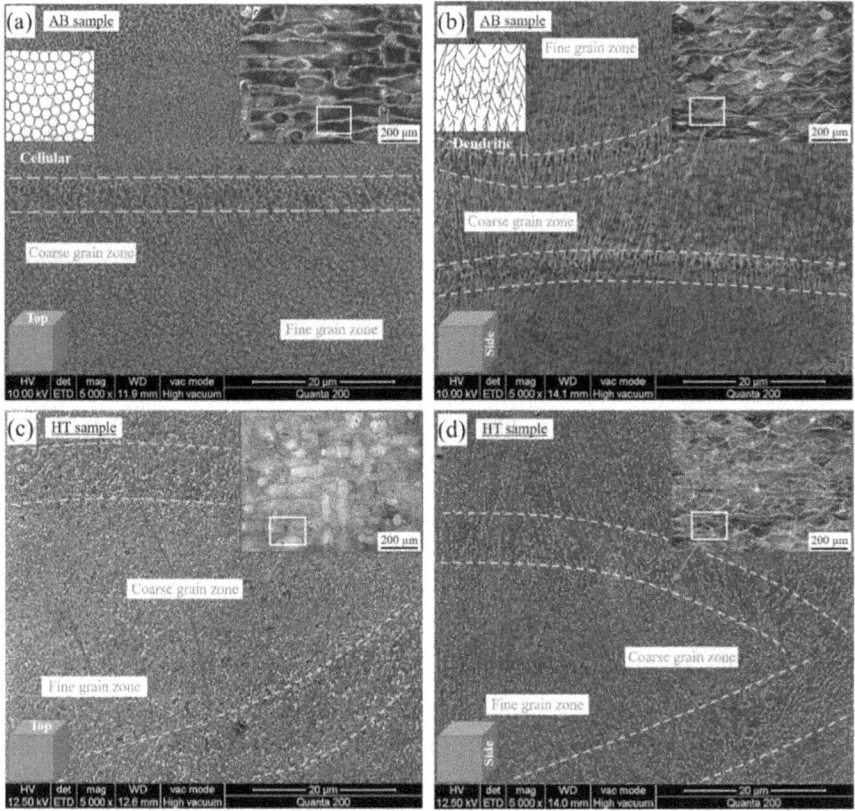

FIGURE 10.3 Microstructure comparison between as-built (AB) and heat treated (HT) microstructures in AlSiMg0.75: (a) AB top surface; (b) AB side surface; (c) HT top surface; (d) HT side surface. Reprinted with permission from Elsevier (Bai et al. 2020).

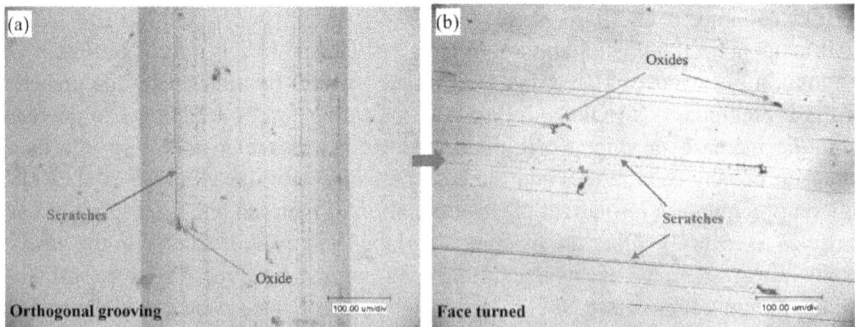

FIGURE 10.4 Optical imaging of the AlSiMg0.75 machined surfaces after: (a) orthogonal micro-cutting (v_c = 20 mm/min, a_p = 20 μm); (b) diamond turning (N = 1,500 rpm, f = 5 μm/rev, a_p = 3 μm). Reprinted with permission from Elsevier (Bai et al. 2020).

dragged along the tool path to produce micro-scratches on the surface degrading the surface roughness to 10.2 nm Ra.

10.1.6 ALTERNATIVE SHAPE GENERATION METHOD 1: ULTRASONIC ELLIPTICAL VIBRATION-ASSISTED MACHINING

Understanding the material characteristics to optimize machining parameters is one method to further improve the surface condition of the diamond-turned optical surface. However, the stringent optical requirements for visible and UV light applications require even lower surface roughness profiles that are typically achieved by MRF and CMP. There are, however, other proposed solutions to augment the fabrication process of an optical surface on the additively manufactured part, such as applying a coating and implementing advanced tooling systems to the ultra-precision machining process.

Ultrasonic elliptical vibration-assisted machining (UEVAM) is a suitable solution to overcome the issue of hard precipitates scratching the machined surfaces (Bai et al. 2020). This is an innovative cutting tool holder that is designed to allow microscopic movements of the tool at a high frequency to induce an alternative chip formation mechanism. These tool motions can occur along up to three dimensions, but are more commonly found in two-axial (2D) applications, where the axes are parallel and perpendicular to the cutting direction (Brehl and Dow 2008). In a 2D scenario, the effective motion of the cutting tool relative to the work material would appear to be elliptically shaped when also accounting for the feed of the tool holder relative to the work piece. In other words, a 2D tool vibrational motion would be circular if the tool holder is stationary, while the additional motion of the tool holder with respect to the workpiece causes the effective vibrational motion to be elliptical, as illustrated in Figure 10.5. Thus, it is common for this process to be referred to as elliptical vibration-assistance rather than vibration-assisted cutting. The vibration motion is induced by piezoelectric or magnetostrictive actuators that are actuated by the input of sinusoidal voltage signals, thus giving the vibration motion harmonic motions at high frequencies in several tens of kHz.

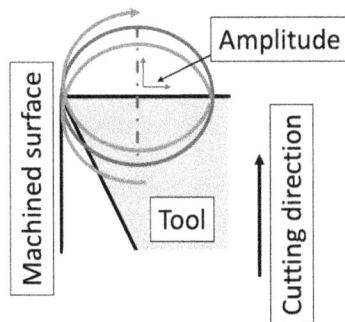

FIGURE 10.5 Tool trajectory in 2D ultrasonic elliptical vibration-assisted machining and its effectiveness at tackling the embedded particle problem.

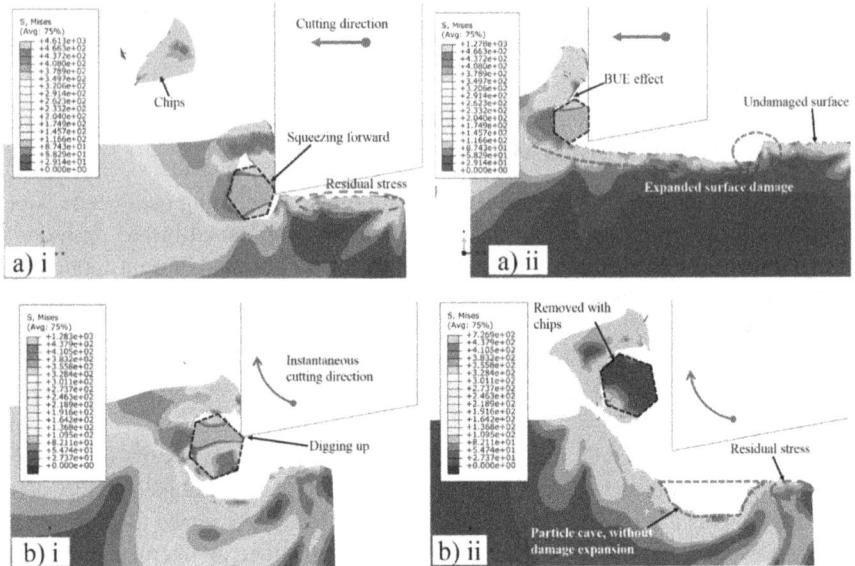

FIGURE 10.6 FEM simulations of the differences in interaction between: (a) a conventional tool motion, and (b) the elliptical tool motion in UEVAM. Reprinted with permission from Elsevier (Bai et al. 2020).

In the case of hard particles, it is believed that UEVAM can effectively remove them from the cutting zone to avoid scratching of the machined surface. Finite element method (FEM) simulations illustrate the benefits of UEVAM to address the particle problem, as shown in Figure 10.6. With a vibrational amplitude and frequency of 4 μm and 38.87 kHz, the surface quality of AlSiMg0.75 alloy could be reduced from 11.03 nm Ra by conventional diamond turning to 5.10 nm Ra with UEVAM (Bai et al. 2020). Turning parameters were set at 1,500 rpm spindle speed, 2 μm/rev feed, and 3 μm cutting depth for this case. Interestingly, the improvement in surface quality with UEVAM was greater than that after heat treatment, where a roughness of 10.07 nm Ra was achieved by conventional diamond-turning after thermal treatment at 300 °C for 3 h. This demonstrated the marginal impact of thermal treatment to improve machinability of the alloy and the potential adoption of UEVAM to shorten the post-processing procedure.

Vibration-assisted machining generally benefits other manufacturing processes in achieving superior surface finishing as well as milling of Ti6Al4V (Su and Li 2022) and burnishing of AlSi10Mg (Teramachi and Yan 2019), but it may not always be the case as reported for vibration-assisted micro-milling of AISI 316L steel (Greco et al. 2021) and the burnishing of maraging steel (Ituarte et al. 2020). Other common benefits of vibration-assistance include the reduction in cutting forces and lower tool wear.

Vibration-assisted machining is, however, limited by a maximum cutting speed defined by Equation 10.4.

$$V_{c,crit} = 2\pi f_{vib} A \tag{10.4}$$

where f_{vib} is the resonant vibration frequency and A is the vibration amplitude. If the effective cutting speed during diamond turning exceeds the critical cutting speed $V_{c,crit}$, tool-workpiece separation will no longer occur and the cutting process may be assumed to be similar to the conventional cutting process where the tool is perpetually in contact with the work material. Therefore, this may cause certain limitations to the overall machining time. A limitation on the cutting speed will limit the spindle speed, and consequently affects the maximum feed rate (Equation 10.1) and the machining time. Aside from this, the benefits of vibration-assisted machining have made the process highly favorable to be included as part of advancing additive manufacturing in hybrid manufacturing (Sealy et al. 2018). Table 10.2 lists several UEVAM parameters employed for conventional metal processing as a guide for further applications on additively manufactured metals.

10.1.7 Alternative Shape Generation Method 2: Replacing Optical Surface with Nickel-Phosphorous Coating

Hilpert et al. (2018) determined that diamond turned AlSi12 could not attain the desirable surface quality of an optical mirror, with the best achievable surface quality of 9.1 nm Sa. Therefore, the alloy surface was electroplated with a 100 μm thick amorphous layer of 11–12 wt.% phosphorous with nickel (i.e., NiP plating). Electroplating is the process using controlled electrolysis to transfer material from an anode to the part to be coated (the cathode). The electrolytic cell comprises an electrical current source, a cathode, an anode, and an aqueous electrolyte (Figure 10.7). The setup is relatively cheap and operable at ambient conditions (i.e., room temperature and pressure).

Electrodeposition of NiP involves the main deposition of nickel ions (Ni^{+2}) with the co-deposition of phosphorous oxyacid onto the cathodic workpiece. Nickel ions are formed in the aqueous solution at the soluble nickel anode in Equation 10.5. Concurrently, an oxyanion such as hypophosphite is reduced by Equation 10.6 to form phosphines that interact with nickel ions to form the nickel-phosphorous material on the cathode in Equation 10.7.

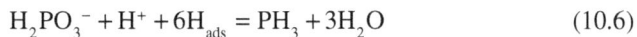

$$Ni - 2e^- = Ni^{2+} \tag{10.5}$$

$$H_2PO_3^- + H^+ + 6H_{ads} = PH_3 + 3H_2O \tag{10.6}$$

$$2PH_3 + 3Ni^{2+} = 3Ni + 2P + 2H^+ \tag{10.7}$$

Phosphorous content in the coating is also an important factor contributing to machinability where higher phosphorous content increases the ductility of the coating and reduces tool wear to improve surface finishing during diamond turning (Pramanik

TABLE 10.2
Ultrasonic Elliptical Vibration-Assisted Machining Parameters for Metals

Material	Amp. (μm)	Freq. (kHz)	Surface roughness	Spindle speed (rpm)	Feed (μm /rev)	Cutting depth (μm)	Ref.
AlSi10Mg	4.0	38.87	5.10 nm Ra	1,500	-	2.0	(Bai et al. 2020)
MB5	1.0	100.0	3.4 nm Ra	-	5.0	1.0	(Xing et al. 2020)
SUS303Se	3.0	40.0	26 nm Rz	90.0	3.0	3.0	(Moriwaki and Shamoto 1991)
SUS42012	2.5	21.8	14 nm Ry	380.0	3.0	3.0	(Shamoto and Moriwaki 1999)
SUS42012	2.0	36.2	16 nm Ra	6.0	-	-	(Zhang et al. 2019)
Stavax	2.0	38.87	10 nm Ra	15.0	2.5	10.0	(Zhang et al. 2011)
Ti6Al4V	6.0	29.75	<30 nm Ra	8.0	5.0	5.0	(Tan et al. 2020)

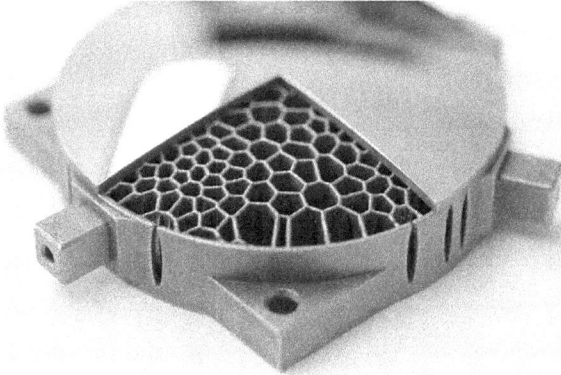

FIGURE 10.7 Additively manufactured AlSi12 optical mirror with NiP plating and lightweight internal structure. Reprinted with permission from Elsevier (Hilpert et al. 2018).

et al. 2008). Despite several studies on tool wear, NiP is considered to be a machinable material with reports of optical surface generation even after 200 km of machining (Pramanik et al. 2009). Generally, better surface finishing through diamond turning and polishing is achievable after heat treatment, but this comes at the expense of cutting tool life (Casstevens and Daugherty 1978).

Thermal treatment of the coating is needed to enhance the mechanical strength (Lelevic and Walsh 2019) and coating-substrate adhesion (Apachitei et al. 2002) of the coating. Treatment temperatures need to exceed 260 °C before observing any changes in the mechanical properties of the coating (Staia et al. 1996). Surface refinement can then be performed after ensuring adhesion of the coating to the aluminum substrate.

Diamond turning of the NiP coating will reshape the optical surface to the target geometry and an optical surface with roughness of between ranging between 2.9 nm Rq (Hilpert et al. 2019) and 5.4 Rq (Hilpert et al. 2018). Figure 10.7 presents an example of the machined optical surface on a NiP coated aluminum alloy constructed with a lightweight internal structure by selective laser melting. The machined surface quality of NiP was reported to be unaffected by the cutting speed but increases together with feed rate and cutting depth (Chon and Namba 2010). However, the achievable surface deteriorated as the feed and cutting depth reduced further, due to relative tool sharpness (Yu et al. 2021). Table 10.3 lists several parameters used in diamond turning nickel phosphorous optical surfaces.

10.1.8 FINISHING PROCESSES

The machined surface after diamond turning with or without UEVAM will inevitably leave behind feed marks, as shown in Figure 10.8(a), which will result in light diffraction and scattering – particularly in X-ray applications. This requires additional processes such as polishing to further reduce the surface roughness of less than 0.3 nm Rq (Figures 10.8(b) and 10.8(c)) for shorter wavelength applications (Namba et al.

TABLE 10.3
Diamond Turning Parameters for NiP

%P	Surface roughness	Spindle speed (rpm)	Feed rate (mm/min)	Feed (µ/rev)	Nose radius (mm)	Cutting depth (µm)	Reference
8.5	3.5 nm Rq	-	-	2.54	25.4	-	(Casstevens and Daugherty 1978)
10	2.54 nm Ra	-	-	1.0	3.0	1.0	(Chon, Namba, and Yoon 2006)
8.5	5 nm Ra	1,000	-	2.0	2.0	4.0	(Pramanik et al. 2003)
13	6.8 nm Rq	1,000	-	2.54	0.8	0.64	(Taylor et al. 1986)
10–11	~8 nm Ra	1,000	-	10.0	2.0	2.0	(Pramanik et al. 2009)
13	3 nm Rq	1,000	2.54	2.5	-	2.54	(Syn, Taylor, and Donaldson 1986)
13	10 nm Ra	2,000	-	1.0	0.5	3.0	(Zhou, Huang, and Nguyen 2017)
11	25 nm Ra	2,000	-	1.0	0.5	3.0	
11	10–14 nm Ra	1,000	-	10.0	-	2.0–4.0	(Pramanik et al. 2008)
-	4.2 nm Ra	2,000	-	-	1.0	2.0	(Bai et al. 2018)
-	2.11 nm Rq	1,500	2.0	-	-	2.0	(Wang et al. 2021)

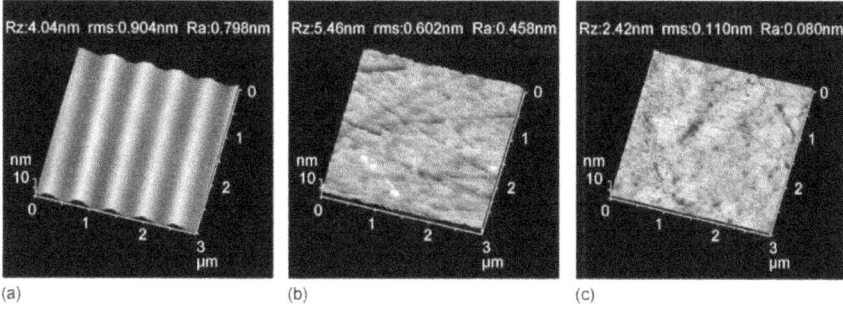

FIGURE 10.8 Machined NiP surfaces measured by scanning probe microscopy: (a) diamond-turning; (b) cloth-polishing; (c) float-polishing. Reprinted with permission from Elsevier (Namba et al. 2008).

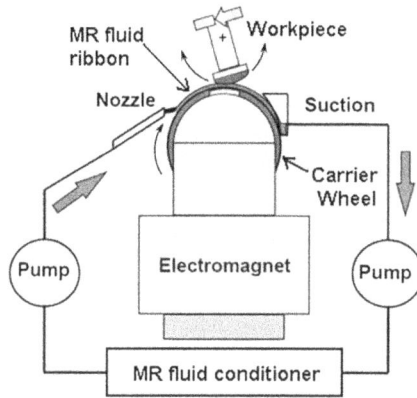

FIGURE 10.9 Schematic of the MRF process. Reprinted with permission from Elsevier (Jain 2009).

2008). A popular choice is through magnetorheological finishing (MRF) due to its unique ability to smoothen the spatial frequencies of aspherical and free-form optics (Jeon et al. 2017; Dumas, Golini, and Tricard 2005), and the higher material removal rates compared to traditional pad polishing processes (Golini et al. 1999).

MRF utilizes a magnetorheological (MR) fluid to be fed to the polishing zone as a ribbon on a rotating carrier wheel (Figure 10.9). A nozzle feeds the fluid to the polishing zone and the fluid is then collected in a suction cup to be deposited into a fluid mixer. This close-looped configuration allows for fluid mixture maintenance and its reuse. As the wheel rotates delivering the MR fluid ribbon, the workpiece also sweeps across the ribbon for surface smoothening by the converging gap between the workpiece and the carrier. Beneath the carrier wheel is an electromagnet designed to direct a magnetic field to the polishing zone and stiffen the MR fluid while conforming to the local curvature or topography of the part being polished

TABLE 10.4
Ranking of MRF Parameters on the Different Geometrical Removal
(Yang, Peng, and Hu 2022)

Desirable outcome	Parameter ranking (More important on the left)			
Affected area	Working depth	Field intensity	Polishing wheel speed	Fluid ribbon flow rate
Volume removal efficiency	Field intensity	Working depth	Polishing wheel speed	Fluid ribbon flow rate

at that position. Material is then removed by plastic shear force exerted by the conformal fluid ribbon as it converges and gets clamped through the gap. It is evident that the important parameters in MRF include the wheel speed, fluid flow, magnetic field strength, and the working depth to determine the effectiveness of the fluid ribbon to polish the workpiece.

The MR fluid comprises non-magnetic polishing abrasives (e.g., diamond, cerium oxide, aluminum oxide, silicon carbide, etc.), magnetic particles, and a carrier liquid mixed with additives or stabilizers (e.g., grease, glycerol, oleic acid, etc.) (Hashmi et al. 2022). The fluid alone involves multiple considerations such as the type and size of abrasive and magnetic particles, and the carrier fluid concentration and viscosity (Sidpara 2014).

Over the years, the MRF process has been adapted with multiple variations to accommodate complex geometries beyond the basic planar surface, such as the magnetorheological abrasive flow finishing (MRAFF) (Jha and Jain 2004) to tackle internal surfaces, MR jet finishing method (Kordonski, Shorey, and Tricard 2004) for freeform geometries and steep cavities, and ball-end MRF (Chen et al. 2016) for concave or spherical surfaces. While the main parameters remain mostly the same for each of the different processes, the range of each parameter for selection can largely vary. Thus, it is not advisable to prescribe a range of parameter selections. It is, however, useful to understand the influence of each parameter on the machining process. Yang, Peng, and Hu (2022) concluded the importance of parameters with respect to the different desirable outcomes listed in Table 10.4. As a gauge, processing parameters for their work included the fluid ribbon flow rate (55–75 liters/h), polishing wheel speed (200–280 rpm), magnetic field intensity current (2–4 A), and the working depth (0.1–0.3 mm). With these parameters and the use of silicon dioxide as the abrasive in the mixture, an optical surface roughness of 1.897 nm Ra was achievable on NiP. A superior NiP surface of 0.851 nm Ra can be achievable by MRF using an unconventional fluid with 1.5 μm iron magnetic particles at a concentration of 37% and 0.2% of nano-alumina abrasives (Bai et al. 2018).

Carbonyl iron particles (CIP) influence the MR effect under the magnetic field due to the elongation of CIP particle chains. Together with the increase in concentration and size of abrasive particles, material removal rates can be increased to reduce the overall machining time. However, the concentration of these particles together influences the overall viscosity of the MR fluid, which would affect the fluid circulation through the fluid delivery system. Thus, a balance must be struck between these few parameters regarding the MR fluid composition. An increase in wheel rotational speed also increases material removal rates due to the increase in shear forces acting at the tool-workpiece gap.

Despite MRF designed to remove diamond turning feed marks that result in spatial frequencies observed in the power spectrum, MRF too is a unidirectional process that would produce surfaces with smaller scales of these characteristic spatial frequencies. Hence, an additional multidirectional polishing process is required to reduce these effects (Moeggenborg et al. 2006). Hilpert et al. (2018) advised to employ chemical mechanical polishing (CMP) as the final finishing process to achieve a final surface quality of 0.4 nm Sa on AlSi12. It is still important to employ MRF after diamond turning to reduce the surface roughness before engaging polishing techniques to avoid long polishing periods.

CMP is a material removal process that involves the chemical reaction of the surface to form an oxidized layer that will be mechanically removed by slurry abrasives in a polishing pad at low pressures (1–3 psi). The combined chemical and mechanical processes are optimal for relatively higher material removal rates and lower subsurface damage compared to traditional polishing processes such as lapping. However, these polishing processes are still limited by the inability to process curved surfaces (Sidpara 2014).

The benefits of these two processes led to the evolution of a combined variation known as chemo-mechanical magnetorheological finishing (CMMRF) (Jain et al. 2010). The concept for material removal remains the same as CMP where a chemical reaction occurs to modify the surface layer for easy mechanical removal, except that the method for mechanical removal in CMMRF follows the MRF process, as illustrated in Figure 10.10. MR fluid compositions will then have to be recalibrated to be equipped with the chemical reactivity of the workpiece. Table 10.5 lists the various constituents of a typical MR fluid. In the case of aluminum, the MR fluid will comprise de-ionized water as the carrier fluid with a mixture of ammonia and citric acid for chemical interaction (Kumar and Singh 2021). A surface roughness of 0.6 nm Ra was achievable on Al7075-T6 by CMMRF (Figure 10.11) with a unique finding that

TABLE 10.5
Constituents of MR Fluid (Kumar and Singh 2021; Ghai et al. 2018; Hashmi et al. 2022)

Constituent	Examples
Base fluid	Deionized water, oil
Abrasive particles	Aluminum oxide, silicon carbide, boron carbide, cerium oxide, zirconium dioxide, diamond
Magnetic particles	Carbonyl iron particles, electrolytic iron powder, iron-cobalt alloy powder
Agglomeration control	Glycol, glycine, glycerol, oleic acid
Passive layer formation	Citric acid, ammonia

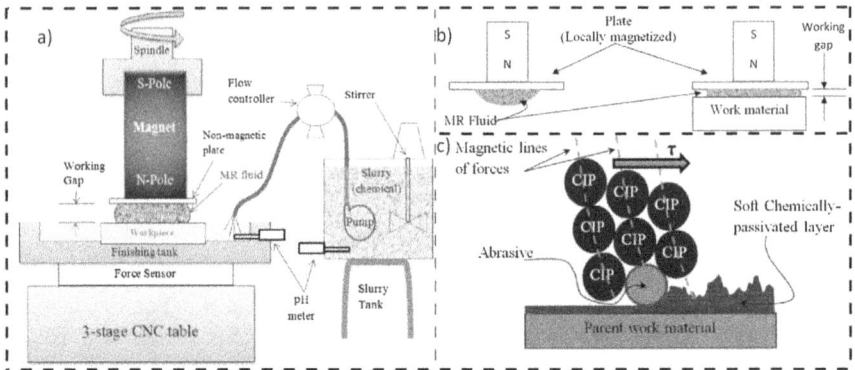

FIGURE 10.10 Schematic of the CMMRF process: (a) Setup; (b) MRF component; (c) interaction of MR fluid with chemically modified surface layer on work material. Reprinted with permission from Elsevier (Ranjan, Balasubramaniam, and Jain 2017).

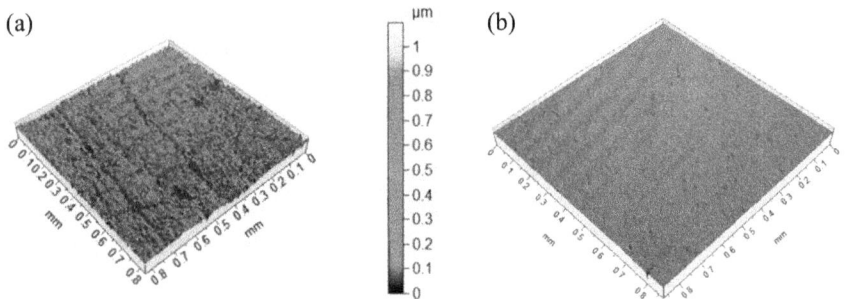

FIGURE 10.11 Comparison of Al7075-T6 surfaces: (a) as lapped (58.9 nm Ra) and (b) after CMMRF (0.597 nm Ra). Reprinted with permission from Elsevier (Ghai et al. 2018).

superior surface finishing was only possible with an alkaline solution in contrast to a more acidic solution for other metals (Ghai et al. 2018). To date, there has yet to be a report on the employment of CMMRF on NiP coatings, which might be attributed to the easy machinability of NiP with MRF and CMP as standalone technologies.

10.1.9 SUMMARIZED WORKFLOW

Figure 10.12 sums up the discussion on the process chain for the post-processing of additively manufactured aluminum alloy optical mirrors. The initial stages of post-processing encompass the typical part handling measures of support structure removal and part alignment on machine tools. Conventionally, diamond turning is performed for desired shape generation where good control of the feed enables the fabrication of relatively good surface quality (up to 10 nm Ra). Heat treatment may

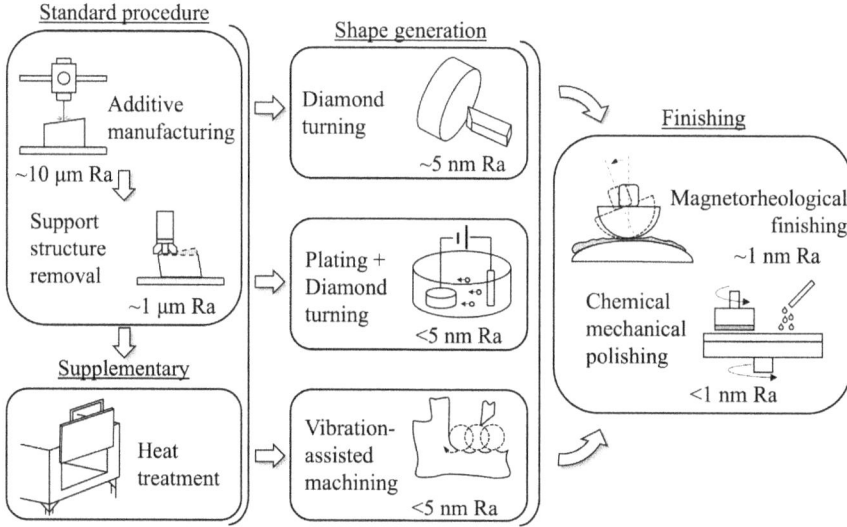

FIGURE 10.12 Variations of process flow to produce optical surface on additively manufactured aluminum alloys.

be employed to remove residual stresses but also generates new microstructures and precipitation that hinder the machined surface quality. Two alternative solutions can avoid such issues –to either replace the optical surface with a nickel-phosphorus coating to be diamond turned or adopt advanced tooling systems such as ultrasonic elliptical vibration-assisted machining. While both methods can improve the surface roughness to <5 nm Ra, turning feed marks remain an issue hindering lower wavelength applications. Thus, further polishing by magnetorheological finishing (MRF), chemical mechanical polishing (CMP), or the hybridized chemo-mechanical magnetorheological finishing (CMMRF) would be necessary to achieve atomic-level surface finishing.

10.2 INTRAVASCULAR STENT PRODUCTION

10.2.1 APPLICATION

Cardiovascular diseases cause millions of deaths annually. A large portion of the deaths are caused by the blockage of blood vessels due to the build-up of plaque. A surgical procedure to tackle this issue is placing a stent (i.e., a tubular mesh) inside the blood vessel to keep the vessel open and smoothen the blow flow. Stents can be made of polymers and metals such as stainless steels, titanium alloys, cobalt chromium alloys, nitinol (a metal alloy comprising nickel and titanium with shape memory properties), and others.

Conventional manufacturing methods for metallic stents include laser cutting and braiding. In laser cutting, a hollow metal tube rotates as the laser cuts out the gaps on the tube surface to form a tubular stent. Laser cutting is only limited to the formation of tubular stents. The process will lead to sharp corners and an oxidized layer that will require secondary processing for improved surface quality. In braiding, more complex stents can be generated using expensive fixtures and tools to braid metal wires into specially designed stents. These fixtures and tools are normally developed for one stent design and changes of the stent design would require the development of new equipment, which drives up the cost of this method. Furthermore, braiding is a labor-intensive operation.

Employing AM to fabricate stents has several attractive advantages. First, AM allows great freedom for customization for stent designs that can have added functions. Moreover, the short production cycle and capability for on-site fabrication allow point-of-care product delivery which is useful and convenient for hospitals and clinics to produce personalized stents for patients without worrying about inventory and delivery. However, there are disadvantages of the AM process that hinder this application in practical. For example, poor surface finish and inadequate dimensional accuracy require post-processing technologies such as chemical etching (Tyagi et al. 2019), magnetic abrasive finishing (Zhang, Chaudhari, and Wang 2019), vibration-assisted conformal finishing (Zhang et al. 2018), and others to come into the picture. In this case study, the design, printing, and post-processing of AM stents will be discussed.

10.2.2 Process Chain

The manufacturing process chain for stents is presented in Figure 10.13. Conventionally, raw materials comprise tubes, wires, sheets, and bars to be processed by laser cutting and braiding into the designed stent geometries. AM employs metal powders as the raw material to produce stents with customized designs and complex geometries followed by post-processing such as heat treatment, surface treatment, inspection, quality assurance, and packaging. Using AM to fabricate stents may pose new challenges for post-processing.

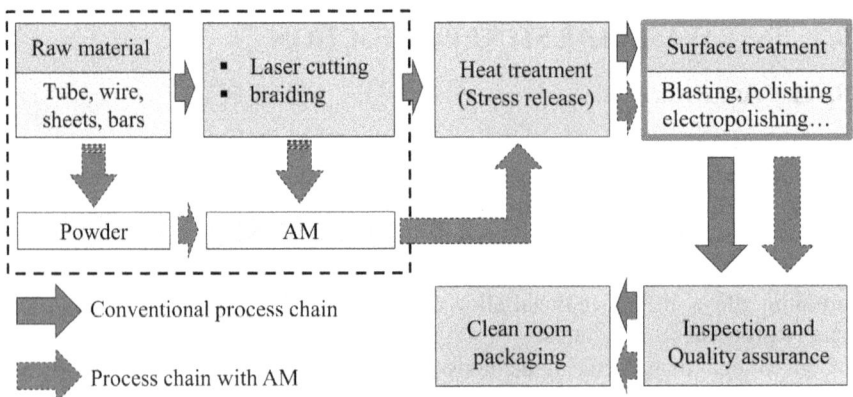

FIGURE 10.13 Manufacturing process chain for stents.

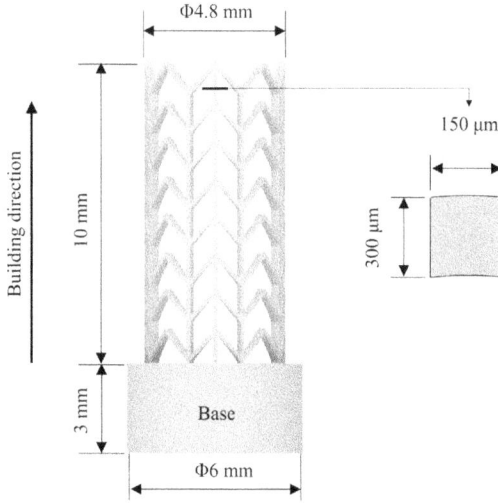

FIGURE 10.14 Stent design with the top view of a cross-sectioned strut (Zhang 2019).

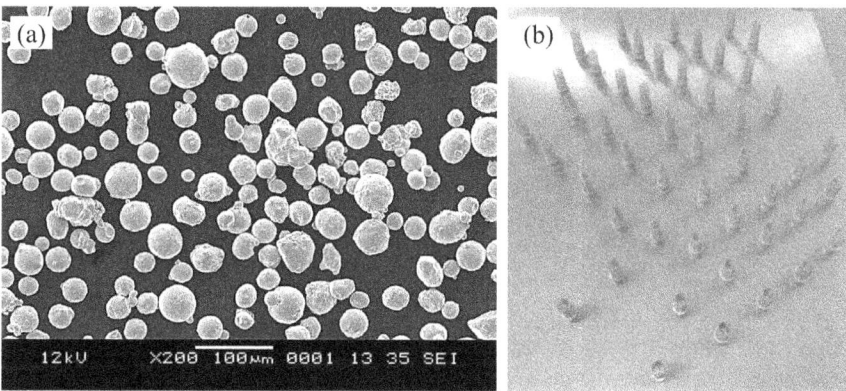

FIGURE 10.15 (a) SEM image of the 316L powders used for printing and (b) the printed stents on the substrate.

10.2.3 DESIGN AND FABRICATION

Figure 10.14 shows a typical example of a stent designed with an outer diameter (OD) and inner diameter (ID) of 4.8 mm and 4.2 mm, respectively. Each strut is part of a ring measuring 150 μm in width and 300 μm in thickness. A 6 mm-diameter and 3 mm-tall bases is designed for clamping during post-processing and can be subsequently removed. Such a stent structure made of 316L stainless steel can be fabricated by LPBF with parameters listed in Table 10.6 for good mechanical properties. Examples of stents producible by LPBF on an EOS 290M SLM printer with stainless steel powder (Figure 10.15(a)) are shown in Figure 10.15(b). The stents

TABLE 10.6
Selective Laser Melting Parameters to Fabricate
the Designed Stents

Parameters	Values
Laser power	195 W
Laser power for support	100 W
Scanning speed	960 mm/s
Scanning speed for support	850 mm/s
Hatch distance	0.09 mm
Layer thickness	0.04 mm
Powder material	0.02 mm
Powder diameter	316 L
Scan strategy	Mixed

were constructed vertically to eliminate the need for support structures inside the stent. Support structures are necessary in the printing base where a 2 mm thick support was constructed between the base of the stent and the building platform. Different printing parameters should be employed to aid the easy removal of the stents. The materials of the stent can also be biocompatible alloys such as nickel titanium (Nitinol) and CoCr.

Defects will arise from the additive manufacturing process – such as partially bonded particles, balling effect, staircase effect, and ripple effect – which greatly deteriorate the dimensional accuracy and surface finish of the fabricated parts. As discussed in earlier chapters of the book, there are many conventional finishing techniques such as machining, mechanical polishing, and electrochemical polishing that have been applied for post-processing. However, the complex geometry of stents requires post-processing methods with high accessibility and material removal efficiency. Mechanical and electrochemical material removal mechanisms will be discussed as these two methods can employ media to reach the internal surfaces. For instance, micro-blasting employs micro-sized abrasives that can flow through the stent and mechanically abrade the internal surfaces. Chemical reactants can also involve solid particles that flow and contact the internal surfaces but features a different material removal mechanism.

10.2.4 Mechanical-Based Finishing

Blasting is a common practice to remove the partially bonded particles on the as-printed parts and to provide a particle-free surface for further processing. Micro-blasting employs micrometer-scale spheres and a miniature nozzle to produce controllable particle flow to process small and intricate parts. The concept of micro-blasting is illustrated in Figure 10.16(a) where the stent is mounted onto a motor to be rotated while a nozzle blasts spherical ceramic abrasive balls that are 150–200 μm in size (Figure 10.16(b) to process the stent. The important parameters of micro-blasting

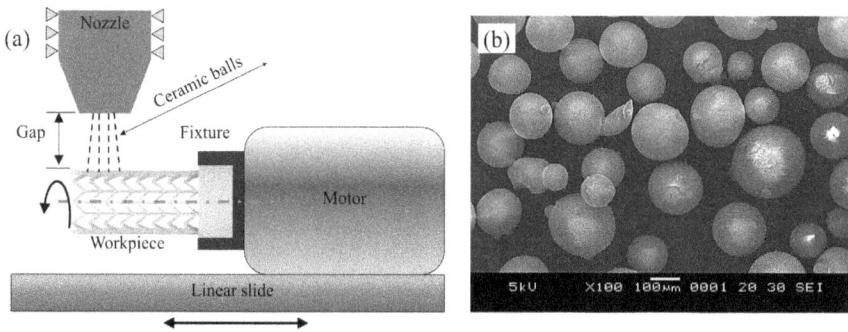

FIGURE 10.16 (a) Schematic of developed micro-blasting setup and (b) SEM imaging of the ceramic abrasive balls (Zhang 2019).

TABLE 10.7
Micro-Blasting Parameters for the SLM Stents

Parameters	Values
Nozzle diameter	5 mm
Media	Ceramic balls with a diameter of 150–200 μm
Spindle rotating speed	30 rpm
Linear slide velocity	1 mm/s
Gap	18, 27, 36 mm
Air pressure	1, 2, 3 bar
Coverage	100, 200, 300%

are listed in Table 10.7, which include the blast pressure, workpiece-nozzle gap distance, and coverage. The blast pressure is controlled by an air valve. A higher blast pressure results in a higher particle speed and a larger material removal rate. Too high blast pressure may lead to abrasive embedding into the workpiece. The work-nozzle gap distance controls the processing area and the impact energy imposed on the workpiece, thus affecting the material removal rate. A small gap distance enables a large material removal rate on a limited region, while a large gap distance can process a large area with low efficiency.

The coverage indicates the process cycles – e.g., 100% coverage means the whole surface of the sample was micro-blasted once, and 300% coverage means another two process cycles are repeated. More materials will be removed after more coverages, which, however, does not necessarily lead to a better surface finish since the surface roughness will saturate after a certain coverage.

10.2.5 Performance of Micro-Blasting

The printed 316L stents were processed with the parameters listed in Table 10.7. with a 3^3 full-factorial experiment design considering the effect of coverage, gap, and

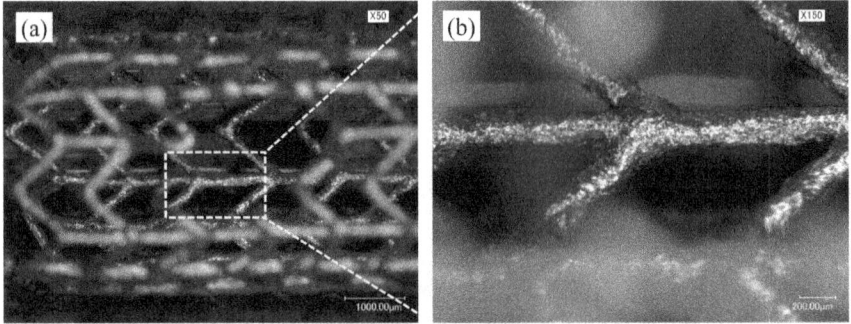

FIGURE 10.17   Finishing of inner surfaces: (a) overall image of the stent and (b) enlarged image of internal surface.

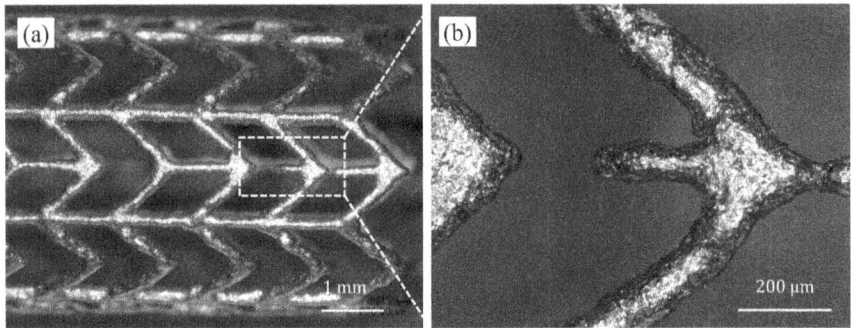

FIGURE 10.18   Strut damage after micro-blasting (Sample #6): (a) overall image and (b) zoom-in image (Zhang 2019).

blast pressure. After micro-blasting, the outer diameter and strut widths of the stent lattices were reduced by an average of 83 µm and 32 µm, respectively. This reduction is mainly caused by the removal of partially bonded particles. Analysis of variance (ANOVA) revealed that the gap distance has the most influence on mass loss and strut width reduction. As seen in Figure 10.17, the micro-blasting has the capability of processing the inner surfaces of the stents. However, high pressures in micro-blasting may also lead to strut damage shown in Figure 10.18. Thus, pressures should be high enough for efficient material removal but limited to prevent damage to the fragile structures generated by AM. The surface roughness can reach below 2 µm Ra from the initial 6–10 µm within several minutes.

In short, one can conclude that micro-blasting is a suitable rough finishing process for SLM stents. It can efficiently remove most of the partially bonded particles and prepare a relatively smooth surface for the following finishing process.

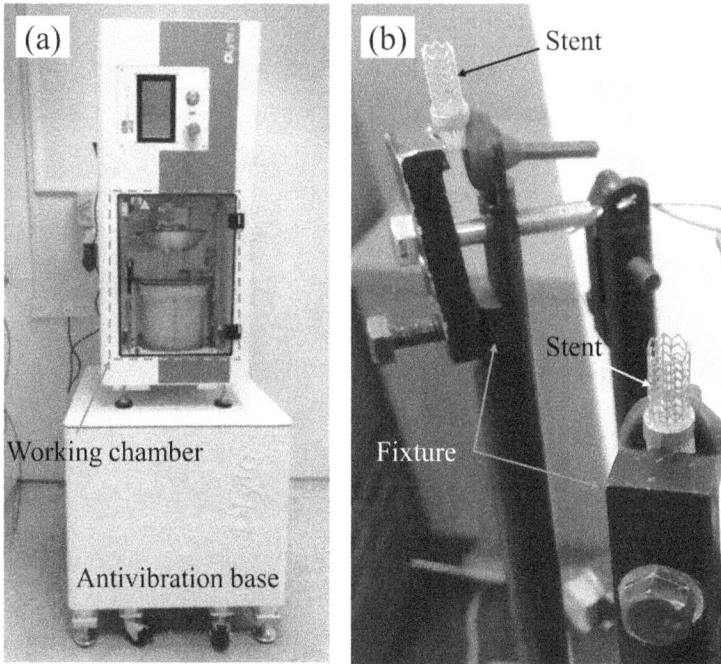

FIGURE 10.19 Dry mechanical electrochemical polishing (DMECP) of SLM stents: (a) the DMECP machine (GPA Innovation DLyte 1I+Ti) and (b) stents mounting on the fixtures for DMECP.

10.2.6 PERFORMANCE OF ELECTROCHEMICAL POLISHING

Chemical etching and electrochemical polishing can remove the partially bonded particles and produce smooth surfaces on AM lattice structures (Pyka et al. 2012; Demir and Previtali 2017). In this case study, the dry mechanical electrochemical polishing (DMECP) concept is used for comparison with micro-blasting considering that both processes only employ solid particles to contact the internal surfaces during the polishing process. A DMECP setup is shown in Figure 10.19(a) where the stents are secured onto fixtures (Figure 10.19(b)) to be embedded inside the polishing media of polymer spheres (DRYLYTE-DLYTE-02). After polishing for 1 h with a voltage set at 30V, the stent surface can be polished to a roughness of 4.14 μm by DMECP alone as presented in Figure 10.20. However, partially bonded particles at the corners between two adjacent struts remain as it can be difficult for the polishing media to access these regions.

A combination of micro-blasting and DMECP works best where partially bonded particles are more easily removed during micro-blasting followed by the polishing action of DMECP. This procedure can achieve surfaces with a roughness of ~60 nm Ra as given in Figure 10.20(d). However, local deformation can still occur in the form of a circular shape as highlighted by the yellow lines in Figure 10.20(d), due to the continued etching of stuck polymer spheres.

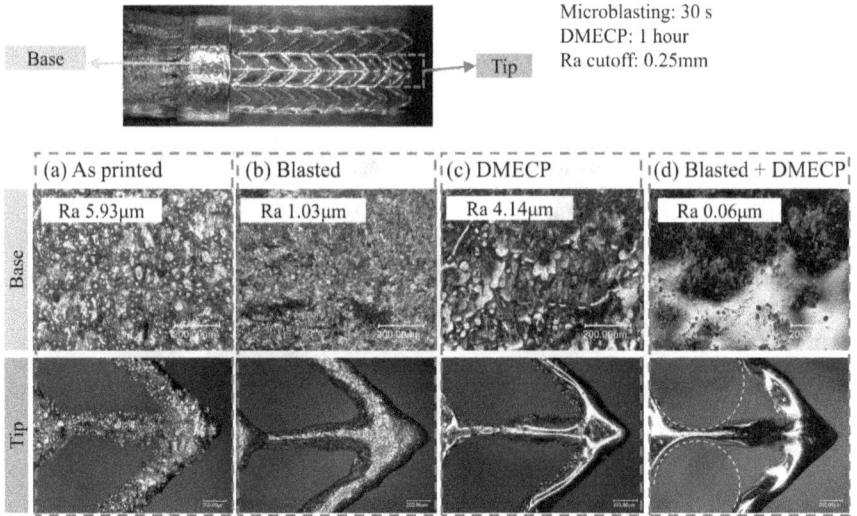

FIGURE 10.20   Hybrid process for stents. (a) as printed, (b) micro-blasted, (c) electropolished and (d) micro-blasted followed by electropolishing.

10.2.7   SUMMARIZED WORKFLOW

Post-processing surface quality is critical for AM stents. Despite the variety of available processes that can reach the internal surfaces, it is evident that there are rough polishing and finish polishing processes. Micro-blasting can serve as the rough finishing method to effectively remove most of the surface defects, such as partially bonded particles, balling, and staircase effect, and achieve a surface with < 5 μm Ra. Caution should be taken to avoid failure and severe deformation of fragile structures. After micro-blasting, conventional electrochemical polishing (EP) or dry mechanical electrochemical polishing (DMECP) can be employed as the secondary finishing method to enhance the surface quality (sub-micrometer scale). Post-cleaning may be needed to remove the residual chemicals for further testing.

REFERENCES

Apachitei, I, F D Tichelaar, J Duszczyk, and L Katgerman. 2002. "The Effect of Heat Treatment on the Structure and Abrasive Wear Resistance of Autocatalytic NiP and NiP–SiC Coatings." *Surface and Coatings Technology* 149 (2): 263–278. doi:10.1016/S0257-8972(01)01492-X.

Atkins, Carolyn, Charlotte Feldman, David Brooks, Stephen Watson, William Cochrane, Melanie Roulet, Emmanuel Hugot, et al. 2018. "Topological Design of Lightweight Additively Manufactured Mirrors for Space." In *Proc.SPIE*. Vol. 10706. doi:10.1117/12.2313353.

Bai, Y., Z. Shi, Y.J. Lee, and H. Wang. 2020. "Optical Surface Generation on Additively Manufactured AlSiMg0.75 Alloys with Ultrasonic Vibration-Assisted Machining." *Journal of Materials Processing Technology* 280. doi:10.1016/j.jmatprotec.2020.116597.

Bai, Yang, Zhiyu Zhang, Donglin Xue, and Xuejun Zhang. 2018. "Ultra-Precision Fabrication of a Nickel-Phosphorus Layer on Aluminum Substrate by SPDT and MRF." *Applied Optics* 57 (34). OSA: F62–F67. doi:10.1364/AO.57.000F62.

Barnes, W P. 1972. "Hexagonal vs Triangular Core Lightweight Mirror Structures." *Applied Optics* 11 (12). Optica Publishing Group: 2748–2751. doi:10.1364/AO.11.002748.

Blackley, W.S., and R.O. Scattergood. 1991. "Ductile-Regime Machining Model for Diamond Turning of Brittle Materials." *Precision Engineering* 13 (2): 95–103. doi:10.1016/0141-6359(91)90500-I.

Brehl, D. E., and T. A. Dow. 2008. "Review of Vibration-Assisted Machining." *Precision Engineering* 32 (3): 153–172. doi:10.1016/j.precisioneng.2007.08.003.

Cardenas, Nelson, Matthew Kyrish, Daniel Taylor, Margaret Fraelich, Oscar Lechuga, Richard Claytor, and Nelson Claytor. 2015. "Quantification of Microscopic Surface Features of Single Point Diamond Turned Optics with Subsequent Chemical Polishing." In *Proc. SPIE*. Vol. 9374. doi:10.1117/12.2080228.

Casstevens, J. M., and C. E. Daugherty. 1978. "<title>Diamond Turning Optical Surfaces On Electroless Nickel</Title>." In *Precision Machining of Optics*, 0159:109–113. doi:10.1117/12.956844.

Chabot, Tristan, Denis Brousseau, Hugues Auger, and Simon Thibault. 2019. "Diamond Turning of Aluminum Image Slicers for Astronomical Applications." In *Proc.SPIE*. Vol. 11175. doi:10.1117/12.2536870.

Chen, Mingjun, Henan Liu, Yinrui Su, Bo Yu, and Zhen Fang. 2016. "Design and Fabrication of a Novel Magnetorheological Finishing Process for Small Concave Surfaces Using Small Ball-End Permanent-Magnet Polishing Head," no. 26: 823–834. doi:10.1007/s00170-015-7573-5.

Cheng, Yuan-Chieh, Wei-Yao Hsu, Khaled Abou-El-Hossein, Oluwole Olufayo, and Timothy Otieno. 2014. "Investigation of Diamond Turning: Of Rapidly Solidified Aluminum Alloys." In *Proc.SPIE*. Vol. 9192. doi:10.1117/12.2060176.

Cheng, Yuan-Chieh, Wei-Yao Hsu, Ching-Hsiang Kuo, Khaled Abou-El-Hossein, and Timothy Otieno. 2015. "Investigation of Rapidly Solidified Aluminum by Using Diamond Turning and a Magnetorheological Finishing Process." In *SPIE Optical Engineering+ Applications*, 9575:957519. International Society for Optics and Photonics. doi:10.1117/12.2186725.

Cheung, C F, and W B Lee. 2000. "A Theoretical and Experimental Investigation of Surface Roughness Formation in Ultra-Precision Diamond Turning." *International Journal of Machine Tools and Manufacture* 40 (7). Elsevier: 979–1002. doi:10.1016/S0890-6955(99)00103-0.

Chon, Kwon Su, and Yoshiharu Namba. 2010. "Single-Point Diamond Turning of Electroless Nickel for Flat X-Ray Mirror." *Journal of Mechanical Science and Technology* 24 (8): 1603–1609. doi:10.1007/s12206-010-0512-3.

Chon, Kwon Su, Yoshiharu Namba, and Kwon-Ha Yoon. 2006. "Precision Machining of Electroless Nickel Mandrel and Fabrication of Replicated Mirrors for a Soft X-Ray Microscope." *JSME International Journal Series C Mechanical Systems, Machine Elements and Manufacturing* 49 (1): 56–62. doi:10.1299/jsmec.49.56.

Demir, Ali Gökhan, and Barbara Previtali. 2017. "Additive Manufacturing of Cardiovascular CoCr Stents by Selective Laser Melting." *Materials and Design* 119. Elsevier: 338–350. doi:10.1016/j.matdes.2017.01.091.

Dumas, Paul, Don Golini, and Marc Tricard. 2005. "Improvement of Figure and Finish of Diamond Turned Surfaces with Magneto-Rheological Finishing (MRF)." In *Proc.SPIE*. Vol. 5786. doi:10.1117/12.603967.

Ghai, Viney, Prabhat Ranjan, Ajay Batish, and Harpreet Singh. 2018. "Atomic-Level Finishing of Aluminum Alloy by Chemo-Mechanical Magneto-Rheological Finishing (CMMRF) for Optical Applications." *Journal of Manufacturing Processes* 32: 635–643. doi:10.1016/j.jmapro.2018.03.032.

Golini, Donald, William I Kordonski, Paul Dumas, and Stephen J Hogan. 1999. "Magnetorheological Finishing (MRF) in Commercial Precision Optics Manufacturing." In *Optical Manufacturing and Testing III*, edited by H. Philip Stahl, 3782:80–91. International Society for Optics and Photonics. doi:10.1117/12.369174.

Greco, Sebastian, Katja Klauer, Benjamin Kirsch, and Jan C Aurich. 2021. "Vibration-Assisted Micro Milling of AISI 316L Produced by Laser-Based Powder Bed Fusion." *Journal of Manufacturing Processes* 71: 298–305. doi:10.1016/j.jmapro.2021.09.020.

Gubbels, Guido P. H., Bart W. H. van Venrooy, Albert J. Bosch, and Roger Senden. 2008. "Rapidly Solidified Aluminium for Optical Applications." In *Advanced Optical and Mechanical Technologies in Telescopes and Instrumentation*, 7018:70183A. International Society for Optics and Photonics. doi:10.1117/12.788766.

Han, Xiao, Nan Kang, Jianchao Jiao, and Chao Wang. 2019. "Research for Surface Characteristics of Aluminum Mirrors Made by Laser Additive Manufacturing." In *Proc. SPIE*. Vol. 11333. doi:10.1117/12.2548108.

Hashmi, Abdul Wahab, Harlal Singh Mali, Anoj Meena, Irshad Ahamad Khilji, Chaitanya Reddy Chilakamarry, and Siti Nadiah binti Mohd Saffe. 2022. "Experimental Investigation on Magnetorheological Finishing Process Parameters." *Materials Today: Proceedings* 48: 1892–1898. doi:10.1016/j.matpr.2021.09.355.

Hilpert, Enrico, Johannes Hartung, Henrik von Lukowicz, Tobias Herffurth, and Nils Heidler. 2019. "Design, Additive Manufacturing, Processing, and Characterization of Metal Mirror Made of Aluminum Silicon Alloy for Space Applications." *Optical Engineering* 58 (9): 1–9. doi:10.1117/1.OE.58.9.092613.

Hilpert, Enrico, Johannes Hartung, Stefan Risse, Ramona Eberhardt, and Andreas Tünnermann. 2018. "Precision Manufacturing of a Lightweight Mirror Body Made by Selective Laser Melting." *Precision Engineering* 53 (July): 310–317. doi:10.1016/j.precisioneng.2018.04.013.

Ituarte, Iñigo Flores, Mika Salmi, Suvi Papula, Juha Huuki, Björn Hemming, Eric Coatanea, Seppo Nurmi, and Iikka Virkkunen. 2020. "Surface Modification of Additively Manufactured 18% Nickel Maraging Steel by Ultrasonic Vibration-Assisted Ball Burnishing." *Journal of Manufacturing Science and Engineering* 142 (7). doi:10.1115/1.4046903.

Jain, V. K. 2009. "Magnetic Field Assisted Abrasive Based Micro-/Nano-Finishing." *Journal of Materials Processing Technology* 209 (20): 6022–6038. doi:10.1016/j.jmatprotec.2009.08.015.

Jain, V K, P Ranjan, V K Suri, and R Komanduri. 2010. "Chemo-Mechanical Magneto-Rheological Finishing (CMMRF) of Silicon for Microelectronics Applications." *CIRP Annals* 59 (1): 323–328. doi:10.1016/j.cirp.2010.03.106.

Jeon, Min-Woo, Sangwon Hyun, Byeong-Joon Jeong, I-Jong Kim, and Geon-Hee Kim. 2017. "Removal of Diamond Turning Marks with Magneto-Rheological Finishing." In *Proc. SPIE*. Vol. 10371. doi:10.1117/12.2274028.

Jha, Sunil, and V.K. Jain. 2004. "Design and Development of the Magnetorheological Abrasive Flow Finishing (MRAFF) Process." *International Journal of Machine Tools and Manufacture* 44 (10): 1019–1029. doi:10.1016/j.ijmachtools.2004.03.007.

Kamiya, Masatsugu, Takao Yakou, Tomohiro Sasaki, and Yoshiki Nagatsuma. 2008. "Effect of Si Content on Turning Machinability of Al-Si Binary Alloy Castings." *MATERIALS TRANSACTIONS* 49 (3): 587–592. doi:10.2320/matertrans.L-MRA2007886.

Kikuchi, Masafumi, Masatoshi Takahashi, Toru Okabe, and Osamu Okuno. 2003. "Grindability of Dental Cast Ti-Ag and Ti-Cu Alloys." *Dental Materials Journal* 22 (2): 191–205. doi:10.4012/dmj.22.191.

Komanduri, R, N Chandrasekaran, and L.M. Raff. 1998. "Effect of Tool Geometry in Nanometric Cutting: A Molecular Dynamics Simulation Approach." *Wear* 219 (1): 84–97. doi:10.1016/S0043-1648(98)00229-4.

Kordonski, William I, Aric B Shorey, and Marc Tricard. 2004. "Magnetorheological (MR) Jet Finishing Technology," November, 77–84. doi:10.1115/IMECE2004-61214.

Kumar, Yogendra, and Harpreet Singh. 2021. "Chemomechanical Magnetorheological Finishing: Process Mechanism, Research Trends, Challenges and Opportunities in Surface Finishing." *Journal of Micromanufacturing*, August. SAGE Publications India, 25165984211038880. doi:10.1177/25165984211038878.

Lelevic, Aleksandra, and Frank C Walsh. 2019. "Electrodeposition of NiP Alloy Coatings: A Review." *Surface and Coatings Technology* 369: 198–220. doi:10.1016/j.surfcoat.2019.03.055.

Moeggenborg, Kevin, Tamara Vincer, Stanley Lesiak, and Roman Salij. 2006. "Super-Polished Aluminum Mirrors through the Application of Chemical Mechanical Polishing Techniques." In *Proc.SPIE*. Vol. 6288. doi:10.1117/12.681043.

Moriwaki, T, and E Shamoto. 1991. "Ultraprecision Diamond Turning of Stainless Steel by Applying Ultrasonic Vibration." *CIRP Annals* 40 (1): 559–562. doi:10.1016/S0007-8506(07)62053-8.

Namba, Y, T Shimomura, A Fushiki, A Beaucamp, I Inasaki, H Kunieda, Y Ogasaka, and K Yamashita. 2008. "Ultra-Precision Polishing of Electroless Nickel Molding Dies for Shorter Wavelength Applications." *CIRP Annals* 57 (1): 337–340. doi:10.1016/j.cirp.2008.03.077.

Pramanik, A., K. S. Neo, M. Rahman, X. P. Li, M. Sawa, and Y. Maeda. 2009. "Ultraprecision Turning of Electroless Nickel: Effects of Crystal Orientation and Origin of Diamond Tools." *International Journal of Advanced Manufacturing Technology* 43 (7–8): 681–689. doi:10.1007/s00170-008-1748-2.

Pramanik, A, K S Neo, M Rahman, X P Li, M Sawa, and Y Maeda. 2003. "Cutting Performance of Diamond Tools during Ultra-Precision Turning of Electroless-Nickel Plated Die Materials." *Journal of Materials Processing Technology* 140 (1): 308–313. doi:10.1016/S0924-0136(03)00751-9.

Pramanik, A, K S Neo, M Rahman, X P Li, M Sawa, and Y Maeda. 2008. "Ultra-Precision Turning of Electroless-Nickel: Effect of Phosphorus Contents, Depth-of-Cut and Rake Angle." *Journal of Materials Processing Technology* 208 (1): 400–408. doi:10.1016/j.jmatprotec.2008.01.006.

Pyka, Grzegorz, Andrzej Burakowski, Greet Kerckhofs, Maarten Moesen, Simon Van Bael, Jan Schrooten, and Martine Wevers. 2012. "Surface Modification of Ti6Al4V Open Porous Structures Produced by Additive Manufacturing." *Advanced Engineering Materials* 14 (6). Wiley Online Library: 363–370. doi:10.1002/adem.201100344.

Rahman, M. Azizur, M. Raihan Amrun, Mustafizur Rahman, and A. Senthil Kumar. 2017. "Variation of Surface Generation Mechanisms in Ultra-Precision Machining Due to Relative Tool Sharpness (RTS) and Material Properties." *International Journal of Machine Tools and Manufacture* 115 (June 2016): 15–28. doi:10.1016/j.ijmachtools.2016.11.003.

Rahman, M Azizur, Mustafizur Rahman, and A Senthil Kumar. 2018. "Influence of Relative Tool Sharpness (RTS) on Different Ultra-Precision Machining Regimes of Mg Alloy." *The International Journal of Advanced Manufacturing Technology* 96 (9): 3545–3563. doi:10.1007/s00170-018-1599-4.

Ranjan, Prabhat, R Balasubramaniam, and V K Jain. 2017. "Analysis of Magnetorheological Fluid Behavior in Chemo-Mechanical Magnetorheological Finishing (CMMRF) Process." *Precision Engineering* 49. Elsevier Inc.: 122–135. doi:10.1016/j.precisioneng.2017.02.001.

Revel, Philippe, Hatem Khanfir, and René-Yves Fillit. 2006. "Surface Characterization of Aluminum Alloys after Diamond Turning." *Journal of Materials Processing Technology* 178 (1): 154–161. doi:10.1016/j.jmatprotec.2006.03.169.

Santos, Mário C., Alisson R. Machado, Wisley F. Sales, Marcos A. S. Barrozo, and Emmanuel O. Ezugwu. 2016. "Machining of Aluminum Alloys: A Review." *The International Journal of Advanced Manufacturing Technology* 86 (9–12). Springer London: 3067–3080. doi:10.1007/s00170-016-8431-9.

Sealy, Michael P, Gurucharan Madireddy, Robert E Williams, Prahalada Rao, and Maziar Toursangsaraki. 2018. "Hybrid Processes in Additive Manufacturing." *Journal of Manufacturing Science and Engineering* 140 (6). doi:10.1115/1.4038644.

Shamoto, Eiji, and Toshimichi Moriwaki. 1999. "Ultaprecision Diamond Cutting of Hardened Steel by Applying Elliptical Vibration Cutting." *Annals of the CIRP* 48 (1): 441–444. doi:10.1016/S0007-8506(07)63222-3.

Siddique, Shafaqat, Muhammad Imran, Eric Wycisk, Claus Emmelmann, and Frank Walther. 2015. "Influence of Process-Induced Microstructure and Imperfections on Mechanical Properties of AlSi12 Processed by Selective Laser Melting." *Journal of Materials Processing Technology* 221: 205–213. doi:10.1016/j.jmatprotec.2015.02.023.

Sidpara, Ajay M. 2014. "Magnetorheological Finishing: A Perfect Solution to Nanofinishing Requirements." *Optical Engineering* 53 (9): 1–8. doi:10.1117/1.OE.53.9.092002.

Staia, M H, E J Castillo, E S Puchi, B Lewis, and H E Hintermann. 1996. "Wear Performance and Mechanism of Electroless Ni-P Coating." *Surface and Coatings Technology* 86–87: 598–602. doi:10.1016/S0257-8972(96)03086-1.

Su, Yongsheng, and Liang Li. 2022. "Surface Integrity of Ultrasonic-Assisted Dry Milling of SLM Ti6Al4V Using Polycrystalline Diamond Tool." *The International Journal of Advanced Manufacturing Technology* 119 (9): 5947–5956. doi:10.1007/s00170-022-08669-4.

Syn, C. K., J. S. Taylor, and R. R. Donaldson. 1986. "Diamond Tool Wear vs Cutting Distance on Electroless Nickel Mirrors." In *Proc. SPIE 0676, Ultraprecision Machining and Automated Fabrication of Optics.* doi:10.1117/12.939527.

Tan, Rongkai, Xuesen Zhao, Shusen Guo, Xicong Zou, Yang He, Yanquan Geng, Zhenjiang Hu, and Tao Sun. 2020. "Sustainable Production of Dry-Ultra-Precision Machining of Ti–6Al–4V Alloy Using PCD Tool under Ultrasonic Elliptical Vibration-Assisted Cutting." *Journal of Cleaner Production* 248. Elsevier: 119254. doi:10.1016/j.jclepro.2019.119254.

Taylor, John S, Choi K Syn, Theodore T Saito, and Robert R Donaldson. 1986. "Surface Finish Measurements Of Diamond-Turned Electroless-Nickel-Plated Mirrors." *Optical Engineering* 25 (9): 1013–1020. doi:10.1117/12.7973947.

Teramachi, Akinori, and Jiwang Yan. 2019. "Improving the Surface Integrity of Additive-Manufactured Metal Parts by Ultrasonic Vibration-Assisted Burnishing." *Journal of Micro and Nano-Manufacturing* 7 (2). doi:10.1115/1.4043344.

Tyagi, Pawan, Tobias Goulet, Christopher Riso, Robert Stephenson, Nitt Chuenprateep, Justin Schlitzer, Cordell Benton, and Francisco Garcia-Moreno. 2019. "Reducing the Roughness of Internal Surface of an Additive Manufacturing Produced 316 Steel Component by Chempolishing and Electropolishing." *Additive Manufacturing* 25. Elsevier: 32–38. doi:10.1016/j.addma.2018.11.001.

Wang, H., S. To, C. Y. Chan, C. F. Cheung, and W. B. Lee. 2011. "Dynamic Modelling of Shear Band Formation and Tool-Tip Vibration in Ultra-Precision Diamond Turning." *International Journal of Machine Tools and Manufacture.* doi:10.1016/j.ijmachtools.2011.02.010.

Wang, S J, X Chen, S To, X B Ouyang, Q Liu, J W Liu, and W B Lee. 2015. "Effect of Cutting Parameters on Heat Generation in Ultra-Precision Milling of Aluminum Alloy 6061." *The International Journal of Advanced Manufacturing Technology* 80 (5): 1265–1275. doi:10.1007/s00170-015-7072-8.

Wang, Yifan, Jun Yu, Zhengxiang Shen, Qiushi Huang, Bin Ma, Zhong Zhang, and Zhanshan Wang. 2021. "Machining Process of Lightweight AlSi10Mg Optical Mirror Based on Additive Manufacturing Substrate." In *Proc.SPIE*. Vol. 12073. doi:10.1117/12.2603970.

Xing, Yintian, Yue Liu, Chao Yang, and Changxi Xue. 2020. "Roughness Model of an Optical Surface in Ultrasonic Assisted Diamond Turning." *Applied Optics* 59 (31). OSA: 9722–9734. doi:10.1364/AO.402613.

Yang, Qilin, Xiaoqiang Peng, and Hao Hu. 2022. "Study on Optimization of a NiP Layer Polishing Process on Magnetorheological Finishing for Metal Substrate." In *Proc.SPIE*. Vol. 12166. doi:10.1117/12.2604879.

Yu, Qian, Tianfeng Zhou, Yupeng He, Peng Liu, Xibin Wang, and Jiwang Yan. 2021. "Effects of Relative Tool Sharpness on Surface Generation Mechanism of Precision Turning of Electroless Nickel-Phosphorus Coating." *Journal of Mechanical Science and Technology* 35 (7): 3113–3121. doi:10.1007/s12206-021-0633-x.

Zhang, Jianguo, Junjie Zhang, Andreas Rosenkranz, Norikazu Suzuki, and Eiji Shamoto. 2019. "Frictional Properties of Surface Textures Fabricated on Hardened Steel by Elliptical Vibration Diamond Cutting." *Precision Engineering* 59: 66–72. doi:10.1016/j.precisioneng.2019.06.001.

Zhang, Jiong. 2019. "Micro-Blasting of 316L Tubular Lattice Manufactured by Laser Powder Bed Fusion." In *European Society for Precision Engineering and Nanotechnology, Conference Proceedings – 19th International Conference and Exhibition, EUSPEN 2019*, 254–255. Bilbao, ES. www.euspen.eu/knowledge-base/ICE19107.pdf.

Zhang, Jiong, Akshay Chaudhari, and Hao Wang. 2019. "Surface Quality and Material Removal in Magnetic Abrasive Finishing of Selective Laser Melted 316L Stainless Steel." *Journal of Manufacturing Processes* 45. Elsevier: 710–719. doi:10.1016/j.jmapro.2019.07.044.

Zhang, Jiong, Alvin You Xiang Toh, Hao Wang, Wen Feng Lu, and Jerry Ying Hsi Fuh. 2018. "Vibration-Assisted Conformal Polishing of Additively Manufactured Structured Surface." *Proceedings of the Institution of Mechanical Engineers, Part C: Journal of Mechanical Engineering Science* 233 (12). IMECHE: 4154–4164. doi:10.1177/0954406218811359.

Zhang, Peirong, Zhanqiang Liu, Jin Du, Guosheng Su, Jingjie Zhang, and Yujing Sun. 2020. "Correlation between the Microstructure and Machinability in Machining Al–(5–25) Wt% Si Alloys." *Proceedings of the Institution of Mechanical Engineers, Part B: Journal of Engineering Manufacture* 234 (9). IMECHE: 1173–1184. doi:10.1177/0954405420911275.

Zhang, Xinquan, A. Senthil Kumar, Mustafizur Rahman, Chandra Nath, and Kui Liu. 2011. "Experimental Study on Ultrasonic Elliptical Vibration Cutting of Hardened Steel

Using PCD Tools." *Journal of Materials Processing Technology* 211 (11): 1701–1709. doi:10.1016/j.jmatprotec.2011.05.015.

Zhao, T, J M Zhou, V Bushlya, and J E Ståhl. 2017. "Effect of Cutting Edge Radius on Surface Roughness and Tool Wear in Hard Turning of AISI 52100 Steel." *The International Journal of Advanced Manufacturing Technology* 91 (9): 3611–3618. doi:10.1007/s00170-017-0065-z.

Zhou, Y J, Z H Huang, and T T Nguyen. 2017. "Rapid Electrolytic Deposition of Diamond Turnable Nickel Phosphorus Alloy Coating." *Surface and Coatings Technology* 320: 23–27. doi:10.1016/j.surfcoat.2017.02.017.

11 Future of Post-Processing

11.1 REVIEW ON POST-PROCESSING TECHNOLOGIES

As this book draws to a close, it is fitting to summarize the various post-processes discussed to evaluate the technological readiness of post-processing technologies for additive manufacturing. Various types of material removal processes have been covered in this book, which were categorized based on the type of energy employed (i.e., mechanical, abrasive, thermal, and chemical). However, material removal technologies to achieve precision surface finishing often come with a hefty price tag and require highly sophisticated equipment and environment conditions. Table 11.1 compiles short descriptions of several process with key advantages and disadvantages for consideration. Table 11.2 lists each of the post-processes that have been covered in this book and a comparison of the achievable surface quality alongside the different material removal rates and cost comparisons.

Innovative techniques employed for conventional machining can be proposed to improve the material removal efficacies of additively manufactured metals, which will be described in Section 11.2. Last, the idealized future of hybrid manufacturing will be discussed in Section 11.3.

As the maturity of understanding metal additive manufacturing technology is far from complete, research on the machinability of additively manufactured metals continues to expand with three key areas of focus:

(1) The machinability of other metal alloy formulations that will be fabricated by additive manufacturing.
(2) Machinability of special structures such as thin walls, thin columns, and complex geometrical structures with low stiffness.
(3) Design of new cutting tools and surface processing strategies to accommodate difficult-to-cut materials.

11.2 SURFACE PROCESSING INNOVATIONS

This book has shown that post-processing of additively manufactured (AM) components does not only involve considerations for material removal mechanisms but also microstructural characteristics that determine the mechanical properties of

DOI: 10.1201/9781003272601-11

TABLE 11.1
Types of Post-Processing Technologies

Process	Description
Micro-turning	A diamond cutting tool driven into a workpiece along a designated tool path remove material by chip formation; (+) Easy design of cheap cutting tools, precise tool path generation with nanometric resolution; (-) Susceptible to high tool wear due to constant tool-workpiece contact, lower material removal rates compared to multi-point cutting tools.
Micro-milling	A process employing a multi-point cutting tool that rotates to intermittently remove material by chip formation. (+) High material removal rates for large area machining, relatively lower tool wear; (-) Challenging to produce micro-milling tools, intermittent cutting leading to fluctuating loads and potential chatter.
Plasma arc machining	Channeling a high-velocity jet of ionized gas to produce a plasma arc that melts and remove material. (+) Effective against almost any metal at high material removal rates without chatter due to the contactless nature; (-) High-cost equipment, high temperatures leading to oxidation, typical reports of tapered surfaces.
Ion beam machining	Accelerating a channel of electrically charged ions at high energies to displace workpiece atoms by transfer of energy and momentum (i.e., sputtering); (+) Precise control of target positioning and material removal rates on almost any material to achieve superior surface quality; (-) Expensive equipment with very low material removal rates.
MRF	Delivering a magnetorheological (MR) fluid, which conforms to the geometry of the polishing area under magnetic-field excitation, to remove material by shearing. (+) Extremely accurate and stable to achieve precision optics; (-) Expensive MR fluid and unsuitable for cylindrical components.
CMP	Hybrid application of chemical etching delivered by the slurry and abrasive polishing achieved by a pad; (+) Capable of achieving uniform planar surfaces; (-) Work material subject to corrosion and non-uniform material removal of multi-material components.
Micro-EDM	Applying an electric voltage to a cathode and anode in proximity within a dielectric fluid to create an electric arc (spark) that melts work material; (+) Capable of machining materials regardless of hardness without tool-workpiece contact to avoid chatter and mechanical stress on the workpiece; (-) Unsuitable for non-conductive metals, material subject to thermal residual stress.
Surface grinding	A grinding wheel covered with diamond grits as a multi-edged cutting tool; (+) Self-sharpening abrasive action during low pressure work; (-) High-cost tools that are constantly degrading while having low material removal rates.
MDIF	A spherical magnet bonded with abrasives to rotate and polish internal surfaces under the influence of an external magnet; (+) Capable of reaching internal surfaces inaccessible by conventional methods; (-) Presently limited to circular internal structures.

TABLE 11.1 (Continued)
Types of Post-Processing Technologies

Process	Description
Sandblasting	A stream of abrasives directed to erode material under high pressures; (+) Easy to perform and effective for most materials; (-) Potential environmental pollution requiring safety equipment and precautions for operators.
AFM	A high-pressure stream of viscous abrasive-laden medium to erode material along its path. (+) No tool-workpiece contact to avoid chatter or vibration, easy implementation to process thin sections; (-) Low material removal rates, unsuitable for soft metals, non-recyclable abrasives, potential environmental pollution.
MAF	Magnetic abrasive particles forming a non-rigid polishing brush for material removal under the influence of a magnetic field; (+) Easy to perform with adaptable tool geometry; (-) Time consuming to setup resulting in high cost of production, unsuitable for mass production.
CAF	Integrating hydrodynamic cavitation erosion (collapse of a fluid cavitation bubble) with microparticle abrasion; (+); (-)
Laser polishing	Utilizing laser technology, which may include pulsed, continuous, or combinations of both to melt and resolidify a thin layer of surface material. (+) Contactless process without tools and capable of machining high aspect ratios; (-) Thermally induced residual stresses and oxidation on surface.
Electron beam polishing	Accelerating a high-velocity electrons through a beam with high energy to heat and vaporize work material. (+) No pressure or force acting on workpiece during precision machining, capable of precise targeting and high aspect ratio machining; (-) Expensive and low material removal rates.

the part. These aspects are inseparably connected with regard to the surface finishing quality. It is, therefore, of interest to further innovate solutions to augment conventional machining processes that attend to the as-built material state prior to other surface processing.

While several of these solutions have been explored on AM metals, there is still a vast variety of innovations that have proven concepts on the material removal processes for conventional materials and it is meaningful to review these implementations. It should be noted that the techniques mentioned in this book are non-exhaustive and designed to demonstrate the available possibilities and potentially give readers inspiration to explore other non-conventional post-processing techniques.

The first two sub-sections 11.2.1 and 11.2.2 will provide short descriptions on the available innovations that have been successfully explored for various materials processed by single-point diamond turning, where the fundamentals of material removal can be applied to most other conventional machining processes. To date, these techniques have yet to be explored on AM metals and therefore

TABLE 11.2
General Comparison Between Post-Processing Technologies

Process	Expected surface quality (Ra)	Material removal rate	Cost
Micro-turning	<0.01 μm		
Micro-milling	0.05–0.4 μm		
PA machining	0.4–3.0 μm		
IB machining	0.001–0.008 μm		
MRF	0.3–1.0 nm		
CMP	0.2–1.0 nm		
Micro-EDM	0.2–1.0 μm		
Surface grinding	0.025 – 0.4 μm		
MDIF	0.1–1.0 μm		
Sandblasting	0.7–1.5 μm		
AFM	0.3–5.0 μm		
MAF	0.15–0.7 μm Ra		
CAF			
Laser polishing	0.2–4.5 μm		
EB polishing	0.7–1.2 μm		
ECM			
PC machining			

PA: Plasma arc; IB: Ion beam; MRF: Magnetorheological finishing; CMP: Chemical-mechanical polishing; EDM: Electric discharge machining; MDIF: Magnetic-driven internal finishing; AFM: Abrasive flow machining; MAF: Magnetic abrasive finishing; CAF: Cavitation abrasive finishing; EB: Electron beam; ECMM: Electro-chemical machining; PC: Photochemical

outlays the potential to further improve the machinability of AM parts during post-processing. Section 11.2.3 will then discuss the challenges of post-processing internal surfaces.

11.2.1 SURFACE EFFECTS

The nanoscopic characteristics of the surfaces desired in precision manufacturing is relatively close to atomic scale interactions on the surface of materials. These geometrical similarities tend to increase the magnitude of influence from surrounding factors that would substantially affect micrometric material removal processing. These effects were classified into three categories (Figure 11.1) – physical, chemical, and physicochemical.

The physicochemical effect involves interatomic chemical interaction coupled with the mechanical stimulus during material removal. It employs simple surfactants including household items such as inks, glue, and alcohols (Udupa et al. 2018), which can substantially lower micro-cutting forces when applied before cutting (Chaudhari et al. 2018). The surfactant or surface-active medium (SAM) supposedly degrades the mechanical stability of the surface and reduces the strength (Shchukin et al. 1978)

PHYSICOCHEMICAL   PHYSICAL   CHEMICAL

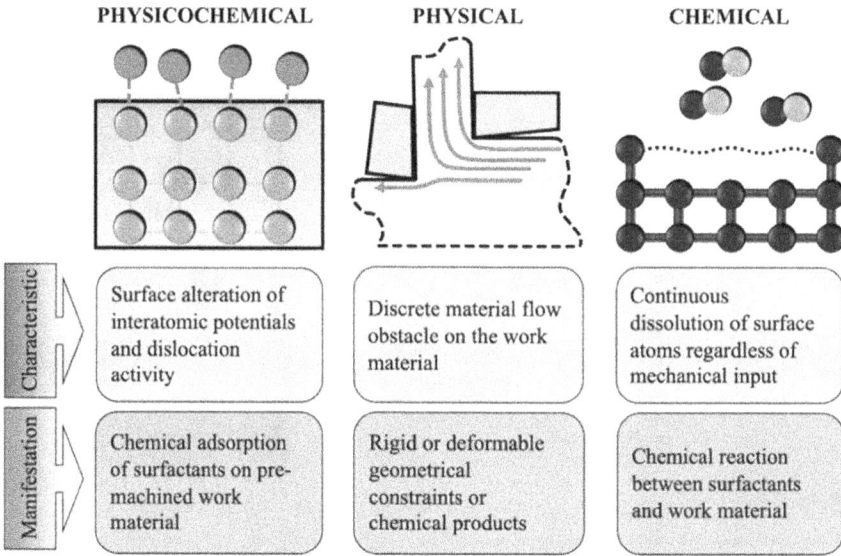

FIGURE 11.1   Categorization of surface effects in microscopic material removal processing. Reprinted with permission from Elsevier (Lee and Wang 2020).

resulting in embrittlement (Rehbinder and Shchukin 1972). This is the change in surface energy of the work material due to favorable bonding between the SAM and the work material atoms (Lee and Wang 2020). Multiple benefits have been reported in applying physiochemical effects in micro-cutting of metals such as the reduction in cutting forces (Zhang, Lee and Wang 2020), improved machined surface quality (Zheng et al. 2021), microstructural changes (Zhang, Lee, and Wang 2021a; Zhang, Lee, and Wang 2021b), and suppressed diamond tool wear (Lee, Shen, and Wang 2020).

The definition of a physical surface effect requires the physical presence of a matter on the work material to disrupt the conventional flow of material during removal. This physical matter may be in the rigid or deformable form for both solids and aqueous solutions.

Extrusion-cutting (de Chiffre 1976) is a rigid physical surface effect where a constraint tool is applied ahead of a cutting tool (Figure 11.2). During chip formation, the cutting tool and the constraint remain fixed at a certain distance from each other, while the work material flows through the gap between the two rigid tools. Hence, the process was coined as an extrusion-like process. It was recently rebranded as large-strain extrusion machining (LSEM) in view of the concept to concentrate the strain in the primary deformation zone (Cai et al. 2015) and reduce the instabilities caused during chip formation, which result in the uniformity of the machined surface quality (Sagapuram et al. 2015). The improvements in machined surface quality may be controlled by the chip compression ratio (i.e., the ratio between the deformed chip thickness and the undeformed chip thickness) to obtain the optimal cutting forces and

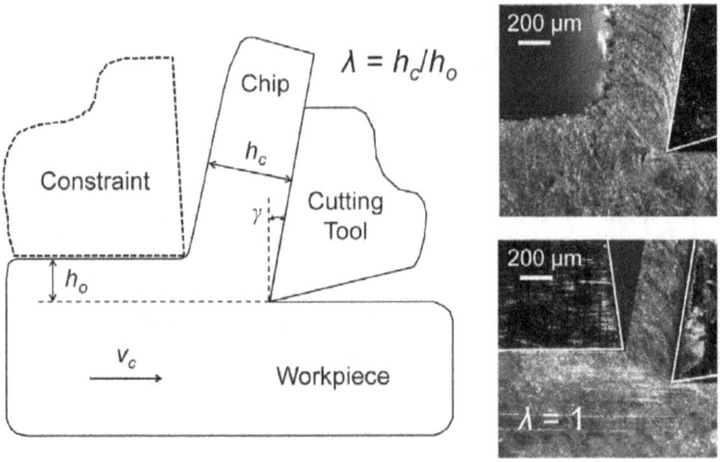

FIGURE 11.2 Illustration of an extrusion-cutting process with a constraint ahead of the cutting tool and the corresponding comparison in chip formation with conventional cutting. Reprinted with permission from Elsevier (Sagapuram et al. 2015).

maintain shear deformation during the cutting process without going into complex extrusion, which induces undesirable subsurface deformation and high energy consumption (Guo, Chen, and Saleh 2020).

The physical surface effect may also manifest in the form of a high-pressured liquid exerting hydrostatic stresses on the work material, which has been reported to affect the deformability of various materials ranging across plastics (Pugh et al. 1971; Pae and Bhateja 1975), metals (Sakata and Okubo 1970; Sakata, Aoki, and Tsujimoto 1971), and ceramics (Bridgman 1947). The concept is to submerge the work material in a chamber of pressurized liquid while performing deformation tests such as scratching (Yoshino et al. 2001) and micro-cutting (Yoshino 2016). At present, the integration of this innovation in micro-cutting has been largely focused on brittle materials due to the desirable notions of the external pressure suppressing crack formation and propagation (Sakata and Aoki 1973). The theory dwells in the fracture toughness (i.e., the criterion for crack propagation) (Equation 11.1), which will require additional stresses under a hydrostatic pressure for crack propagation (Equation 11.2).

$$K_I = \sigma_t \sqrt{\pi c} \qquad (11.1)$$

$$K_I' = \left(\sigma_t - p_1\right)\sqrt{\pi c} \qquad (11.2)$$

where σ_t is the tensile stress acting on a crack tip, c is the length of a crack, and p_1 is the externally applied hydrostatic pressure.

While the theory on fracture toughness may not seem directly beneficial for application in metal cutting, the implementation of hydrostatic pressure has an influence

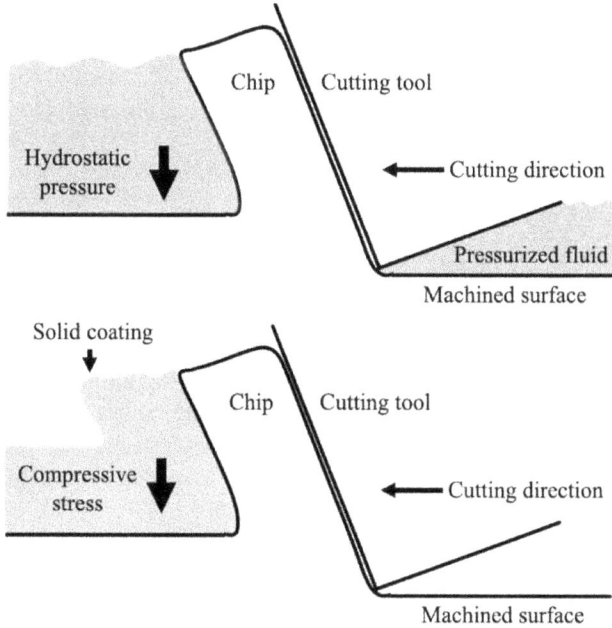

FIGURE 11.3   Illustration of the believed similarities in the concept of applying hydrostatic pressure and a coating during cutting. Reprinted with permission from Elsevier (Lee and Wang 2020).

on the dislocation density that can further vary the flow stress between large and small grains (Yuan and Wu 2014). The hydrostatic pressure can also affect the transformations of the microstructure (Krawczynska et al. 2019) as heat builds up during material removal (due to severe plastic deformation) and cooling follows in both cutting chips and machined surfaces. Thus, it is admissible that the implementation of hydrostatic pressure can affect the material removal process of AM metals.

In the middle of the solid rigid tool and the high-pressure fluid is the application of a coating to induce a deformable physical surface effect. The concept is believed to liken the application of hydrostatic pressure with a difference in the surface stress developing from the resistance to deformation of the coating as illustrated in Figure 11.3 (Lee et al. 2019; Lee and Wang 2021; Lee, Kumar, and Wang 2021).

11.2.2   FIELD-ASSISTED MACHINING

A magnetic field is understood to affect ferromagnetic materials by inducing magnetic forces of attraction and repulsion. This has led to the implementation of eddy current damping effects to suppress the detrimental effects of tool-tip vibration in micro-cutting (Yip and To 2017a). The theory follows that an eddy current is generated as a conductive material moves through a magnetic field, which then creates its own magnetic field with opposing directions relative to the external magnetic field. The opposing force is known as the Lorentz force, which then serves as a repulsion to the motion of the conductive material that is perpendicular to the magnetic field orientation. The Lorentz force is calculated based on Equation 11.3, which indicates its proportional relationship with the conductor velocity and the magnetic field.

$$F_{Lorentz} = E \times q + q \times V_e \times B \qquad (11.3)$$

where E is the electric field, q is the conductor charge, V_e is the velocity of the conductor, and B is the magnetic field intensity. It was also believed that conversion of kinetic energy into heat energy would occur as part of the eddy current damping effect during machining (Yip and To 2019), which would aid in the suppression of tool wear and improve the surface finishing of titanium alloys (Yip and To 2017b). Enhanced plasticity in the form of elongated slip band area was also reported during cutting of steel (el Mansori and Mkaddem 2007; Mkaddem and el Mansori 2012; Dehghani et al. 2017).

While this theory is well-received due to the association between a magnetic field and ferromagnetic materials, various research works have also identified significant evidence on the influence of a magnetic field on paramagnetic and diamagnetic materials. This is referred to as the magneto-plastic effect, which has been implemented to improve the cutting performance of copper (Guo, Lee, Zhang, and Wang 2022) and even a brittle ceramic (Guo, Lee, Zhang, Sorkin, et al. 2022). It was proposed that the magnetic field affects the spin conversion in radical pairs of dislocations and stoppers (Figure 11.4) and reduces the pinning energy to improve dislocation mobility (Golovin 2004). The magneto-plastic effect allowed for the reduction in cutting forces, improvements in surface quality (i.e., reduction in surface roughness), reduced sub-grain formation and crystallographic retention, and the enhanced plasticity for delayed ductile–brittle transition of brittle materials. The intriguing findings on the influence of the weak magnetic field on paramagnetic and

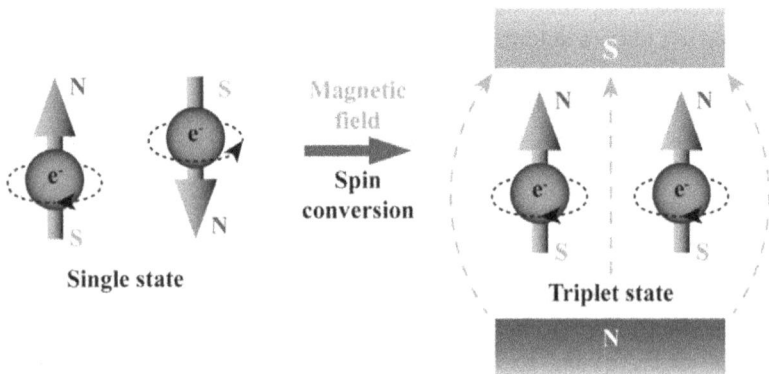

FIGURE 11.4 Spin conversion theory under a magnetic field. Reprinted with permission from Elsevier (Guo, Lee, Zhang, and Wang 2022).

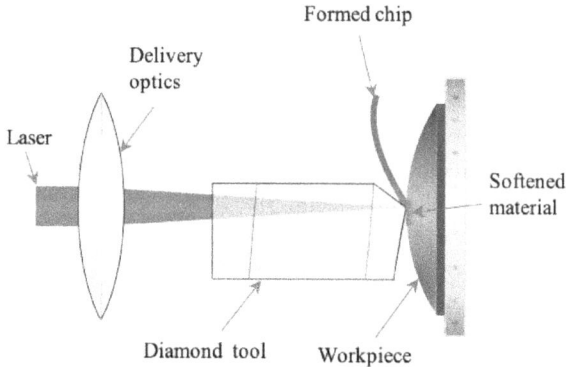

FIGURE 11.5 Concept of a localized laser beam as the heat source during micro-cutting. Reprinted with permission from Elsevier (Navare et al. 2020).

diamagnetic materials during precision cutting would be highly attractive to the post-processing of a wide range of additively manufactured metals and even other materials such as ceramics.

Another multi-physical phenomenon is the application of heat to the cutting region, which has been explored in the localized and generic forms. The latter indicates the overall heating of the work material (Lee 2019), while the latter employs a targeted heat source such as a laser to transfer heat to the material. Laser-assisted machining has been developed over the years to overcome the brittleness of ceramics and glasses during machining and is achieved by focusing a laser beam through a monocrystalline diamond tool to concurrently heat the material during cutting (Figure 11.5). The reason for administering heat to the cutting zone is to soften the work material for enhanced plasticity and mitigate crack formation during cutting. The innovation was proven successful for the generation of infrared-optic surfaces on various materials such as zinc sulfide (Navare et al. 2020), germanium (Shahinian et al. 2020), fused silica (Shahinian et al. 2019), tungsten carbide (Kang et al. 2019), and calcium fluoride (Bodlapati et al. 2021).

The fundamental cause for such an occurrence boils down to the excitation of atomic kinetic energy. The vibration of atoms increases the tendency for interatomic bonding to be broken during deformation. This manifests on a larger scale as slip deformation where a row of atoms is granted easier mobility to glide across another plane of atoms resulting in macroscopic plasticity. Fundamentally, this phenomenon occurs on the primary slip systems (i.e., a network of slip planes and slip directions for atomic displacement), but the larger input of heat energy that is converted into atomic kinetic energy calls for secondary slip systems to also be activated, which can further enhance the plasticity to improve machinability of brittle materials. From the perspective of cutting metals, promoting plasticity may not necessarily correspond well with superior surface finishing as is the case with Ti-Al alloys, which reports the major influence of the ductile-induced microstructure on the poor machined surface quality (Zhang et al. 2022). However, the randomness of the grain orientations and its influence on the material removal performance can be potentially resolved through the implementation of heat as reported by the potential reduction in anisotropic characteristics of single-crystalline materials (Wang et al. 2016; Wang, Kumar, and Riemer 2018). At present, thermal-assisted cutting of metals is widely studied but little is still known about the effectiveness of such a technique on as-built metals in additive manufacturing, particularly the interaction of heat with microstructural features such as the melt pool boundaries and different material phases emerging from the building process.

11.2.3 Internal Finishing

In the meantime, internal surface finishing remains a challenge to be resolved despite the great advantages of AM to produce parts as a single component according to internal surface designs. It may seem like achieving desirable surface quality on AM metal surfaces is seemingly easy with the various innovations that have yet to be explored, but the inaccessibility of most tools in reaching internal surfaces creates the bottleneck for the uptake of AM in the manufacturing industry.

For instance, internal cooling channels are essential for cooling of molds and dies, which would extend the life of the molds by delaying the onset of thermal fatigue cracking, plastic deformation, and corrosion. While AM offers a revolutionary solution of producing molds with specially designed internal cooling channels in a one stop process (compared to drilling of cooling channels), the poor roughness of these internal surfaces can severely affect the cooling performance (Yamaguchi et al. 2021). Thus, internal finishing remains one of the most challenging problems to resolve in the research on metal AM. Chapter 5 discussed various abrasive based material removal processes that have been applied to address this issue, but there has yet to be an economical solution toward optimizing the manufacturing process for precision surface generation on an industrial scale.

11.3 HYBRID ADDITIVE MANUFACTURING

The crux of hybrid AM is to have a single solution to integrate the full manufacturing process from build to finish while having an emphasis on efficient management of energy consumption and material wastage (Campatelli et al. 2020; Wippermann et al. 2020). On the side lines, the implementation of hybrid AM aims to overcome the limitations of both AM technology (e.g., inadequate surface roughness, support structure removal, limited build size, etc.) and traditional material removal processes (e.g., challenges in internal surface fabrication), to create an effective solution for mass production (Strong et al. 2018). Several successful integrations of these processes were covered in Chapter 9 demonstrating the careful planning of inventory and the cost effectiveness of a technologically feasible solution that accommodates a wider range of product design specifications. A typical example centers on the integration of DED to build the component with a milling process to achieve the desirable surface quality, all within a single-clamping setup to reduce machine tool errors and fabrication time.

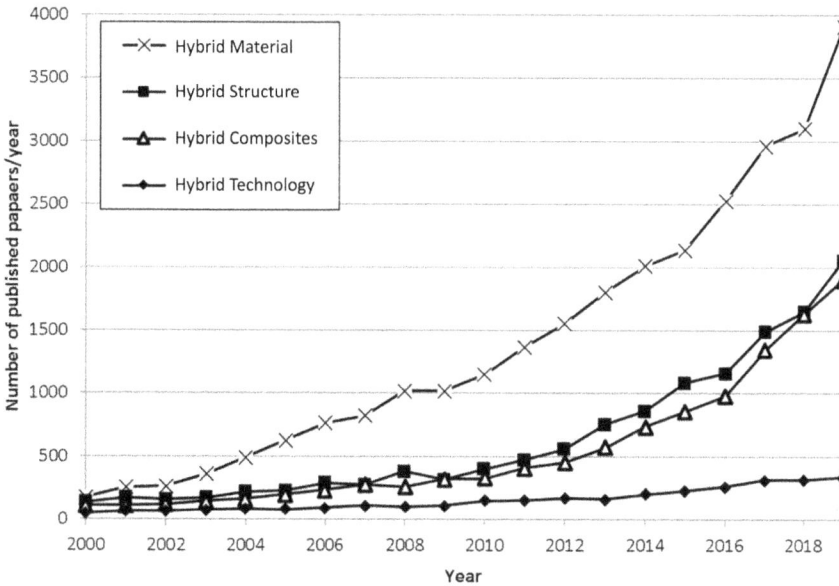

FIGURE 11.6 Comparison of published articles in the different areas of AM research. Reprinted with permission from EDP Sciences (Popov and Fleisher 2020).

The integrative technologies further revolutionize manufacturing methodologies promoting strong suits that make it attractive as a highly investable solution for the future of manufacturing. Further development in this technology is postulated to emerge as the one-stop manufacturing solution for any product type regardless of geometry and material. Research trends in hybridization (Figure 11.6) show that the status of the technology is still in its infancy stages with great untapped potential for further development, and it is essential to map out the areas to be explored for the enhancements for hybrid AM.

In the interests of precision manufacturing with hybrid AM technology, other considerations aside from the additive and subtractive process management include determining optimal build and tooling paths based on the assessment of the digitalized part geometry, detection of defect formation during build/post-processing, and corrective action decision-making and planning. With these considerations in mind, an idealized concept of hybrid AM for precision manufacturing can be realized, which combines the different aspects surrounding the integration of the additive and subtractive processes with smart features.

11.3.1 Artificial Intelligence

Artificial intelligence (AI) has been proven to boost both the performance and product quality in both additive and subtractive manufacturing processes (Chuo et al. 2022; Wang et al. 2020). In additive manufacturing, various machine learning techniques have been developed to optimize the part design, optimize the process parameters, and monitor in-process defect formations. In subtractive manufacturing, intelligent algorithms have been established to monitor the tool wear, optimize process parameters, and predict the surface quality. An example for applied AI in post-processing is the use of a broad echo state learning system to predict the material removal rates in magnetically driven internal finishing based on the input of polishing forces, which presented a prediction error smaller than that of physics-based models (Zhang et al. 2022). It is believed that artificial intelligence will enhance the hybrid additive and subtractive manufacturing processes where optimization of part properties, process planning, and manufacturing efficiency are of major concern.

11.3.2 FIELD-ASSISTED INTEGRATIONS

External excitation fields such as the magnetic, thermal, and ultrasonic, have also presented benefits to enhance machinability. These field-assisted techniques also have the potential to be applied during the building process. For instance, the magnetic field has been reported to fine tune the microstructure of Inconel 625 during wire arc additive manufacturing, which aid to obtain the desirable mechanical properties of the fabricated components (Wang et al. 2021). However, this area of employing field-assistance during additive manufacturing remains largely unexplored and extensive investigations will be required to study the interaction between the applied fields, the control parameters of the energy fields, and metal plasticity especially where microstructure is considered.

This book concludes having touched base with a short review on additive manufacturing technologies and the workflows, followed by a detailed overview on the various post-processing technologies employed for additively manufactured metals before ending off with a summarized comparison of these processes.

REFERENCES

Bodlapati, Charan, Di Kang, Jayesh Navare, Robert Turnbull, Yuxiang Zhong, and Hossein Shahinian. 2021. "Cutting Performance of Laser Assisted Diamond Turning of Calcium Fluoride with Different Crystal Orientations." *Applied Optics* 60 (9). OSA: 2465–2470. doi:10.1364/AO.415265.

Bridgman, P W. 1947. "The Effect of Hydrostatic Pressure on the Fracture of Brittle Substances." *Journal of Applied Physics* 18 (2): 246–258. doi:10.1063/1.1697610.

Cai, S L, Y Chen, G G Ye, M Q Jiang, H Y Wang, and L H Dai. 2015. "Characterization of the Deformation Field in Large-Strain Extrusion Machining." *Journal of Materials Processing Technology* 216: 48–58. doi:10.1016/j.jmatprotec.2014.08.022.

Campatelli, Gianni, Filippo Montevecchi, Giuseppe Venturini, Giuseppe Ingarao, and Paolo C Priarone. 2020. "Integrated WAAM-Subtractive Versus Pure Subtractive Manufacturing Approaches: An Energy Efficiency Comparison." *International Journal of Precision Engineering and Manufacturing-Green Technology* 7 (1): 1–11. doi:10.1007/s40684-019-00071-y.

Chaudhari, Akshay, Zhi Yuan Soh, Hao Wang, and A. Senthil Kumar. 2018. "Rehbinder Effect in Ultraprecision Machining of Ductile Materials." *International Journal of Machine Tools and Manufacture* 133 (May). Elsevier Ltd: 47–60. doi:10.1016/j.ijmachtools.2018.05.009.

Chuo, Yu Sung, Ji Woong Lee, Chang Hyeon Mun, In Woong Noh, Sina Rezvani, Dong Chan Kim, Jihyun Lee, Sang Won Lee, and Simon S Park. 2022. "Artificial Intelligence Enabled Smart Machining and Machine Tools." *Journal of Mechanical Science and Technology* 36 (1): 1–23. doi:10.1007/s12206-021-1201-0.

de Chiffre, L. 1976. "Extrusion-Cutting." *International Journal of Machine Tool Design and Research* 16 (2): 137–144. doi:10.1016/0020-7357(76)90032-9.

Dehghani, Alireza, Sasan Khalaj Amnieh, Alireza Fadaei Tehrani, and Aminollah Mohammadi. 2017. "Effects of Magnetic Assistance on Improving Tool Wear Resistance and Cutting Mechanisms during Steel Turning." *Wear* 384–385 (February). Elsevier B.V.: 1–7. doi:10.1016/j.wear.2017.04.023.

el Mansori, Mohamed, and Ali Mkaddem. 2007. "Surface Plastic Deformation in Dry Cutting at Magnetically Assisted Machining." *Surface and Coatings Technology* 202 (4–7): 1118–1122. doi:10.1016/j.surfcoat.2007.07.104.

Golovin, Yu I. 2004. "Magnetoplastic Effects in Solids." *Physics of the Solid State* 46 (5): 789–824. doi:10.1134/1.1744954.

Guo, Yang, Jisheng Chen, and Amr Saleh. 2020. "In Situ Analysis of Deformation Mechanics of Constrained Cutting toward Enhanced Material Removal." *Journal of Manufacturing Science and Engineering* 142 (2). doi:10.1115/1.4045613.

Guo, Yunfa, Yan Jin Lee, Yu Zhang, Anastassia Sorkin, Sergei Manzhos, and Hao Wang. 2022. "Effect of a Weak Magnetic Field on Ductile–Brittle Transition in Micro-Cutting of Single-Crystal Calcium Fluoride." *Journal of Materials Science & Technology* 112: 96–113. doi:10.1016/j.jmst.2021.09.049.

Guo, Yunfa, Yan Jin Lee, Yu Zhang, and Hao Wang. 2022. "Magneto-Plasticity in Micro-Cutting of Single-Crystal Copper." *Journal of Materials Science & Technology* 124 (October). Elsevier: 121–134. doi:10.1016/J.JMST.2022.03.003.

Kang, Di, Jayesh Navare, Yang Su, Dmytro Zaytsev, Deepak Ravindra, and Hossein Shahinian. 2019. "Observations on Ductile Laser Assisted Diamond Turning of Tungsten Carbide." In *Optical Design and Fabrication 2019 (Freeform, OFT)*, JT5A.11. OSA Technical Digest. Washington, DC: Optica Publishing Group. doi:10.1364/FREEFORM.2019.JT5A.11.

Krawczynska, Agnieszka Teresa, Przemyslaw Suchecki, Boguslawa Adamczyk-Cieslak, Barbara Romelczyk-Baishya, and Malgorzata Lewandowska. 2019. "Influence of High Hydrostatic Pressure Annealing on the Recrystallization of Nanostructured Austenitic Stainless Steel." *Materials Science and Engineering: A* 767 (November). Elsevier: 138381. doi:10.1016/J.MSEA.2019.138381.

Lee, Yan Jin. 2019. "Thermal Expansion Control in Heat Assisted Machining of Calcium Fluoride Single Crystals." In *19th International Conference of the European Society for Precision Engineering and Nanotechnology (Euspen)*. Bilbao, Spain.

Lee, Yan Jin, Jing Yi Chong, Akshay Chaudhari, and Hao Wang. 2019. "Enhancing Ductile-Mode Cutting of Calcium Fluoride Single Crystals with Solidified Coating." *International Journal of Precision Engineering and Manufacturing-Green Technology* 7 (6): 1019–1029. doi:10.1007/s40684-019-00126-0.

Lee, Yan Jin, A Senthil Kumar, and Hao Wang. 2021. "Beneficial Stress of a Coating on Ductile-Mode Cutting of Single-Crystal Brittle Material." *International Journal of Machine Tools and Manufacture* 168: 103787. doi:10.1016/j.ijmachtools.2021.103787.

Lee, Yan Jin, and Hao Wang. 2020. "Current Understanding of Surface Effects in Microcutting." *Materials & Design* 192: 108688. doi:10.1016/j.matdes.2020.108688.

Lee, Yan Jin, and Hao Wang. 2021. "Characterizing Crack Morphology toward Improving Ductile Mode Cutting of Calcium Fluoride." *Ceramics International* 47 (20): 28543–28556. doi:10.1016/j.ceramint.2021.07.012.

Lee, Yan Jin, Yung-Kang Shen, and Hao Wang. 2020. "Suppression of Polycrystalline Diamond Tool Wear with Mechanochemical Effects in Micromachining of Ferrous Metal." *Journal of Manufacturing and Materials Processing* 4 (3): 81. doi:10.3390/jmmp4030081.

Mkaddem, Ali, and Mohamed el Mansori. 2012. "Extension of the Slip Band Area under Magnetization during Steel Machining." *Materials and Manufacturing Processes* 27 (10): 1073–1077. doi:10.1080/10426914.2011.654163.

Navare, Jayesh, Di Kang, Dmytro Zaytsev, Charan Bodlapati, Deepak Ravindra, and Hossein Shahinian. 2020. "Experimental Investigation on the Effect of Crystal Orientation of Diamond Tooling on Micro Laser Assisted Diamond Turning of Zinc Sulfide." *Procedia Manufacturing* 48 (January). Elsevier: 606–610. doi:10.1016/J.PROMFG.2020.05.088.

Pae, K D, and S K Bhateja. 1975. "The Effects of Hydrostatic Pressure on the Mechanical Behavior of Polymers." *Journal of Macromolecular Science, Part C* 13 (1). Taylor & Francis: 1–75. doi:10.1080/15321797508068145.

Popov, Vladimir V, and Alexander Fleisher. 2020. "Hybrid Additive Manufacturing of Steels and Alloys." *Manufacturing Rev.* 7. doi:10.1051/mfreview/2020005.

Pugh, H Ll. D, E F Chandler, L Holliday, and J Mann. 1971. "The Effect of Hydrostatic Pressure on the Tensile Properties of Plastics." *Polymer Engineering & Science* 11 (6). John Wiley & Sons, Ltd: 463–473. doi:10.1002/pen.760110605.

Rehbinder, P A, and E D Shchukin. 1972. "Surface Phenomena in Solids during Deformation and Fracture Processes." *Progress in Surface Science* 3 (2): 97–188. doi:10.1016/0079-6816(72)90011-1.

Sagapuram, Dinakar, Ho Yeung, Yang Guo, Anirban Mahato, Rachid M'Saoubi, W Dale Compton, Kevin P Trumble, and Srinivasan Chandrasekar. 2015. "On Control of Flow Instabilities in Cutting of Metals." *CIRP Annals* 64 (1): 49–52. doi:10.1016/j.cirp.2015.04.059.

Sakata, M., and S. Aoki. 1973. "Torsional Strength of Glass under Hydrostatic Pressure." *Journal of Engineering Materials and Technology* 95 (2): 83–86. doi:10.1115/1.3443136.

Sakata, Masaru, Shigeru Aoki, and Takashi Tsujimoto. 1971. "An Experimental Study on the Deformation of Materials under High Hydrostatic Pressures: 2nd Report, on Torsional Deformation." *Bulletin of JSME* 14 (74): 737–744. doi:10.1299/jsme1958.14.737.

Sakata, Masaru, and Tetsuo Okubo. 1970. "An Experimental Study on the Deformation of Materials under High Hydrostatic Pressures: 1st Report, on the Deformation of Cylinders under Internal and External Pressures." *Bulletin of JSME* 13 (63): 1059–1065. doi:10.1299/jsme1958.13.1059.

Shahinian, Hossein, Kang Di, Jayesh Navare, Charan Bodlapati, Dmytro Zaytsev, and Deepak Ravindra. 2020. "Ultraprecision Laser-Assisted Diamond Machining of Single Crystal Ge." *Precision Engineering* 65 (September). Elsevier: 149–155. doi:10.1016/J.PRECISIONENG.2020.04.020.

Shahinian, Hossein, Jayesh Navare, Charan Bodlapati, Dmytro Zaytsev, Di Kang, and Deepak Ravindra. 2019. "Micro-Laser Assisted Single Point Diamond Turning of Fused Silica Glass." In *Proc.SPIE.* Vol. 11175. doi:10.1117/12.2536282.

Shchukin, E D, Yu. V Goryunov, N V Pertsov, and L S Bryukhanova. 1978. "Development of Investigations of the Adsorption Reduction of the Strength of Solids in the Works of P. A. Rebinder and His School." *Soviet Materials Science: A Transl. of Fiziko-Khimicheskaya Mekhanika Materialov / Academy of Sciences of the Ukrainian SSR* 14 (4): 341–346. doi:10.1007/BF01154708.

Strong, Danielle, Michael Kay, Brett Conner, Thomas Wakefield, and Guha Manogharan. 2018. "Hybrid Manufacturing – Integrating Traditional Manufacturers with Additive Manufacturing (AM) Supply Chain." *Additive Manufacturing* 21 (May). Elsevier: 159–173. doi:10.1016/J.ADDMA.2018.03.010.

Udupa, Anirudh, Koushik Viswanathan, Mojib Saei, James B. Mann, and Srinivasan Chandrasekar. 2018. "Material-Independent Mechanochemical Effect in the Deformation of Highly-Strain-Hardening Metals." *Physical Review Applied* 10 (1). American Physical Society: 1. doi:10.1103/PhysRevApplied.10.014009.

Wang, C, X P Tan, S B Tor, and C S Lim. 2020. "Machine Learning in Additive Manufacturing: State-of-the-Art and Perspectives." *Additive Manufacturing* 36: 101538. doi:10.1016/j.addma.2020.101538.

Wang, Hao, O. Riemer, K. Rickens, and E. Brinksmeier. 2016. "On the Mechanism of Asymmetric Ductile-Brittle Transition in Microcutting of (111) CaF2 Single Crystals." *Scripta Materialia* 114. Elsevier B.V.: 21–26. doi:10.1016/j.scriptamat.2015.11.030.

Wang, Hao, A. Senthil Kumar, and Oltmann Riemer. 2018. "On the Theoretical Foundation for the Microcutting of Calcium Fluoride Single Crystals at Elevated Temperatures." *Proceedings of the Institution of Mechanical Engineers, Part B: Journal of Engineering Manufacture* 232 (6): 1123–1129. doi:10.1177/0954405416666907.

Wang, Yangfan, Xizhang Chen, Qingkai Shen, Chuanchu Su, Yupeng Zhang, S Jayalakshmi, and R Arvind Singh. 2021. "Effect of Magnetic Field on the Microstructure and Mechanical Properties of Inconel 625 Superalloy Fabricated by Wire Arc Additive Manufacturing." *Journal of Manufacturing Processes* 64: 10–19. doi:10.1016/j.jmapro.2021.01.008.

Wippermann, A., T. G. Gutowski, B. Denkena, M. A. Dittrich, and Y. Wessarges. 2020. "Electrical Energy and Material Efficiency Analysis of Machining, Additive and Hybrid Manufacturing." *Journal of Cleaner Production* 251 (April). Elsevier: 119731. doi:10.1016/J.JCLEPRO.2019.119731.

Yamaguchi, Mitsugu, Tatsuaki Furumoto, Shuuji Inagaki, Masao Tsuji, Yoshiki Ochiai, Yohei Hashimoto, Tomohiro Koyano, and Akira Hosokawa. 2021. "Internal Face Finishing for a Cooling Channel Using a Fluid Containing Free Abrasive Grains." *The International Journal of Advanced Manufacturing Technology* 114 (1): 497–507. doi:10.1007/s00170-021-06893-y.

Yip, W. S., and S. To. 2017a. "An Application of Eddy Current Damping Effect on Single Point Diamond Turning of Titanium Alloys." *Journal of Physics D: Applied Physics* 50 (43). doi:10.1088/1361-6463/aa86fc.

Yip, W. S., and S. To. 2017b. "Tool Life Enhancement in Dry Diamond Turning of Titanium Alloys Using an Eddy Current Damping and a Magnetic Field for Sustainable Manufacturing." *Journal of Cleaner Production* 168: 929–939. doi:10.1016/j.jclepro.2017.09.100.

Yip, W S, and S To. 2019. "Reduction of Tool Tip Vibration in Single-Point Diamond Turning Using an Eddy Current Damping Effect." *The International Journal of Advanced Manufacturing Technology*. Springer, 1–11. doi:10.1007/s00170-019-03457-z.

Yoshino, M, T Aoki, T Shirakashi, and R Komanduri. 2001a. "Some Experiments on the Scratching of Silicon: In Situ Scratching inside an SEM and Scratching under High External Hydrostatic Pressures." *International Journal of Mechanical Sciences* 43 (2): 335–347. doi:10.1016/S0020-7403(00)00019-9.

Yoshino, Masahiko, Takayuki Aoki, Taisuke Sugishima, and Takahiro Shirakashi. 2001b. "Scratching Test on Hard-Brittle Materials under High Hydrostatic Pressure." *Journal of Manufacturing Science and Engineering* 123 (10): 231–239. doi:10.2493/jjspe.65.1481.

Yoshino, Masahiko. 2016. "Effects of External Hydrostatic Pressure on Orthogonal Cutting Characteristics." *Procedia Manufacturing* 5: 1308–1319. doi:10.1016/j.promfg.2016.08.102.

Yuan, Fuping, and Xiaolei Wu. 2014. "Hydrostatic Pressure Effects on Deformation Mechanisms of Nanocrystalline Fcc Metals." *Computational Materials Science* 85 (April). Elsevier: 8–15. doi:10.1016/J.COMMATSCI.2013.12.047.

Zhang, J, W Tian, F Zhao, X Mei, G Chen, and H Wang. 2022. "Material Removal Rate Prediction Based on Broad Echo State Learning System for Magnetically Driven Internal Finishing." *IEEE Transactions on Industrial Informatics*, 1–10. doi:10.1109/TII.2022.3204003.

Zhang, Jiayi, Yan Jin Lee, and Hao Wang. 2020. "Surface Texture Transformation in Micro-Cutting of AA6061-T6 with the Rehbinder Effect." *International Journal of Precision Engineering and Manufacturing-Green Technology* 8: 1151–1162. doi:10.1007/s40684-020-00260-0.

Zhang, Jiayi, Yan Jin Lee, and Hao Wang. 2021a. "Microstructure Evaluation of Shear Bands of Microcutting Chips in AA6061 Alloy under the Mechanochemical Effect." *Journal of Materials Science & Technology* 91: 178–186. doi:10.1016/j.jmst.2021.02.044.

Zhang, Jiayi, Yan Jin Lee, and Hao Wang. 2021b. "Mechanochemical Effect on the Microstructure and Mechanical Properties in Ultraprecision Machining of AA6061 Alloy." *Journal of Materials Science and Technology* 69. doi:10.1016/j.jmst.2020.08.024.

Zhang, Yu, Yan Jin Lee, Shuai Chang, Yuyong Chen, Yuchao Bai, Jiong Zhang, and Hao Wang. 2022. "Microstructural Modulation of TiAl Alloys for Controlling Ultra-Precision Machinability." *International Journal of Machine Tools and Manufacture* 174: 103851. doi:10.1016/j.ijmachtools.2022.103851.

Zheng, Zhongpeng, Yan Jin Lee, Jiayi Zhang, Xin Jin, and Hao Wang. 2021. "Ultra-Precision Micro-Cutting of Maraging Steel 3J33C under the Influence of a Surface-Active Medium." *Journal of Materials Processing Technology* 292: 117054. doi:10.1016/j.jmatprotec.2021.117054.

Index

For Product Safety Concerns and Information please contact our EU
representative GPSR@taylorandfrancis.com
Taylor & Francis Verlag GmbH, Kaufingerstraße 24, 80331 München, Germany

9 781032 224480